Traveling Cultures and Plants

Studies in Environmental Anthropology and Ethnobiology

General Editor: **Roy Ellen**, FBA
Professor of Anthropology, University of Kent at Canterbury

Interest in environmental anthropology has grown steadily in recent years, reflecting national and international concern about the environment and developing research priorities. This major new international series, which continues a series first published by Harwood and Routledge, is a vehicle for publishing up-to-date monographs and edited works on particular issues, themes, places or peoples which focus on the interrelationship between society, culture, and environment. Relevant areas include human ecology, the perception and representation of the environment, ethno-ecological knowledge, the human dimension of biodiversity conservation, and the ethnography of environmental problems. While the underlying ethos of the series will be anthropological, the approach is interdisciplinary.

Traveling Cultures and Plants

*The Ethnobiology and Ethnopharmacy of
Human Migrations*

Edited By
Andrea Pieroni and Ina Vandebroek

Berghahn Books
New York • Oxford

First edition published in 2007 by

Berghahn Books
www.berghahnbooks.com

©2007, 2009 Andrea Pieroni and Ina Vandebroek
First paperback edition published in 2009

Library of Congress Cataloging-in-Publication Data
Traveling cultures, plants, and medicine : the ethnobiology and ethnopharmacy
of human migrations / edited by Andrea Pieroni and Ina Vandebroek. -- 1st ed.
    p. cm. -- (Studies in environmental anthropology and ethnobiology ; v. 7)
  Includes bibliographical references.
  ISBN 978-1-84545-373-2 (hbk) -- ISBN 978-1-84545-679-5 (pbk)
  1. Ethnobiology. 2. Traditional medicine. 3. Emigration and immigration--
Social aspects. 4. Urban anthropology. I. Pieroni, Andrea. II. Vandebroek, Ina.

  GN476.7.T73 2007
  306.4'5--dc22

                                2007024373

British Library Cataloguing in Publication Data
A catalogue record for this book is available from the British Library

Printed in the United States on acid-free paper

        ISBN 978-1-84545-373-2   hardback
        ISBN 978-1-84545-679-5   paperback

Contents

List of Tables and Figures

Tables

Figures

⌘ Introduction

Andrea Pieroni and Ina Vandebroek

Ethnobiology as a Contested Discipline: From the Exotic and Remote to Urban Studies in Our Own Backyard

The research area that represents the lynchpin of this book is situated between *transcultural health studies* and what we call today *urban ethnobotany* (Balick et al. 2000). In the late 1990s, ethnobotanists became interested in urban ethnobotany when they began to realize that biocultural diversity (Maffi 2001) is not restricted to dispersed and marginal communities of the planet; it is also very much a part of metropolitan areas. Young ethnobotanists often talk of their dream to "boldly go" in the tradition of Star Trek, where no man (or woman) has gone before, to some remote and exotic tropical place where they can study peoples and their knowledge, beliefs, and practices, including their use of plants. Many young ethnobotanists would prefer to work with indigenous tribes in tropical rainforests, because there is a notion deeply rooted in Western public opinion that a wealth of ethnobotanical knowledge is hidden away in some far-off jungle just waiting to be uncovered (Voeks 2004). They do not seem to be aware of the fact that ethnobiology and ethnobotany can be studied just as effectively in immigrant communities who live, literally speaking, in our own backyard. Indeed, no passport or airplane ticket is required to study exotic plants and the traditional beliefs surrounding their use in multicultural, urban settings. One could stubbornly insist, of course, that this kind of urban ethnobotany will probably be "minimal" compared to the "original" wealth of knowledge held by remote indigenous forest tribes. It is true that urban ethnobotany is still in its infancy, and as yet no extensive studies have been conducted that would allow us to compare the breadth of plant knowledge in rural, tropical study areas with urban temperate study areas. Or to put it another way, there still is very limited data available to make comparisons *in situ* (in immigrants' country of origin) with *ex situ* (in immigrants' host countries).[1]

Nevertheless, the dynamic and challenging subfield of urban ethnobotany is rapidly gaining attention worldwide, and on-going studies are already providing a surprising amount of ethnobiological and ethnopharmaceutical data from immigrant communities in urban settings (e.g., Balick et al. 2000; Corlett et al. 2003; Pieroni et al. 2005; Sandhu and Heinrich 2005; Johnson et al. 2006; Waldstein 2006).

Migrants' Health

The growing interest of medical anthropologists and other social scientists in ethnicity and health studies in multicultural societies parallels an increasing awareness that the meaning of health in multicultural societies is often broader and more complex than what is understood in classical Western biomedical terms. In order to understand human well-being, we need to take into consideration *emic* or insiders' health perceptions, beliefs, and practices (Balarajan 1996), instead of focusing merely on the *etic* or outsiders' approach. This is significant given the contested nature of the concepts "ethnicity" and "health" (Anand 1999; Bradby 2003). *Emic* approaches to migrants' health studies represent important turning points in public health discourses aimed at improving interventions devoted to ethnic "minority" groups in Western countries (Trevino 1999; Mackenbach 2006). For instance, it has been shown that immigration to Western metropolitan areas has a significant impact on migrants' experience and meaning of illness, and on their health care-seeking strategies (Papadopoulos et al. 2004; Owusu-Daaku and Smith 2004; Gilgen et al. 2005; Belliard and Ramírez-Johnson 2005).

Research on transcultural health also includes studies conducted on migrants' dietary habits (Jonsson et al. 2002; Burns 2004; Cwiertka 2005), as nutrition and health are closely intertwined (Etkin 2006; Pieroni and Price 2006, and chapters in this volume).

Open Questions

Several scientific questions remain open for investigation in the field of ethnobotany and transcultural health. They include the following:

- Do migrants still depend on their own health care strategies within the domestic domain, including the continued use of food and medicinal plants brought over from their home countries or purchased in local shops in the host society for common, chronic, and/or culturally important health conditions? If so, why is this the case?
- In what ways do migrants' health care-seeking strategies change over time, in response to "internal" dynamics of identity and representation within the migrant community, and to external environmental, cultural, social, and political changes in the host country, including public health care policies?

- What are the existing articulations between migrants' own health care systems and the biomedical system?
- To what extent are institutional health actors in the host country aware of these strategies?

Mobility in Urban Environments

Mobility is a key factor in shaping human history (Sanjek 2003). Over the past several decades, national and transnational migrations have gained momentum so that they now take place on a large scale in an increasingly urbanized and "globalized" world. Plural societies are as much a part of our reality today as the Internet, with some 140 million persons living outside their country of birth. Ethnic diversity is a hallmark of large metropolitan areas such as New York City and London. Furthermore, by 2007, the majority of the world's population will live in urban areas; hence migration is now considered to be a primarily urban phenomenon (Galea et al. 2005).

Migration and cities appear to be intrinsically linked within most immigrants' countries of origin, where people tend to move away from rural areas toward urban centers. Transnational migration is responding to the same tendency, and usually involves settlement in large urban areas upon arrival in the host country. Immigration is also intertwined with health and health care, as many immigrants often lack health insurance coverage, or face linguistic, cultural, or legal barriers to biomedical care. The usual result is in an underutilization of the official, biomedical health care system in the host country. On the other hand, people who cross borders and arrive in new societies do not come empty handed. They bring along their own traditions, lifestyles, world and health views, as well as their own support systems, which include knowledge about plants for health care and nutrition. These attitudes and practices are maintained to varying extents in the host society (e.g., Gomez-Beloz and Chavez 2001, Nguyen 2003).

This volume is a compilation of studies that deal with transnational movements, urban living, health and health care, biocultural adaptation, and past and present interactions between people and plants for health, food, well-being, and identity. Each chapter explores different dimensions that exist between immigrants, health, well-being, disease and illness, plants, and/or identity by drawing examples from different continents at different moments in time, and addressing the dynamics that occur between immigrant groups and host societies.

Strengthening or Adapting Cultural Identities?

Biocultural adaptation, cultural negotiation, and identity are key issues for anthropological discourses on displacement and migrations (Janes and Pawson 1986; Belliard and Ramírez-Johnson 2005). Research in culturally homogenous and/or non-urban environments has shown that to follow the pattern of change in traditional knowledge and use of plants among migrants implies the analysis of acculturation processes (Bodeker et al. 2005; Nesheim et al. 2006). Acculturation has been identified as one of the explanatory concepts that encompass the complexity of migrant adaptation (Janes and Pawson 1986). Acculturation has been discussed in communication sciences as being the result of two simultaneous processes: one involving deculturation from the original culture, and the other enculturation toward the host culture (Kim 2001). The old model of cultural adaptation is quite problematic, since it is highly unlikely that a culture, after moving, simply "adapts" to the new autochthonous culture. There are at least two reasons for this:

- Adaptation does not represent a sort of "destiny". On the contrary, it is only one of many diverse possibilities that migrants have in their interface with the host culture. Adaptation is, in fact, the result of cultural negotiations.
- The host environment is not always culturally homogenous. Indeed, Western metropolitan milieus are never culturally homogenous.

Often migrants may choose other strategies, those aimed at strengthening their cultural identities, for example, in a process that is similar to what other scholars in ethnoecology define as "resilience" at ecological and cultural edges (Turner et al. 2003). Strengthening their own identities in the midst of autochthonous/host populations means that migrants may want deliberately to retain their traditional knowledge and practices, in order to affirm their distinct cultural identity. In other words, they may wish to say, "I am different from you, and I am proud of it!" The strategy of "strengthening identities" and the cultural adaptation model are theoretical opposites; what migrants do in reality probably lies somewhere in between the two. The exact location depends very much on the dynamics in the migrants' changing interface with their host cultural context. Moreover, since ethnicity is the complex result of social processes too (Barth 1969; Baumann 1999), and cultural boundaries are very dynamic (and may even be seen as constructs that are created by our own processes of representation, see Clifford and Marcus 1986; Marcus 1998), plant uses, and especially their representations, may change rapidly in response to continuously shifting cultural negotiations.

The Chapters in This Book

This volume is partially based on papers presented by the panel, "Ethnopharmacy and Migration", at the 9th International Congress of Ethnobiology (organized by the International Society of Ethnobiology at the Department of Anthropology, University of Kent at Canterbury, UK, 14–17 June 2004). It is an attempt to offer an overview of diverse approaches and research questions, all of which focus on "traveling" plants for food and medicine, and transcultural health and diet studies. This book is organized according to immigrant group and continent. The first three chapters look into Latino immigrants in New York City; chapters 4 and 5 give an account of Asians in the United States and Europe; chapters 6 to 9 deal with immigrants in Western Europe, ranging from Surinamese immigrants in Amsterdam, to Latino, Kurdish, and Somali immigrants in London; and the last three chapters focus on plant uses and historical migrations and/or cultural exchanges in Northern and Southern Europe, and Northwestern Africa, respectively.

Latino Immigrants in New York

In the first chapter, Andreana Ososki and coworkers present an overview of ethnobotanical research on medicinal plant knowledge pertaining to women's health conditions among rural and urban communities in the Dominican Republic, and Dominican healers in New York City. The work of Ososki et al. shows that Dominican traditional medicine is not confined within the borders of the Dominican Republic. On the contrary, Dominican healers in New York City have retained and adapted their plant knowledge according to the local prevalence of women's health conditions and the availability of plants. The authors also demonstrate that selected Dominican rural study sites hold a greater amount of general medicinal plant knowledge than urban sites for most of the women's health conditions that were surveyed, with the exception of uterine fibroids (benign tumors that develop within the wall of the uterus or attached to it). The rural-urban disparity in plant knowledge for uterine fibroids in the Dominican Republic is further pronounced by a transnational comparison between the Dominican Republic and New York City. In the latter environment, a higher number of distinct plant species was reported as compared to the urban and rural Dominican Republic study sites. This may be because women in urban settings are more familiar with the diagnosis of uterine fibroids, especially in a metropolis with modern screening facilities like New York City. Furthermore, in New York City, several plant species were substituted. Sometimes, but not always, these plant species were closely related to the species used in the Dominican Republic. This example nicely illustrates how the dynamics of medicinal plant knowledge and use are triggered by migration. The authors call for continued research that addresses cultural variation and change by examining the distribution of intracultural plant knowledge across rural, urban, and transnational landscapes.

Chapter 2 presents the results of a survey conducted by Ina Vandebroek and coworkers with Dominican immigrants in New York City, who self-medicate with medicinal plants (non-specialists or laypeople). Plant remedies reported by Dominican participants as "used in New York City only" (and hence not used in the Dominican Republic, nor used in both countries) were compared with Dominican and Caribbean plant-use data from scientific literature. The results demonstrate that there exists a high correspondence of plant remedies used for flu, common cold, and respiratory disorders between New York City and the Dominican Republic. On the other hand, plants used in New York City to treat high cholesterol and diabetes are not corroborated by the reviewed literature from the Dominican Republic. These results lead to a new research question: Are plant remedies for high cholesterol and diabetes "exclusive" to New York City because of a higher prevalence of these health conditions or better screening facilities, or did former inventories in the Dominican Republic simply not record data on diabetes and cholesterol?

In the third chapter, Anahí Viladrich approaches Latino immigrants' healing practices in New York City from a different angle. She describes New York City *botánicas* (ethnic healing stores annex plant pharmacies that sell religious and ritual items, and/or dried and fresh medicinal plants) as focal points that maintain connections between a dispersed immigrant community of practitioners (healers) and followers (clients) of spiritual, religious, and physical healing. These *botánicas* are culturally grafted entities that resemble *botánicas* in immigrants' home countries. However, the multicultural urban setting in New York City allows immigrant healers to experiment, interchange, disseminate, and gain profit from their knowledge on a larger scale than in their country of origin. At the same time, Viladrich points out that the widening and blending of different spiritual and religious healing practices such as *Santería, Espiritismo,* and Yoruba are compromised by serious limitations on plant availability in the host country. The United States government strictly regulates plant species allowed into the country; hence rituals that originally require 100 to 200 different plants have been adapted so that only 21 herbs are needed. In New York City, healing has shifted on a continuum from informal exchange towards a profit making enterprise. Plants are not only remedies for clients' relief of spiritual, emotional, or physical complaints, but also a concrete means of survival for healers.

Asian Migrants in the United States and Sweden

In Chapter 4, Usha Palaniswamy presents a study that focuses on Indians in the United States and their knowledge, beliefs, and preventive practices in relation to diabetes. Palaniswamy reveals that dietary modifications have occurred within the participating Indian groups, which have different levels of acculturation. Some groups, for example, have reduced their intake of traditional foods that contain plant species with known hypoglycemic potential. Reasons for no longer using these species included a lack of knowledge about these plants' health benefits and non-availability of these plants at neighborhood grocery stores in the United States. Interestingly, neither Indian immigrants who already lived for more than ten years in the United States, nor second-generation immigrants born in the

United States, have abandoned traditional diets completely. Furthermore, the immigrants in this study indicated that they were growing specific plants from India in their home gardens, despite difficulties in procuring plants from India or other sources and/or growing these plants in a temperate climate. Palani-swamy urges that health promotion programs for Indian immigrants incorporate education about the benefits of traditional diets.

In Chapter 5, Pranee Lundberg explores the knowledge, beliefs, and practices of Thai immigrant women in Sweden in relation to medicinal plant use, and identifies the following contexts: (1) Thai immigrant women's belief that certain plants will help users become or remain strong, healthy, and free from disease; (2) medicinal plants can be used in the treatment of common health conditions; and (3) traditional health beliefs and practices such as Thai massage and puer-perium rituals are used in an effort to regain health or avoid ill health. Lundberg found that medicinal plants were administered as condiments and spices, and as teas or health drinks, either in response to illness or for illness prevention. Further-more, the Thai women who were interviewed described how they used plants curatively to treat common illnesses, such as colds, cough, sore throat, and stom-achache, as well as chronic health conditions such as menopause, hyperglycemia, and hypertension. The study also found that Thai immigrant women in Sweden combine biomedical health care with traditional Thai herbal remedies for dis-eases such as cancer. Lundberg's interviewees reported, not only about herbs, but also about *emic* holistic views related to health. Lundberg believes that recording these practices is important for health care providers working with Thai immigrants in Sweden, because the information will enable them to deliver cul-turally appropriate health care.

Surinamese in Amsterdam

In Chapter 6, Tinde van Andel and Charlotte van't Klooster give an account of their urban ethnobotany research on Surinamese immigrants in Amsterdam. Their study shows that a wide variety of Surinamese medicinal plants is offered for sale in Amsterdam, many of which are imported from Surinam where most are har-vested in the wild. Hence, in spite of the fact that in The Netherlands most legal immigrants have health insurance and access to modern health facilities, medicinal plants still play an important role in the well-being of Surinamese im-migrants. The medicinal plants listed during the survey were most frequently used for gynecological problems and rituals for psychosocial ailments. This observation has led the authors to conclude that gynecological and psychosocial ailments must be perceived as culturally important by the Dutch Surinamese community. They point out that it is quite easy to import medicinal plants from Surinam to The Netherlands. No permit is needed for personal use as long as no endangered species are involved. This situation is in sharp contrast with the strict customs regulations described for Colombians in London (chapter 7) and for Latinos in New York (chapters 1 and 2), and might help explain the higher number of local/"exotic" versus common/"globalized" plant species in the urban pharmacopoeias of The Netherlands compared with the United Kingdom and the United States.

Migrants in Urban United Kingdom

In chapter 7, Melissa Ceuterick and colleagues explore the use of home remedies by Latino immigrants in London, and demonstrate that immigrants' actual use of particular plant species for health care and well-being is a practical matter of plant availability in the host country. The United Kingdom's harsh importation laws make it difficult for herbal material from non-European Union countries to enter the country. Ceuterick et al. argue that, as a consequence, Latino immigrants in London restrict their plant use to commonly known and "globalized" species such as garlic, chamomile, mint, eucalyptus, *Aloe vera,* and citrus fruits. The authors also offer practical methodological lessons drawn from their research. For instance, querying participants for "plants to treat disease" yielded very few answers. This may be because in the Latino view of health, self-medication with home remedies is linked to discomfort and the re-establishment of well-being rather than disease. Furthermore, querying participants about "plants" can cause confusion because this term is culturally interpreted as "herbs", hereby ruling out a range of health foods and food medicines.

In chapter 8, Sarah Keeler focuses less on plants and plant uses, and more on local food culture and the articulation of cultural identity. Keeler's study was conducted among the Kurdish community in Hackney, London, and her core question was: "In what ways do the *how, where,* and *with whom* of eating habits contribute to the sense of self and other?" Her paper is based on fourteen months of fieldwork with Kurdish migrants and refugees in a multiethnic London community and looks at food culture from this perspective. In particular, it explores the ways in which food becomes a marker of ethnic difference, and a means of imagining and maintaining identity boundaries within the Kurdish diaspora, and as a consequence of narrating their history and migratory experience, including the changing contours of social relations in the diaspora. Of particular significance for the Kurdish community is the way these processes are played out with regard to their relations with the sizable Turkish community, with whom they coexist in the same inner-city area of London. The study reveals how foodstuffs and cuisine culture become, within these narratives, emblematic of a politico-historical experience in the homeland, in which Kurdish ethnic identity was perceived as being obscured, and aspects of Kurdish culture appropriated by Turkish antagonists, and how this experience is reproduced in local political economic relations in the position of exile. Keeler also looks at how the reclamation of this "stolen" and degraded identity is enacted through foodways that memorialize imaginings of the homeland and assert a continued presence and resistance through embodied practices of food production and consumption. In exploring these questions, the author considers the traditional concepts of use value and exploitation of resources in symbolic rather than material terms, and more specifically, how this exploitation of resources within public space is used as a means of managing expressions of identity. For Kurds embedded within the local ethnic economy of Hackney, and often identified from an external perspective as "Turkish speakers" or even Turks, there is a considerable degree of ambivalence toward the many factors that shape their identification. These factors include a sense of responsibility to and pride in maintaining a distinct ethnic identity; the desire to integrate socially into the local culture; and the very pragmatic economic interests met by the essentializing expressions of ethnicity in the commercial arena. Keeler suggests that it is perhaps because of this ambivalence, and the relative fluidity of ethnic identifications at the local level (particularly within the economy and public space), as familiar social relations are juxtaposed, reconfigured, and rendered unfamiliar in the multicultural metropolis, that foodways become more salient as signifiers of historical experience, ethnic integrity, and purity. In the case of food culture in Hackney's "Little Turkey", the consciousness of ethnic identity and its expression through food is informed by wider economic, political, and social discourses at local and transnational levels.

In chapter 9, Neil Carrier focuses on khat consumption within the Somali diaspora in the United Kingdom. Carrier emphasizes khat's social and cultural importance in the life of Somalis, and also its reception in the West, as well as its glib absorption into debates concerning substances with very different cultures of consumption. The spread of khat and khat consumers to the Western countries has raised major public health concerns for the migrant communities involved. Carrier criticizes the approach the media and diverse stakeholders are taking on this issue and writes of their "demonization" of such practices. He explains that, despite these reactions, khat consumption is probably not going to disappear in the near future, as the use of a plant means much more to an immigrant community than its taste or effect.

Non-urban Historical Migrations in Europe

In chapter 10, Cassandra Quave and Andrea Pieroni offer a cross-cultural comparison of traditional medical practices, botanical remedies, and foods in a small, ethnic Albanian Arbëreshë community and in an autochthonous Italian village in Lucania, southern Italy. Despite the fact that the two communities share similar sociodemographic, economic, and geographic characteristics, they differ considerably in the way they use plants for food and medicine. Not only are the autochthonous Italians familiar with a much larger botanical pharmacopoeia than the Arbëreshë, but they also follow a different medical framework. The authors suggest that the lower number of medicinal plants known to the Arbëreshë is due to their heavy reliance on a ritualistic or spiritually based system of complementary medicine. In the Arbëreshë system, plants serve an underlying role as spiritual objects and are not necessarily associated with being the source of a cure. With regards to wild food botanicals, the Arbëreshë more frequently integrate *liakra,* or wild/weedy greens, into their daily diet. While their cuisine has certainly grown due to an informational flux with neighboring Italian communities, this exchange of culinary knowledge has probably not been a mutual phenomenon, as the authors have not been able to clearly trace Albanian food plants and food plant uses in the surrounding South Italian culinary traditions. The multifaceted picture provided in this study demonstrates the complexity of comparative ethnobiological studies on migratory populations, and how these are influenced by social processes that are very rarely mutual. This study also illustrates how "traveling plants" and their related uses can be more obscure than first impressions would suggest.

In chapter 11, Ingvar Svanberg gives another perspective on historical migration and cultural exchange trajectories in his research on plant knowledge and historical cultural contacts in the Atlantic Fringe. Svanberg's case study focuses on tormentil (*Potentilla erecta* L.), which until the end of the nineteenth century was used as a source of tannins within the household economy of Atlantic islands where trees are lacking or scarce. Svanberg shows that when the local ecology restricted the amount of available tanning plants, as in the case of the Hebrides and the Orkney, Shetland and Faroe Islands, tormentil was found to be a good substitute. Svanberg suggests that the islanders' knowledge of *Potentilla erecta* uses was probably not brought by the Vikings from Norway, but was more likely to have been discovered locally either in the Faroes, Shetland, or Orkney Islands and then spread to the other islands. This has been confirmed using ethnolinguistic analysis and clearly demonstrates how plant knowledge has been exchanged among populations since ancient times, and how the cultural history of plant uses—along with archaeobotany and the history of Materia Medica/pharmacognosy—could represent interesting arenas for studying the dynamics of ethnobiological knowledge in human societies that are fortunate enough to retain a relevant heritage of historical texts or written sources.

Displacements in Northwestern Africa

In Chapter 12, Gabriele Volpato and colleagues focus on the procurement of traditional remedies and the transmission of medicinal plant knowledge among Sahrawi people displaced in southwestern Algerian refugee camps. The authors suggest that Sahrawi refugees have preserved the use and knowledge of traditional medicinal remedies in the camps, and have established a variety of networks in order to obtain these remedies. The authors have recorded mostly wild plants gathered in the part of western Sahara controlled by the *Polisario,* the political and military organization that represents the refugees. They found that soldiers stationed there play an important role in the procurement of these remedies for refugees' families. The detailed study by Volpato et al. shows how the conservation of traditional medicine in this kind of context represents a means of maintaining cultural identity, and that the procurement of remedies from the "liberated territories" in western Sahara allows refugees to maintain ties with their place of origin. The conservation of traditional knowledge and practices also represents resistance to acculturation and a way of alleviating despair. Many refugees report feeling that their lives have been wasted. When they emigrate, they abandoned hope of ever returning to western Sahara. This, and the influence of different host cultures on the Sahrawi who have studied abroad, inevitably leads to younger generations losing out on traditional knowledge and breaking ties with western Sahara.

Concluding Remarks

The chapters in this volume bring to our attention that there exists a challenging dynamism in what migrants consciously or subconsciously choose to retain, abandon, and/or acquire while being exposed to host societies. For instance, migrants may retain the use of particular plant species, health beliefs, or traditional practices in host societies because of their perceived shared cultural importance or because their continued use reinforces migrants' cultural identity. In the latter case, these species, beliefs, or practices become salient markers of *identity* that express "tell me what you eat/consume, or what you use to cure yourself, and I will tell you who you are." On the other hand, the plant uses or practices that people abandon or newly acquire after they migrate may well become a marker of the immigrant *experience.* Shifting health conditions and changing folk pharmacopoeias can inform us about the new realities migrants are confronted with. Health conditions that were prevalent in immigrants' area of origin may become replaced with (a set of) different health conditions that are very much a part of the reality of life in the host society. This dynamic interaction between migrants and host societies may result in shifting pharmacopoeias adapted with substituted plant species. Likewise, changes in pharmacopoeias can be triggered by the availability of particular plant species in the host society, or by contact with different kinds of (traditional) healing systems that coexist within multicultural urban spaces. Further studies may wish to address these themes more in detail, either by looking at the knowledge, beliefs, and practices that have been retained or abandoned by migrants, or alternatively, by focusing on those that are newly acquired. Continued research is also needed on the underlying reasons for these changes. In addition, by comparing healing pharmacopoeias between immigrant groups from diverse cultural backgrounds in different host societies, it may become clear if and to what extent globalization has had an impact on these pharmacopoeias by substituting the original "exotic" plant species by common "globalized" species.

Notes

1. Often the country of origin represents a tropical, or subtropical, developing country and the host country an industrialized country.

References

Anand, S.S. 1999. "Using ethnicity as a classification variable in health research: perpetuating the myth of biological determinism, serving sociopolitical agendas, or making valuable contributions to medical sciences?" *Ethnicity & Health* 4: 241–244.

Balarajan, R. 1996. "Editorial." *Ethnicity & Health* 1: 3–5.

Balick, M.J., F. Kronenberg, A.L. Ososki, M. Reif, A. Fugh-Berman, B. O'Connor, M. Roble, P. Lohr, and D. Atha. 2000. "Medicinal plants used by Latino healers for women's health conditions in New York City." *Economic Botany* 54: 344–357.

Barth, F. 1969. *Ethnic groups and boundaries. The social organization of culture difference.* Long Grove (Illinois): Waveland Press.

Baumann, G. 1999. *The multicultural riddle: Re-thinking national, ethnic and religious identities.* London: Routledge.

Belliard, J.C., and J. Ramírez-Johnson. 2005. "Medical pluralism in the life of a Mexican immigrant woman." *Hispanic Journal of Behavioral Sciences* 27: 267–285.

Bodeker, G., C. Neumann, P. Lall, and Z. Min Oo. 2005. "Traditional medicine use and health worker training in a refugee setting at the Thai-Burma border." *Journal of Refugee Studies* 18: 76–99.

Bradby, H. 2003. "Describing ethnicity in health research." *Ethnicity & Health* 8: 5–13.

Burns, C. 2004. "Effect of migration on food habits of Somali women living as refugees in Australia." *Ecology of Food and Nutrition* 43: 213–229.

Clifford, J., and J. Marcus, eds. 1986. *Writing culture: the poetic and politics of ethnography.* Berkeley: University of California Press.

Corlett, J.L., E.A. Dean, and L.E. Grivetti. 2003. "Hmong gardens: botanical diversity in an urban setting." *Economic Botany* 57: 365–379.

Cwiertka, K.J. 2005. "From ethnic to hip: circuits of Japanese cuisine in Europe." *Food & Foodways* 13: 241–272.

Etkin, N.L. 2006. *Edible medicines. An ethnopharmacology of food.* Tucson: University of Arizona Press.

Galea, S., N. Freudenberg, and D. Vlahov. 2005. "Cities and population health." *Social Science & Medicine* 60: 1017–1033.

Gilgen, D., D. Maeusezahl, C. Salis-Gross, E. Battegay, P. Flubacher, M. Tanner, M.G. Weiss, and C. Hatz. 2005. "Impact of migration on illness experience and help-seeking strategies of patients from Turkey and Bosnia in primary health care in Basel." *Health & Place* 11: 261–273.

Gomez-Beloz, A., and N. Chavez. 2001. "The botánica as a culturally appropriate health care option for Latinos." *Journal of Alternative and Complementary Medicine* 7: 537–546.

Janes, C.R., and I.G. Pawson. 1986. "Migration and biocultural adaptation: Samoans in California." *Social Science & Medicine* 22: 821–834.

Johnson, L., H. Strich, A. Taylor, B. Timmermann, D. Malone, N. Teufel-Shone, R. Drummon, R. Woosley, E. Pereira, and A. Martinez. 2006. "Use of herbal remedies by diabetic Hispanic women in the Southwestern United States." *Phytotherapy Research* 20: 250–255.

Jonsson, I.M., H.R.M. Hallberg, and I.B. Gustafsson. 2002. "Choice of food and food traditions in prewar Bosnia-Herzegovina: focus group interviews with immigrant women in Sweden." *Ethnicity & Health.* 7: 149–161.

Kim, Y. 2001. *Becoming intercultural. An integrative theory of communication and cross cultural adaptation.* Thousand Oaks, California: Sage.

Mackenbach, J.P. 2006. "Migrant health: new challenges for European public health research (Editorial)." *European Journal of Public Health* 16: 345.

Marcus, G. 1998. *Ethnography through thick and thin.* Princeton: Princeton University Press.

Maffi, L., ed. 2001. *On Bio-Cultural Diversity. Linking Language, Knowledge and the Environment.* Washington D.C.: Smithsonian Institution Press.

Nesheim, I., S.S. Dhillion, and K.A. Stolen. 2006. "What happens to traditional knowledge and use of natural resources when people migrate?" *Human Ecology* 34: 99–131.

Nguyen, M.L.T. 2003. "Comparison of food plant knowledge between urban Vietnamese living in Vietnam and in Hawai'i." *Economic Botany* 57: 472–480.

Owusu-Daaku, F.T.K. and F.J. Smith. 2004. "Health-seeking behavior: perspective of Ghanaian women in London and Kumasi." *International Journal of Pharmacy Practice* 13: 71–76.

Papadopoulos, I., S. Lees, M. Lay, and A. Gebrehiwot. 2004. "Ethiopian refugees in the United Kingdom: migration, adaptation, and settlement experiences and their relevance to health." *Ethnicity & Health* 1: 55–73.

Pieroni, A., H. Münz, M. Akbulut, K.H.C. Baser, and C. Durmuskahya. 2005. "Traditional phytotherapy and transcultural pharmacy among Turkish immigrants living in Cologne, Germany." *Journal of Ethnopharmacology* 102: 69–88.

Pieroni, A., and L.L. Price, eds. 2006. *Eating and healing. Traditional food as medicine.* Binghamton, New York: The Haworth Press.

Sandhu, D.S., and M. Heinrich. 2005. "The use of health foods, spices, and other botanicals in the Sikh community in London." *Phytotherapy Research* 19: 633–642.

Sanjek, R. 2003. "Rethinking migration, ancient to future." *Global Networks* 3: 315–336.

Turner, N., I.J. Davidson-Hunt, and M. O'Flaherty. 2003. "Living on the edge: ecological and cultural edges as sources of diversity for social-ecological resilience." *Human Ecology* 31: 439–461.

Trevino, F.M. 1999. "Quality of health care for ethnic/racial minority populations." *Ethnicity & Health* 4: 153–164.

Voeks, R.A. 2004. "Disturbance pharmacopoeias: medicine and myth from the humid tropics." *Annals of the Association of American Geographers* 94: 868–888.

Waldstein, A. 2006. "Mexican migrant ethnopharmacology: pharmacopoeia, classification of medicines, and explanations of efficacy." *Journal of Ethnopharmalogy* 108: 299–310.

Chapter 1

Medicinal Plants and Cultural Variation across Dominican Rural, Urban, and Transnational Landscapes

Andreana L. Ososki, Michael J. Balick,
and Douglas C. Daly

Introduction

Ethnobotanical knowledge evolves as it is exchanged, transferred, and appropriated by people adapting to new surroundings and changing environments (Lee et al. 2001; Voeks and Leony 2004). As people migrate between rural and urban environments, they exchange knowledge, cultural traditions, and medicinal plants. Fixed borders do not exist between rural, urban, and transnational groups, nor do they exist between laypeople and healers, as information is shared through various channels.

Medicinal plant knowledge is important for health care initiatives (Bodeker 1995; Bodeker and Kronenberg 2002) and conservation efforts (King 1996; Balick et al. 2002), yet our understanding of the distribution of plant knowledge within a community or across multiple communities is in its infancy (Campos and Ehringhaus 2003). Ethnobotanical studies tend to understate the variation of plant knowledge by reporting plant use information as homogeneous throughout a cultural group. Ethnobotanical knowledge is diverse and can differ markedly from one individual to another, as well as from one community to another, making it difficult to analyze. Studying cultural variation of ethnomedical knowledge gives insight into the distribution and transmission of plant use information in a community. These studies can help demonstrate the complexities and dynamics of medicinal plant knowledge and offer insight into cultural change.

In addition, a description of these patterns provides a benchmark for future studies to evaluate erosion or acquisition of knowledge over time.

Dominican traditional medicine is not limited to the borders of the Dominican Republic, as New York City has a growing Dominican community that continues to maintain traditional healing practices (Balick et al. 2000; Reiff et al. 2003). When Dominicans immigrate to New York City, they continue to use traditional medical practices. Twenty-four percent of Dominicans surveyed in New York City emergency rooms in 1997 reported using complementary and alternative medicine (CAM) in the form of home remedies or traditional medicine for their presenting complaint (Allen et al. 2000).

In this chapter, we examine the differences and similarities of medicinal plant knowledge in the Dominican Republic and among Dominicans in New York City. Using ethnographic and ethnobotanical fieldwork, we highlight the richness of plant species and herbal therapies used specifically for women's health, the diversity of this knowledge across rural, urban, and transnational landscapes, and the processes affecting cultural variation.

Study Sites

Research was conducted in two provinces in the Dominican Republic and in New York City, the latter as part of an on-going project known as the Urban Ethnobotany Project, which began in 1996. Fieldwork in the Dominican Republic was conducted from September 2000 to August 2001.

New York City

New York City has the second largest Hispanic population and the largest Dominican population (an estimated 424,847 people) of any US city (US Census Bureau 2000). Dominican healers involved in the Urban Ethnobotany Project resided in Washington Heights, Inwood, and the Bronx.

Dominican Republic

The Dominican Republic (48,225 km^2) occupies two-thirds of the island of Hispaniola, with Haiti occupying the remaining one-third. The population of the Dominican Republic is estimated to be slightly over 9 million (US Census Bureau 2005) with 70 percent of the population residing in urban centers. The ethnic origins of the Dominican population are 73 percent mulattos (defined as persons of mixed European and African ancestry), 16 percent of European descent, and 11 percent of African descent (Bolay 1997). The native language is Spanish and is blended with indigenous and African elements specific to the Dominican Republic (Cambeira 1997).

The study sites in the Dominican Republic were located in the provinces of La Vega and San Cristóbal (figure 1.1). Both a rural community and an urban community were selected in each province. La Vega is located in the central region of the country known as the Cibao, and San Cristóbal is located southwest of the capital, Santo Domingo. The study sites located in La Vega were Los Calabazos, a rural community, and the city of La Vega. Within the city, Proyecto Habitacional de San Miguel was selected as the study area. The study sites in the province of San Cristóbal were La Colonia, a rural community, and Proyecto Hacienda Fundación, a section of the city of San Cristóbal. We will refer to the urban sites as San Cristóbal and La Vega. The study sites were selected based on the following criteria: presence of a women's organization, rural and urban sites at comparable distances in both provinces, and agreement of the community to participate in the study.

Figure 1.1. *Map of the Dominican Republic showing the four study sites: Los Calabazos, La Vega city (San Miguel), La Colonia, and San Cristóbal city (Hacienda Fundación)*

Research Methods

Medicinal Plants and Dominicans in New York City

The Urban Ethnobotany Project has been working with Latino healers in New York City since 1996 to learn about their traditional healing practices for women's health conditions (Balick et al. 2000; Reiff et al. 2003; Fugh-Berman et al. 2004). The project has focused on the following health conditions: endometriosis (growth of endometrial tissue outside of the uterus); hot flashes (sudden brief sensations of heat often experienced during menopause); menorrhagia (excessive uterine bleeding); and uterine fibroids (benign connective tissue tumors). Most recently, this study has focused on herbal therapies for uterine fibroids and hot flashes.

Data collected from Dominican healers in New York City were based on multiple patient-healer consultations, with different patients previously diagnosed by a medical physician with the aforementioned health conditions. Following the consultations, both healers and patients were interviewed. Interviews with the healers provided additional insight into healing beliefs and practices (Reiff et al. 2003) and a more thorough understanding of medicinal plant recommendations for the patients (Balick et al. 2000). Accompanied by Dominican healers, we collected plant samples, which was a valuable and necessary component of these interviews. Fresh and dried plant samples were collected at local *botánicas* in New York City to verify their scientific names and to better understand how plants are selected, which parts are used, and how the plants are prepared into medicines.

Botánicas are shops that sell traditional remedies and function as herbal pharmacies providing fresh and dried herbs, mixtures, and tinctures, as well as religious and ritual items such as candles, oils, figurines of saints, and holy water (*agua bendita*) (Fisch 1968; Borello and Mathias 1977; Delgado and Santiago 1998). Voucher specimens, often plant fragments from *botánicas,* were prepared of all of the plants reported during the consultations and are stored at the herbarium of the New York Botanical Garden. In addition to traditional healers, shopkeepers and assistants, who handled the medicinal plants at *botánicas,* shared herbal therapies commonly prepared with Dominican plants.

Based on these initial interviews and plant collecting trips in New York City, we became familiar with common medicinal plants used and sold by Dominicans there. In addition, we conducted a comparative literature review to gather baseline data about Dominican plants used for women's health conditions, which further provided us with a working knowledge of Dominican plants, their uses, and their common names (Ososki et al. 2002). Through our literature review we found only limited discussion and minimal details provided on women's health categories suggesting the need for further investigation in this area, which led us to conduct research in the Dominican Republic and expand the number of health conditions studied to include menstruation, pregnancy, and menopause.

Medicinal Plants and Dominicans in the Dominican Republic

In the Dominican Republic, data were collected from local adult women (generalists) and healers (specialists), using a survey and an interview format. Before conducting the fieldwork, data collection instruments were reviewed and approved by the City University of New York Graduate School Institutional Review Board (IRB # 6-10-06-01). Several measures of ethnobotanical knowledge were evaluated: the number of plant species recorded per interview, the percentage of individuals who knew a remedy for a health condition, and the number of field reports for specific medicinal plants and remedies.

The sampling methods used in the Dominican Republic were structured differently than those used in New York City. The data collected from Dominican healers in New York City were obtained via multiple consultations with different patients, while that collected with Dominican healers in the Dominican Republic involved querying healers about treatments they use for a list of health conditions and did not involve patients. Six women healers participated in the study in New York City. Eleven healers, seven women and four men, were interviewed in the Dominican Republic. Because the sample sizes and data collection methods differ, comparisons can only provide a rough estimate. However, this comparison can give a basis upon which to build further studies to address questions of transmission and acquisition of plant knowledge in different environments.

The survey elicited both qualitative and quantitative information about the diversity of herbal remedies and medicinal plant species known and used for ten women's health conditions. The initial questions generated sociodemographic data such as age, birthplace, length of residence, civil status, number of children, educational level, religious affiliation, affiliation in community organizations, and occupation. A short household inventory estimated socioeconomic background.

The bulk of the survey was about medicinal plant knowledge. Each participant was asked to free-list as many medicinal plants that they knew and used. Then each participant was asked to free-list all the plants or remedies they knew for each of the women's health conditions. As needed during the survey, symptoms were elicited and discussed to further clarify the women's health conditions selected. For those plants reported for the ten health conditions, we asked about plant part used, remedy preparation, form of administration, how it treated the health condition, if it had been used by the person interviewed, how they learned about its use, any precautions when using the remedy, and other uses. Several questions on the survey elicited names of healers, midwives, or other specialists knowledgeable about medicinal plants. Additional questions concerned plant collection and health care data. Questions for this survey were developed from previous ethnobotanical studies (Brett 1994; Alexiades 1996; Balick et al. 2000), health surveys (Aday 1989), and suggestions from our Dominican collaborators. To ensure culturally appropriate survey and interview questions and relevant women's health conditions, we depended on our fieldwork in New York City with Dominican healers and piloted the questions with our collaborators in the Dominican Republic.

Semi-structured interviews were used with healers to collect data about women's health conditions, including descriptions, symptoms, causes, and treatments. Data was also collected about plant collection, preparation, and administration. Interviews took place in healers' homes and were usually followed by a walk to collect plants. Voucher specimens were made of medicinal plants reported during the study and are housed at the herbaria of New York (NYBG) and the Jardín Botánico de Santo Domingo (National Botanical Garden, Santo Domingo) in the Dominican Republic (JBSD). All the interviews were recorded on audiotape except for one healer who did not want to be taped. Informed consent was obtained prior to beginning the interview or survey.

Dominican Women's Health Conditions

To narrow the focus of women's medicine, ten health conditions were selected (table 1.1). The conditions were selected at the initial stages of the field research prior to conducting the surveys and interviews. The term "condition" is used rather than illness because in many cases the conditions are part of a woman's life cycle and therefore are not considered illnesses. The survey focused on ten conditions, but allowed those interviewed to discuss a broad range of plant species and other home remedies. The conditions were chosen based on initial conversations with Dominican women about common health ailments (vaginal infections, menstrual cramps, postpartum care), research conducted in New York City with Dominican healers (Balick et al. 2000), and a literature review of Dominican ethnomedical studies which included infertility, suspended menstruation, and morning sickness (Ososki et al. 2002). Hot flashes, menorrhagia (excessive menstruation), and uterine fibroids were included for comparison with the study in New York City. Plants were not reported for treatment of endometriosis in the New York City study, so it was not included in the Dominican Republic study.

Table 1.1. Women's Health Conditions Selected in the Dominican Republic

English	Spanish
Menstrual cramps	*Dolores menstruales, calambres*
Excessive menstruation	*Derrame de la mujer*
Suspended menstruation	*No llega la menstruación*
Pregnancy prevention	*Prevenir el embarazo*
Morning sickness	*Mal estar de la barriga durante el embarazo*
Postpartum care	*En riesgo después de dar luz*
Infertility	*Infertilidad*
Menopausal hot flashes	*Calores del cambio, menopausia*
Uterine fibroids	*Fibromas*
Vaginal infections	*Infecciones vaginales, infecciones de la mujer*

Pregnancy prevention was also selected to explore preventative family planning measures in Dominican traditional medicine. From a biomedical viewpoint these conditions represent a broad range of health conditions: postpartum care and morning sickness can be classified as obstetric care, while menstruation (excessive, suspended, and cramps), infertility, vaginal infections, hot flashes, uterine fibroids, and pregnancy prevention fall under gynecological care (Venes and Thomas 2001).

Results

To understand the cultural variation of medicinal plant knowledge in rural and urban Dominican Republic and New York City, we examined several measures of ethnomedical knowledge: the plant species and the number of field reports of these species in rural and urban communities, the percentage of rural and urban women who knew a remedy for a health condition, and the number of plant species known in New York City versus the Dominican Republic for selected health conditions. A total of 226 surveys were administered to Dominican women (Los Calabazos (n = 33), La Colonia (n = 34), La Vega (n = 87), San Cristóbal (n = 72)). Healers were not included in these analyses of urban and rural communities because the number of healers per study site was significantly different (G-test, p < 0.0005) and the mean number of distinct plant species known by healers (29.27 ± SE 2.71) was significantly greater than the mean for women (6.74 ± SE 0.42).

Of the 226 women surveyed, 200 (88 percent) named a plant or remedy for one of the ten health conditions. A total of 2,148 field reports of plants were cited by these women, including 187 plant species from 70 different plant families. Plants reported in the surveys that were not collected or verified with a photo were not included in these calculations. The total number of field reports for plant species reported by women in Los Calabazos and La Colonia was 589 and 459, respectively. The total number of field reports for La Vega was 673 and 427 for San Cristóbal. The combined value for rural communities was 1,048 and for urban communities 1,100.

Comparing Medicinal Knowledge across Rural and Urban Landscapes in the Dominican Republic

PLANT SPECIES KNOWN IN RURAL AND URBAN COMMUNITIES

A total of 205 plant species were known by women and healers for the ten women's health conditions in the Dominican Republic. During surveys and interviews, other health conditions were discussed informally and recorded, but they are not included in this analysis. This methodology allowed us to investigate plant species that were mentioned frequently for particular health conditions and those species mentioned commonly in each community.

Several plant species were known in common between rural and urban communities, while other species were mentioned in only one community. A total of thirty-three plant species were known in common in all four communities (18 percent, n = 187). Thirty-nine plant species (21 percent) were known only in Los Calabazos, twenty-two (12 percent) were cited only in La Colonia, nineteen (10 percent) were specific to La Vega, and fourteen (7 percent) were cited only in San Cristóbal. The plant species most frequently mentioned by women were *Matricaria recutita* L. and *Kalanchoe gastonis-bonnieri* Raym.-Hamet & H. Perrier, with 103 and 95 field reports, respectively. Table 1.2 lists all plants in the text and includes family and common names.

Table 1.2. Medicinal Plant Species Reported in Text

Species [Family] {Voucher}[a]	Vernacular Name in Spanish and English* (*between brackets)
Adiantum tenerum Sw. [Pteridaceae] {452}	*cilantrico de pozo, (fan maidenhair)*
Agave antillarum Descourt. [Agavaceae] {57, 62, 241, 378}	*maguey, maguey verde, maguey blanco, (agave, green agave, white agave)*
Ambrosia artemisiifolia L. [Asteraceae] {338, 371, 419}	*altamisa, artemisa, (ragweed)*
Apium graveolens L. [Apiaceae] {SR9}	*apio, (celery)*
Averrhoa carambola L. [Oxalidaceae] {SR12}	*carambola, (starfruit)*
Beta vulgaris L. [Chenopodiaceae] {95, PV168}	*remolacha, (beet)*
Capsicum annuum L. [Solanaceae] {PV164}	*ají, ají dulce (sweet pepper)*
Centrosema pubescens Benth. [Fabaceae] {236}	*diverta caminante, (butterfly pea)*
Chamaemelum nobile (L.) All. [Asteraceae] {5, 30, 32, 65}	*manzanilla, (chamomile)*
Cinnamomum verum J. Presl [Lauraceae] {R133}	*canela, (cinnamon)*
Citrus sp. [Rutaceae] {79, 99, 116}	*naranja, limón, (orange, lime)*
Citrus aurantifolia (Christm.) Swingle [Rutaceae] {440, 471}	*limón, limón agrio, (lime)*
Citrus aurantium L. [Rutaceae] {503}	*naranja agria, (sour orange, bitter orange)*
Citrus sinensis Osbeck [Rutaceae] {213}	*china, naranja, (orange)*
Coffea arabica L. [Rubiaceae] {PV160}	*café, (coffee)*
Dioscorea alata L. [Dioscoreaceae] {SR8}	*ñame, (yam)*
Genipa americana L. [Rubiaceae] {36, 464, R142}	*jagua*
Glycine max (L.) Merr. [Fabaceae] {SR4}	*soya, (soy)*
Hamelia patens Jacq. [Rubiaceae] {417}	*buzunuco, (scarletbush)*

(continued)

Table 1.2. Continued

Helichrysum italicum (Roth) G. Don f. [Asteraceae] {85}	*siempre fresca, (curry plant)*
Illicium verum Hook.f. [Illiciaceae] {R74}	*anís de estrella, (star anise)*
Kalanchoe gastonis-bonnieri Raym.-Hamet & H. Perrier [Crassulaceae] {45, 363, 376}	*mala madre, (palm beachbells)*
Lavandula angustifolia Mill. [Lamiaceae] {R75}	*algucema, alhucema, (lavender)*
Linum usitatissimum L. [Linaceae] {R164}	*lino, (flax)*
Matricaria recutita L. [Asteraceae] {R76}	*manzanilla, (chamomile)*
Momordica charantia L. [Cucurbitaceae] {54, 250, 448}	*cundeamor, sorosí, (bitter melon, balsam pear)*
Opuntia ficus-indica (L.) Mill [Cactaceae] {46, 257, 274}	*alquitira, tuna, tuna de España, (Indian fig, prickly pear)*
Peperomia pellucida (L.) Kunth [Piperaceae] {336}	*siempre fresca (man to man)*
Petiveria alliacea L. [Phytolaccaceae] {89, 90, 136, 252}	*anamú, (guinea henweed)*
Petroselinum crispum (Mill.) Nyman ex A. W. Hill [Apiaceae] {92, PV165}	*perejíl, (parsley)*
Phoenix dactylifera L. [Arecaceae] {113}	*palma, (date palm)*
Pimenta haitiensis (Urb.) Landrum [Myrtaceae] {R89}	*canelilla*
Pisonia aculeata L. [Nyctaginaceae] {434, 500}	*uña de gato, (pullback)*
Plantago major L. [Plantaginaceae] {273, 436}	*llantén, (common plantain)*
Rosmarinus officinalis L. [Lamiaceae] {86, 287}	*romero, (rosemary)*
Roystonea hispaniolana L. H. Bailey [Arecaceae] {540}	*palma, (Hispaniolan royal palm)*
Ruta chalepensis L. [Rutaceae] {37, 339, 407}	*ruda, (rue)*
Saccharum officinarum L. [Poaceae] {42, 321, 450}	*caña, melaza, (sugarcane, molasses)*
Spermacoce assurgens Ruiz & Pav. [Rubiaceae] {128, 142, 178, 390}	*juana la blanca, (woodland false buttonweed)*
Tilia mandshurica Rupr. & Maxim. [Tiliaceae] {34, R108}	*flor de tilo, (linden)*
Uncaria tomentosa (Willd. ex Roem. & Schult.) DC. [Rubiaceae] {81}	*uña de gato, (cat's claw)*
Vaccinium macrocarpon Aiton [Ericaceae] {100}	*(cranberry)*
Zingiber cassumunar Roxb. [Zingiberaceae] {R143}	*jengibre amargo, (Cassumunar ginger)*
Zingiber zerumbet (L.) Sm. [Zingiberaceae] {56}	*jengibre amargo, (bitter ginger)*

[a]All numbers are A. Ososki collections

The six plant species most frequently cited by women in the rural communities were *Agave antillarum* Descourt., *Momordica charantia* L., *Spermacoce assurgens* Ruiz & Pav., *Coffea arabica* L., *Ambrosia artemisiifolia* L., and *Opuntia ficus-indica* (L.) Mill. Of these plants, *Agave antillarum, Momordica charantia,* and *Ambrosia artemisiifolia* were mentioned more frequently in La Colonia, while *Coffea arabica* and *Opuntia ficus-indica* were cited more frequently in Los Calabazos. A similar number of field reports of *Spermacoce assurgens* were mentioned in both rural communities. Those plants cited in rural communities are common in disturbed habitats (*Momordica charantia* and *Spermacoce assurgens*) or cultivated in *patios* (outside area surrounding a home), *conucos* (cultivated plots) or gardens (*Ambrosia artemisiifolia, Coffea arabica,* and *Opuntia ficus-indica*).

In the urban communities, the six species most frequently reported by women were *Matricaria recutita, Kalanchoe gastonis-bonnieri, Saccharum officinarum* L., *Beta vulgaris* L., *Cinnamomum verum* J. Presl, and *Opuntia ficus-indica.* Of these species, *Matricaria recutita, Saccharum officinarum, Beta vulgaris,* and *Cinnamomum verum* are usually purchased at supermarkets, pharmacies, or markets, while *Kalanchoe gastonis-bonnieri* and *Opuntia ficus-indica* are grown in home gardens. *Opuntia ficus-indica* is mentioned frequently in both rural and urban communities. The most frequently reported plants in rural and urban sites are different species, except for *Opuntia ficus-indica;* however, all ten plants were known in all four communities except for *Momordica charantia,* which was not reported in Los Calabazos for women's health conditions, although it was observed growing in this community.

Although some women in urban communities had front or backyard gardens, they tended to report more cultivated species found in supermarkets such as *Apium graveolens* L., *Capsicum annuum* L., and *Linum usitatissimum* L. They also reported *Tradescantia spathacea* Sw., a common ornamental, and *Pimenta haitiensis* (Urb.) Landrum, an allspice endemic to Hispaniola and sold in municipal markets. This is not surprising considering an urban environment has less access to wild plant resources. Although some urban women visited relatives in the countryside to collect medicinal plants, many resorted to plants found in their urban environment or at markets. Rural women were able to collect their plants locally, except for exotic plants such as *Illicium verum* Hook.f., and *Lavandula angustifolia* Mill., which are not in cultivation in the Dominican Republic and need to be purchased at pharmacies. A flip side of this observation is that rural women have less access to plants sold in supermarkets and municipal markets because of the distance of these markets from rural communities. Common market plants such as *Capsicum annuum* and *Averrhoa carambola* L. were not reported for any of the women's conditions in the rural communities. Surprisingly, these plants were not observed in rural home gardens. *Adiantum tenerum* Sw. and *Hamelia patens* Jacq. were only reported in La Colonia and were found commonly along trails or roadsides. *Hamelia patens* was observed growing in Los Calabazos, but it was not reported for women's health. The same was true of *Centrosema pubescens* Benth., which was observed in both rural communities, but reported for women's medicine only in Los Calabazos.

A few plants mentioned in the urban sites merit further discussion. These are *ñame* (yam, *Dioscorea alata* L.), *soya* (soy, *Glycine max* (L.) Merr.), and *lino* (flax, *Linum usitatissimum*). Field reports of these plants are interesting because of their popularity for women's health in the United States and other countries. Soy and flax in particular have been reputed to be sources of phytoestrogens (Adlercreutz and Mazur 1997; Setchell 1998). During the Dominican fieldwork, two women mentioned the use of flax in a plant mixture for vaginal infections. One of them also cited the same flax mixture for infertility. This is not surprising, because the majority of Dominican women interviewed mentioned that infertility is caused by an infection. Two other women mentioned soy for hot flashes; both of them had learned about this remedy from reading magazines. Of these women, one also reported yam for hot flashes. She had also learned about this remedy from a magazine article. Neither healers nor rural women reported using these plants. This illustrates a relatively unheralded way in which new knowledge about medicinal plants is acquired, suggesting the impact of global communication such as the Internet, e-mail, magazines, and television on medicinal plant knowledge. Exchange of herbal therapies is not limited to talking with a relative or neighbor but can occur through various channels and may transcend national borders. In the future, more women in these communities might report soy and flax as important plants for women's health.

A number of medicinal plant species are used in common by both rural and urban women, but each community also uses plants that were not reported in the other communities. Proximity and availability appear to be contributing factors in plant choice, as well as preferences of a person's family. As mentioned above, we observed some of the same plant species (*Centrosema pubescens* and *Hamelia patens*) growing in both rural communities, but they were only reported as medicinal plants in one of the communities. This example illustrates that a person's family or relatives may be a stronger predictor of medicinal plant preference as compared to plant availability and access.

NUMBER OF PLANT SPECIES KNOWN IN RURAL AND URBAN COMMUNITIES

For further comparison, the four study sites were collapsed into two categories of rural and urban. Of 187 plant species from 70 botanical families, there were 82 plant species (44 percent) that were known by both groups (figure 1.2.A). Women in rural communities cited an additional 70 plant species (37 percent) that were not known in the urban sites, while the women living in the urban communities knew an additional 35 plant species (19 percent) that were not known in the rural sites. Of the total plant species, 152 or 81 percent were known in the rural communities and 117 species or 63 percent were known in the urban communities. There is a significant difference between the total number of plant species known by rural and urban women (G-test, $p = 0.012$). Figure 1.2.B shows the number of plant families that were specific to the rural and urban sites and those families that overlapped.

A   B

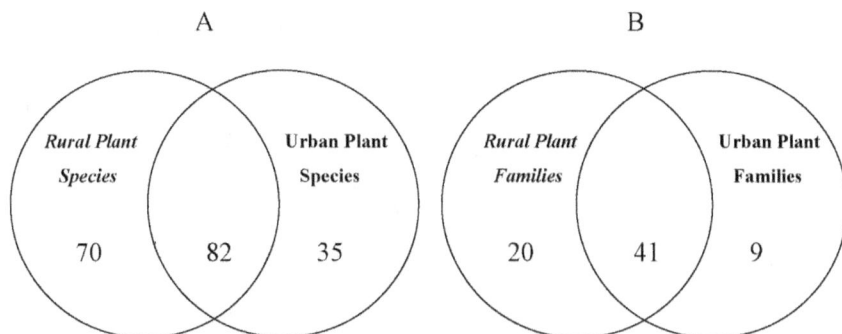

Figure 1.2. *Venn diagrams illustrating the number of medicinal plant species (A) and botanical families (B) reported by women in rural (n = 67) and urban study sites (n = 159) in the Dominican Republic for ten women's health conditions*

For further analysis, the mean number of plant species known per study participant was compared for rural and urban sites. The women in the rural sites reported a greater number of plant species ($9.97 \pm SE\ 1.08$) than in the urban sites ($5.38 \pm SE\ 0.34$). An analysis of variance test confirmed a significant difference in the means of the two groups (F-test = 27.73; $p < 0.0001$). Both rural and urban women knew many of the same plant species, yet a greater number of plant species were known in rural communities than in urban communities.

PERCENTAGE OF RURAL AND URBAN WOMEN
WHO KNEW A REMEDY FOR A HEALTH CONDITION

Figure 1.3 illustrates the percentage of women who knew a remedy for the ten women's health conditions. We used the G-test to compare the percentage of women who reported a remedy in each community for each health condition.

The greatest percentage (50 percent or more) of women in all four communities knew remedies for postpartum care and vaginal infections. These conditions appear to be salient women's health conditions. Every woman interviewed in Los Calabazos knew a remedy for postpartum care, while 82 percent of the women in La Colonia knew a remedy. A greater percentage of women in Los Calabazos knew a remedy for postpartum care than the other three communities (G-test, $p < 0.0001$). San Cristóbal had the lowest percentage of women name a remedy for postpartum care (61 percent).

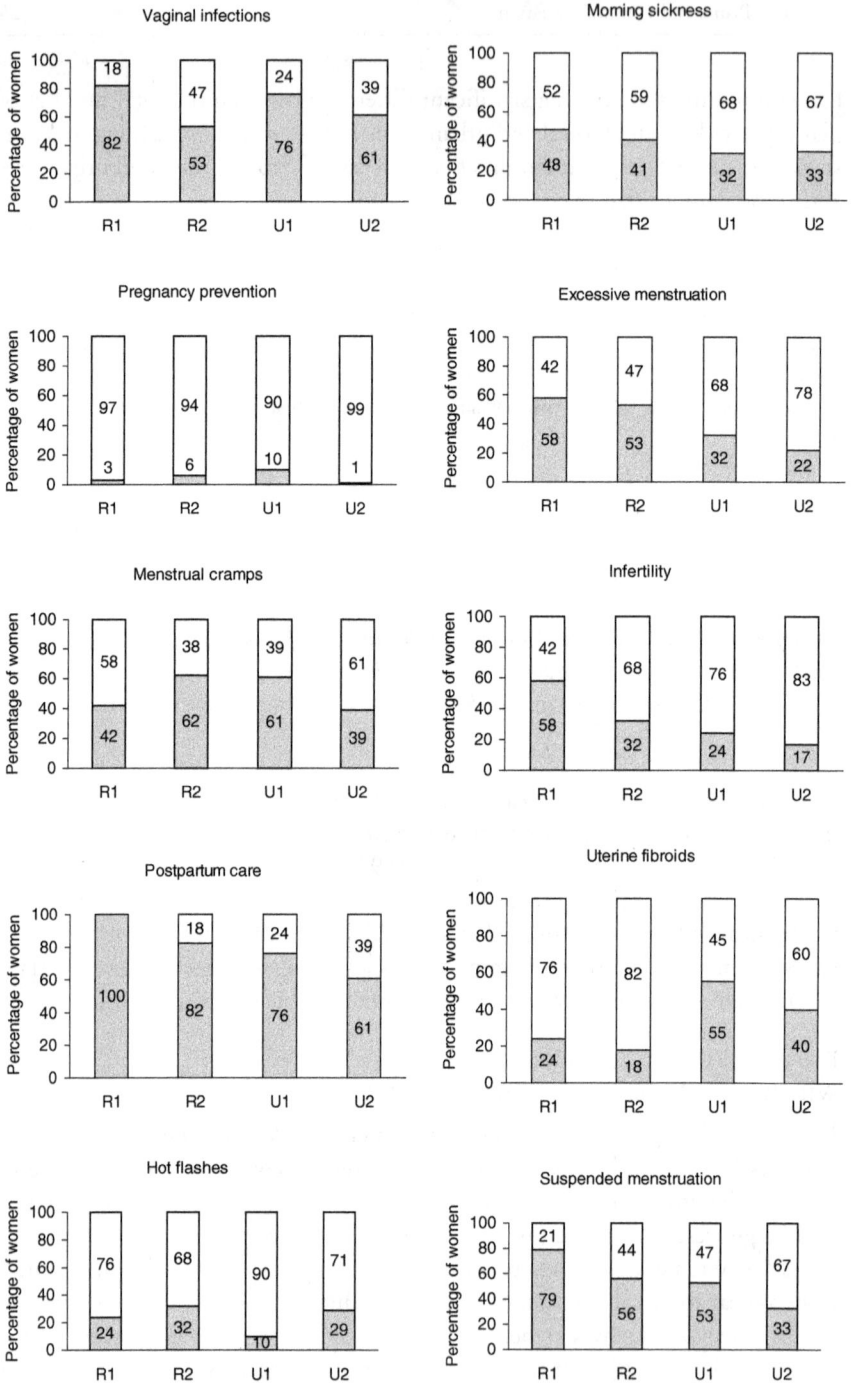

Figure 1.3. *Percentage of Dominican women in Los Calabazos R1 (n = 33), La Colonia R2 (n = 34), La Vega U1 (n = 87), and San Cristóbal U2 (n = 72) who reported a remedy for the ten health conditions. Reports:* shaded; *No reports:* white

The percentage of women who knew a remedy for vaginal infections varied. Los Calabazos (82 percent, $n = 33$) and La Vega (76 percent, $n = 87$), both in the same province, had a greater percentage of women report a remedy for this condition (G-test, $p = 0.0065$) than the communities in the province of San Cristóbal (La Colonia = 53 percent, $n = 34$; San Cristóbal = 61 percent, $n = 72$). In all communities, a small percentage of women (≤ 10 percent) knew a remedy for pregnancy prevention. A significantly greater percentage of women in rural communities (G-test, $p < 0.0001$) mentioned a remedy for excessive menstruation than in urban communities. For this condition, both rural communities had 53 percent or more women who knew such a remedy, while the urban communities showed a maximum of 32 percent of the women.

A greater percentage of women reported traditional remedies for infertility and suspended menstruation in the two rural communities as compared with the urban communities. A significantly greater percentage of women in Los Calabazos knew a remedy for infertility (G-test, $p < 0.0002$) and suspended menstruation (G-test, $p < 0.0005$) than in La Colonia, La Vega, and San Cristóbal. For those same conditions, La Colonia had a significantly greater percentage of women who knew a traditional remedy than San Cristóbal (G-test, $p = 0.01$), but La Colonia and La Vega were not significantly different.

Uterine fibroids was the only condition for which a greater percentage of women in urban communities reported a remedy than rural communities. At least 40 percent of the women interviewed in each urban community reported a remedy for this condition. A maximum of 24 percent of women in the rural communities reported a remedy for treating fibroids. Both La Vega and San Cristóbal, when compared with La Colonia, showed significantly different percentages (G-test, $p < 0.0001$) of women who reported a remedy for uterine fibroids. Los Calabazos and La Vega also had significantly different percentages of women who knew a remedy (G-test, $p = 0.0003$) for this condition; however, the percentages in Los Calabazos and San Cristóbal were not significantly different (G-test, $p = 0.06$).

There are several plausible reasons for these findings. It is possible that the condition of uterine fibroids affects urban women more than rural women. They may also occur more frequently in urban women than rural women, or uterine fibroids may be considered a recently recognized women's condition which has not been incorporated into medicinal plant knowledge in the rural communities. Because uterine fibroids are usually diagnosed with an ultrasound, another explanation may be that fibroids are less likely to be detected in women who live in rural communities where there is less access to modern medical technology; therefore, the condition is diagnosed with less frequency. As a result, urban women might know more about uterine fibroids than rural women. Further studies are needed to understand these differences.

Of the women interviewed, three in the rural sites (4 percent, $n = 67$) mentioned having fibroids and eleven (7 percent, $n = 159$) reported having fibroids in the urban sites. Remedies reported to treat fibroids further support the trend observed in the data above. A popular remedy for this condition was a mixture of *Beta vulgaris* and *Saccharum officinarum* (beets and molasses), which was reported by a greater percentage of women (19 percent, n = 159) in the urban than the rural sites (7 percent, n = 67).

Results from this data suggest that a greater percentage of rural women know medicinal plants and remedies for the ten health conditions. Uterine fibroids were an exception because a greater percentage of urban women reported traditional remedies than rural women. Pregnancy prevention showed a low percentage of women for all communities. It should also be noted that at least one woman in every community was able to report a plant or remedy for each condition.

The issue of differences in health concerns between rural and urban areas could in some cases be a function of cultural differences: one might expect rural women to be more concerned with postpartum care, for instance. Surprisingly, we expected more women to report medicinal plant treatments for pregnancy prevention. One explanation for the low percentage of women who reported a remedy for this condition may be that women did not feel it was appropriate to share this kind of information.

Transnational Patterns in Medicinal Plant Knowledge:
Comparing Plant Knowledge in New York City and the Dominican Republic

In an attempt to uncover transnational patterns of Dominican ethnomedical knowledge, we compared our data from New York City (Balick et al. 2000; Reiff et al. 2003) with the data collected in the Dominican Republic. This study provides preliminary analyses about the distribution of medicinal plant knowledge across transnational landscapes and suggests the degree to which Dominican healers retain traditional knowledge in a new urban setting, distant from their homeland. Three conditions surveyed in the Dominican Republic were the same as those surveyed in New York City: hot flashes, excessive menstruation (menorrhagia), and uterine fibroids.

An ethnobotanical literature review by Ososki and co-authors (2002) provided a data set for further comparisons. This data set consists of plant species reported in thirty ethnobotanical references specific to the Dominican Republic for the above women's conditions and associated symptoms. Table 1.3 shows the number of plant species reported in each data set for a health condition and shows the number of species that were listed in all three data sets for a single health condition (overlap). *Citrus* species were reported in all three data sets for hot flashes, while *Citrus aurantifolia* (Christm.) Swingle (*limón agrio, lime*) was reported in the Dominican Republic, and *C. aurantium* L. and *C. sinensis* Osbeck (*naranja, orange*) were reported in the literature. *Naranja* and *limón agrio* were reported in New York City, but because the specimens were identified only to genus, all *Cit-*

rus species were collapsed and reported as one species that overlapped as seen in table 1.3. For hot flashes, *Genipa americana* L., *Kalanchoe gastonis-bonnieri,* and *Tilia mandshurica* Rupr. & Maxim. were mentioned by healers in New York City and the Dominican Republic but not in the literature. Two plant species, *Plantago major* L. and *Ruta chalepensis* L., were reported in all data sets for menorrhagia. *Petiveria alliacea* L. was reported by both Dominican healers and New York City healers for this condition, but it was not reported in the literature data set. Eight plant species overlapped among the three data sets for uterine fibroids and include *Beta vulgaris, Momordica charantia, Opuntia ficus-indica, Petiveria alliacea, Petroselinum crispum* (Mill.) Nyman ex A. W. Hill, *Rosmarinus officinalis* L., *Ruta chalepensis,* and *Saccharum officinarum.*

Table 1.3. Number of Medicinal Plant Species Reported for Hot Flashes, Menorrhagia, and Uterine Fibroids in Dominican Ethnomedical Literature ($n = 30$) and by Dominican Healers in New York City ($n = 6$) and the Dominican Republic ($n = 11$)

	Number of plant species			
Health condition	Dominican Republic Healers	New York City Healers	Dominican Literature	Overlap
Hot flashes	63	7	17	1
Menorrhagia	24	7	13	2
Uterine fibroids	56	68	57	8

Of the three conditions, uterine fibroids provides the most accurate comparison of plant use between the two study areas because of the comparable sample sizes of healers from New York City ($n = 6$) and the Dominican Republic ($n = 8$) that reported a medicinal plant species. A Venn diagram (figure 1.4) displays the number of plant species that overlapped for uterine fibroids among the three data sets and the number of plant species distinct to each. The total number of

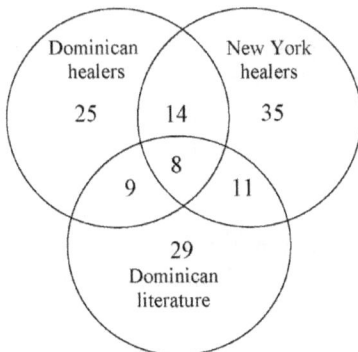

Figure 1.4. *Venn diagram illustrating the number of medicinal plant species reported in Dominican ethnobotanical literature (references,* n = 30*) and by Dominican healers in the Dominican Republic (*n = 8*) and New York City (*n = 6*) for uterine fibroids*

plant species known by Dominican healers in New York City and the Dominican Republic for uterine fibroids was 102, and an additional 29 species were reported in the literature review. Of a total of 131 species, only eight (6 percent) overlapped in all three data sets.

A total of 22 plant species (22 percent, n = 102) were reported in common between Dominican and New York City healers for uterine fibroids. New York City healers knew a greater number of distinct plant species for fibroids (n = 46) than Dominican healers (n = 34), similar to the rural-urban contrast observed among women in the Dominican Republic (figure 1.3). The greater number of plants reported for fibroids in New York City than in the Dominican Republic, and the greater number of women able to report such a remedy in urban communities in the Dominican Republic than in rural communities, suggests that women tend to be more familiar with fibroids in urban environments.

Two plants that were popular treatments for uterine fibroids by New York City healers were *Agave* sp. and *Kalanchoe gastonis-bonnieri* (Balick et al. 2000). These plants were reported in the Dominican Republic for other women's health conditions, but healers did not report these plants for uterine fibroids. A plant species frequently reported for uterine fibroids among healers in the Dominican Republic that was not reported by New York City healers was *Spermacoce assurgens*. It is important to remember that these plants are often used in mixtures rather than in individual preparations.

Beet juice with molasses was a remedy reported by both New York City (Fugh-Berman et al. 2004) and Dominican healers for uterine fibroids. A *botella* was another remedy reported for uterine fibroids in New York City and the Dominican Republic, although it was not always referred to by this name in the New York City study, rather only the plants and preparation were reported. A *botella* often refers to a complex mixture of plants that is prepared as a decoction and usually stored in a bottle and administered orally at room temperature or cold if refrigeration is available. Caramelized sugar, honey, *malta* (malted barley beverage), and alcohol (rum, gin, or wine) may be added as additional ingredients. The preparation of a *botella* varies depending on the healer, the illness being treated, and ingredient availability. It should be noted that Dominicans may also use the term *botella* to refer to a tincture in which plant fragments such as leaves, chopped branches, or roots are placed in a bottle and then alcohol is added; but, this preparation is often referred to as *mama juana*, which was not reported for the ten women's conditions. *Botella* preparations specifically used to cleanse the uterus and to remove any impurities after pregnancy may also be called a *bebedizo*. Perhaps the use of a *botella* for postpartum care has been adapted for fibroids as a way of cleansing and removing fibroids from the uterus. Differences in treatment approaches may be explained by individual preferences; perhaps each healer uses his/her own remedy for a health condition.

Table 1.4 lists 13 plant species reported in New York City for uterine fibroids (Balick et al. 2000) that were not reported in the Dominican Republic for any women's health condition. Several of the plants listed in table 1.4 do not even grow in the Dominican Republic, such as *Uncaria tomentosa* (Willd. ex Roem. & Schult.) DC. and *Vaccinium macrocarpon* Aiton.

Table 1.4. Medicinal Plant Species Reported Exclusively by Dominican Healers in New York City (*n* = 6)

Species [Family] {Voucher}[a]	Vernacular Name in Spanish and English* (*between brackets)
Chamaemelum nobile (L.) All. [Asteraceae] {5, 30, 32, 65}	*manzanilla, (chamomile)*
Coccothrinax argentea (Lodd. ex Schult. & Schult.f.) Sarg. ex Becc. [Arecaceae] {106}	*cana, (silver thatch palm)*
Fevillea cordifolia L. [Cucurbitaceae] {72, 75}	*ayamo, jayamo, (jallamo, antidote caccoon)*
Ficus religiosa L. [Moraceae] {78}	*alamo, (sacred fig, peepul)*
Helichrysum italicum (Roth) G. Don f. [Asteraceae] {85}	*siempre fresca, (curry plant)*
Myrsine sp. [Myrsinaceae] {104}	*palo santo, (colicwood)*
Phoenix dactylifera L. [Arecaceae] {113}	*palma, (date palm)*
Solanum tuberosum L. [Solanaceae] {118}	*papa, (potato)*
Tabebuia impetiginosa (Mart. ex DC.) Standl. [Bignoniaceae] {35}	*palo de arco, (pau d'arco)*
cf. *Taraxacum officinale* Weber [Asteraceae] {61}	*diente de león, (dandelion)*
Uncaria tomentosa (Willd. ex Roem. & Schult.) DC. [Rubiaceae] {81}	*uña de gato, (cat's claw)*
Vaccinium macrocarpon Aiton [Ericaceae] {100}	*(cranberry)*
Zingiber zerumbet (L.) Sm. [Zingiberaceae] {56}	*jengibre amargo, (bitter ginger)*

[a]All numbers are A. Ososki collections

Of these thirteen *taxa* reported exclusively in New York City for uterine fibroids, five have the same common name as plants reported in the Dominican Republic (table 1.5). Three of the plants exclusive to New York City but sharing common names with Dominican Republic medicinal plants are also in the same botanical families. In the Dominican Republic, the chamomile used was *Matricaria recutita*, while in New York City Dominican healers used *Chamaemelum nobile* (L.) All., both Asteraceae. In the Dominican Republic *siempre fresca* was identified as *Peperomia pellucida* (L.) Kunth in the Piperaceae family, while *siem-*

pre fresca in New York City was *Helichrysum italicum* (Roth) G. Don f. in the Asteraceae family. The *palmas* of New York City and of the Dominican Republic are both palms, *Phoenix dactylifera* L. and *Roystonea hispaniolana* L. H. Bailey, respectively. In contrast, the *uña de gato* used in New York City was *Uncaria tomentosa*, which is usually imported from Peru and is a different plant family than *Pisonia aculeata* L. used in the Dominican Republic. The two plants known as *jengibre amargo, Zingiber cassumunar* Roxb. in the Dominican Republic and *Zingiber zerumbet* (L.) Sm. in New York City, are in the genus *Zingiber.* Table 1.5 illustrates the importance of collecting voucher specimens for ethnobotanical studies because common names can be the same but the plant species are different.

Table 1.5. Plant Species Used by Dominican Healers in New York City and the Dominican Republic That Share the Same Vernacular Name

New York Plant Species Species [Family] {Voucher}[a]	Dominican Plant Species Species [Family] {Voucher}[a]	Vernacular Name in Spanish and English* (*between brackets)
Chamaemelum nobile (L.) All. [Asteraceae] {5, 30, 32, 65}	*Matricaria recutita* L. [Asteraceae] {R76}	*manzanilla, (chamomile)*
Helichrysum italicum (Roth) G. Don f. [Asteraceae] {85}	*Peperomia pellucida* (L.) Kunth [Piperaceae] {336}	*siempre fresca, (curry plant, man to man)*
Phoenix dactylifera L. [Arecaceae] {113}	*Roystonea hispaniolana* L. H. Bailey [Arecaceae] {540}	*palma, (date palm, Hispaniolan royal palm)*
Uncaria tomentosa (Willd. ex Roem. & Schult.) DC. [Rubiaceae] {81}	*Pisonia aculeata* L. [Nyctaginaceae] {434, 500}	*uña de gato, (cat's claw, pullback)*
Zingiber zerumbet (L.) Sm. [Zingiberaceae] {56}	*Zingiber cassumunar* Roxb. [Zingiberaceae] {R143}	*jengibre amargo, (bitter ginger, cassumunar ginger)*

[a]All numbers are A. Ososki collections

The results in tables 1.4 and 1.5 signify that people seeking medicinal plants in new settings find substitutes for commonly used species to which access is no longer possible. Homonyms may be applied in cases where the substitute is related to the original, or resembles the original (e.g., the thorns of *Uncaria tomentosa* and of *Pisonia aculeata*), or is perceived as serving the same purposes. The United States government strictly regulates plant species allowed into the country; therefore, some species that are commonly used in the Dominican Republic may not qualify for importation, such as fresh fruits which may be used medicinally but also may carry pathogens that could harm agriculture in the United States. As a result, New York City healers may be forced to rely on other species. Different species that are substituted may be selected because of their efficacy or use for similar conditions in other pharmacopoeias in New York City. In some cases, plants with similar physical characteristics and morphology may be substituted. For example, the flowers of *Chamaemelum nobile* and *Matricaria recutita* or the rhizomes of *Zingiber zerumbet* and *Zingiber cassumunar* are similar in appearance. Other plants may be selected because they share a common name even though their morphology is quite different such as *siempre fresca*, which refers to *Peperomia pellucida*, a succulent plant, and can also refer to a non-succulent, *Helichrysum italicum*. Further studies are needed to better understand how plant species are substituted in new settings.

Discussion

Several patterns were observed from our findings. Rural women showed a tendency to know more medicinal plants for women's health conditions than urban women except for uterine fibroids. More urban women and healers in the Dominican Republic and New York City knew medicinal plants for uterine fibroids than their rural counterparts. Some plants were known specifically in one community while a number of plants were known in all five study sites. Understanding these rural, urban, and transnational similarities and differences can help further untangle the complexities of medicinal plant knowledge.

Cultural Similarities and Differences

Rural and urban communities shared species and knowledge but there were also plants unique to the pharmacopoeia of each community. Both rural and urban women were able to report home remedies and plants for all ten health conditions. While the communities shared many plant species in common, it was demonstrated that some plants were preferred over others in each community.

In some cases, a plant species was found growing in more than one community but was only reported in one site. This is illustrated by a couple of plant walks during which the first author would occasionally point out a plant that was reported in another study site to see if they had a use for it in the currently visited site. For these few cases the women responded that the plant was considered a "weed" and not used. In future studies, it would be interesting to combine the lists of reported plants from several communities and use the checklist or plant interview technique as described by Boom (1987) to query individuals.

In the Dominican Republic, rural women reported more plant species than women living in urban communities. These results suggest an erosion of plant knowledge as people migrate to urban centers. These results, perhaps in part, are due to the limited access and exposure in urban environments to natural resources including plants and increased preference for other forms of treatment such as pharmaceuticals. Other studies (Milliken and Albert 1996; Alexiades 1999: 353) have reported a similar pattern of increased use of pharmaceuticals and a decreased use of plant medicine among indigenous groups as they become acculturated.

The Dominican Republic is culturally diverse due to historical events and migrations of various cultural groups. It should be noted that in addition to plant species distribution across the Dominican Republic, cultural differences regionally may also help to explain varying plant use in the provinces of La Vega and San Cristóbal. The percentages of women who reported a remedy for vaginal infections and hot flashes in figure 1.3 show that regional differences between the provinces, in addition to urban and rural dichotomies, may further suggest patterns to explain cultural variation of Dominican medicinal plant knowledge. San Cristóbal is considered to have stronger African origins than La Vega and the Cibao region (Cambeira 1997, B. Peguero pers. comm.). In addition, San Cristóbal is more urbanized as it is closer to the capital than La Vega, which is composed of more *campesinos* (peasants). Future longitudinal studies in these same communities could compare plant knowledge over time to see how their pharmacopoeias evolve and to better understand the acquisition of cultural plant knowledge. For example, it would be interesting to see if flax and soy become more widely reported by women in these communities as these herbs increase in popularity in urban centers, or if more rural women become familiar with remedies for uterine fibroids.

Plants for Women's Medicine

The variety of plants used for the ten women's health conditions is extensive and includes cultivated, wild-collected, and purchased plants. Individuals in urban environments tended to purchase their plants or grow them in home gardens. A few species, such as *Plantago major* and *Spermacoce assurgens,* are wild-collected because they are readily available in empty lots, along roadsides, and in disturbed habitats. Rural women tend to rely on plants that they have access to in their home gardens and *conucos* or on plants that can be wild-collected. Rural women reported few plants that needed to be purchased.

Many plants are known in all five study sites, but we did note some plants that were specific to a community. For example, *Adiantum tenerum* was specific to La Colonia. *Adiantum tenerum* was reported by nine different women in La Colonia and was used primarily for postpartum care. It was also reported for suspended menstruation, infertility, and vaginal infections. Medicinal plants used for uterine fibroids in New York City and in the Dominican Republic also showed similarity and variation. Species were reported in New York City that were not reported in the Dominican Republic, and other species were mentioned in the Dominican Republic that were not cited in New York City. Transnational comparisons of Dominican traditional medicine provide an additional perspective on the evolution of healing practices as people migrate to new environments, and they offer a lens to see how plant knowledge is exchanged through a multitude of networks.

Traditional Knowledge of Women's Health Conditions

The higher percentage of rural women who reported a remedy for the ten selected conditions may be because there is a greater reliance on natural resources in rural settings as compared to urban environments. Distance from health centers, costs of transportation, and high costs of medications may also contribute to the reliance on traditional medicine, manifested as an increased level of medicinal plant knowledge in rural areas. Women in urban communities may rely on allopathic medicine perhaps because it is more readily available and accessible to them than medicinal plants. Education and social pressures may also encourage urban women to use biomedical resources more than medicinal plants. However, this may change with the spreading popularity of herbal use in the United States and other countries. In the future, perhaps Dominican women living in urban areas will report more herbal medicines, incorporating those from their families as well as new uses acquired from the media or neighbors from other regions or even, perhaps, from abroad.

Conclusion

Plant knowledge for Dominican women's medicine both differs and coincides across Dominican rural, urban, and transnational landscapes. The groups are not exclusive, and much of the ethnomedical knowledge overlaps. Some plant species and healing knowledge may be group-specific because of cultural preferences, access to plant resources, and influences of media and technology, while other species and healing knowledge are shared between groups.

The rural communities tended to hold an overall greater knowledge base about women's traditional medicine than the urban communities. This holds true for the number of plant species known and the percentage of women who reported a remedy for different health conditions. In conclusion, the greater knowledge demonstrated in rural environments probably stems from close contact with natural resources and reliance on these resources for health care. People in urban communities, in contrast, have access to a diverse array of treatment approaches and less contact with natural resources and diverse plant populations. Yet this increased access to biomedicine by individuals in urban settings such as La Vega, San Cristóbal, and New York City has not caused them to lose their knowledge about traditional medicine; instead, the two are used in parallel.

Ethnobotanical studies are needed that address cultural variation and change. In addition, studies are needed to further elucidate the transmission of plant knowledge and to highlight factors that affect the acquisition of plant knowledge. How is ethnomedical knowledge appropriated, transferred, and exchanged in different settings? What mechanisms affect the loss and gain of this knowledge? Further examination of pharmacopoeias that are specific to rural, urban, and transnational communities may provide insight on factors that affect the transmission of cultural knowledge. A deeper grasp of the distribution of plant knowledge in communities may provide useful insight to help explore these questions.

Acknowledgments

We would like to thank the healers and community members in New York City and the Dominican Republic who participated in these studies. We also would like to acknowledge our Dominican collaborators who facilitated this work: Milcíades Mejía, Daisy Castillo, Ricardo García, and Brígidio Peguero of the Jardín Botánico Nacional de Santo Domingo, Librada Dionicio of Confederación Nacional de Mujeres del Campo (CONAMUCA), and Aurelio Suero of Colectivo Mujer y Salud. Fredi Kronenberg and Marian Reiff of Columbia University, Jolene Yukes of The New York Botanical Garden, Edward Kennelly, Dwight Kincaid, and Juan Carlos Saborío of Lehman College, and Claudia Scholz of John Hopkins also contributed at different stages of these projects. We would also like to acknowledge the following sources for funding the New York City fieldwork, as well as the support of the NYBG Institute of Economic Botany: Arum Ltd., The Educational Foundation of America, Edward P. Bass of The Philecology Trust, Center for Environmental Research and Conservation at Columbia University, Fetzer Institute, MetLife Foundation, National Geographic Society, National Institutes of Health, Nature's Herbs, Nu Skin International, Prospect Hill Foundation, Sony Entertainment, and Maxell Corporation; and the Dominican fieldwork: Allen-Ososki, The City University of New York Graduate School, Fulbright-Hays Research Award, Garden Club of Allegheny County, and The New York Botanical Garden.

References

Aday, L.A. 1989. *Designing and conducting health surveys: A comprehensive guide.* San Francisco: Jossey-Bass Publishers.

Adlercreutz, H., and W. Mazur. 1997. "Phyto-oestrogens and Western disease." *Annals of Medicine* 29: 95–120.

Alexiades, M.N., ed. 1996. *Selected guidelines for ethnobotanical research: A field manual.* Advances in Economic Botany 10. Bronx: New York Botanical Garden.

Alexiades, M.N. 1999. "Ethnobotany of the Ese Eja: Plants, health, and change in an Amazonian society." Ph.D. Dissertation. New York: City University of New York Graduate Center.

Allen, R., L.F. Cushman, S. Morris, J. Feldman, C. Wade, D. McMahon, M. Moses, and F. Kronenberg. 2000. "Use of complementary and alternative medicine among Dominican emergency department patients." *American Journal of Emergency Medicine* 18: 51–54.

Balick, M.J., R. Arvigo, G. Shropshire, J. Walker, D. Campbell, and L. Romero. 2002. "The Belize Ethnobotany Project: Safeguarding medicinal plants and traditional knowledge in Belize." In *Ethnomedicine and drug discovery,* ed. J.C. Wootton and M.W. Iwu. New York: Elsevier Science, 267–281.

Balick, M.J., F. Kronenberg, A.L. Ososki, M. Reiff, A. Fugh-Berman, B. O'Connor, M. Roble, P. Lohr, and D. Atha. 2000. "Medicinal plants used by Latino healers for women's health conditions in New York City." *Economic Botany* 54: 344–357.

Bodeker, G. 1995. "Traditional health systems: Policy, biodiversity, and global interdependence." *Journal of Alternative and Complementary Medicine* 1: 231–243.

Bodeker, G., and F. Kronenberg. 2002. "A public health agenda for traditional, complementary, and alternative medicine." *American Journal of Public Health* 92: 1582–1592.

Bolay, E. 1997. *The Dominican Republic: A country between rain forest and desert.* Weikersheim: Margraf Verlag.

Boom, B. 1987. *Ethnobotany of the Chácobo Indians, Beni, Bolivia.* Advances in Economic Botany 4. Bronx: New York Botanical Garden.

Borello, M.A., and E. Mathias. 1977. "Botánicas: Puerto Rican folk pharmacies." *Natural History* 86: 65–73.

Brett, J.A. 1994. "Medicinal plant selection criteria among the Tzeltal Maya of Highland Chiapas, Mexico." Ph.D. dissertation. San Francisco: University of San Francisco.

Cambeira, A. 1997. *Quisqueya la Bella: The Dominican Republic in historical and cultural perspective.* Amonk: M.E. Sharp.

Campos, M.T., and C. Ehringhaus. 2003. "Plant virtues are in the eyes of the beholders: A comparison of known palm uses among indigenous and folk communities of Southwestern Amazonia." *Economic Botany* 57: 324–344.

Delgado, M., and J. Santiago. 1998. "HIV/AIDS in a Puerto Rican/Dominican community: A collaboration project with a botanical shop." *Social Work* 43: 183–186.

Fisch, S. 1968. "Botánicas and spiritualism in a metropolis." *Milbank Memorial Fund Quarterly* 46: 377–388.

Fugh-Berman, A., M.J. Balick, F. Kronenberg, A.L. Ososki, B. O'Connor, M. Reiff, M. Roble, P. Lohr, B.J. Brosi, and R. Lee. 2004. "Ethnomedical studies on the treatment of fibroids: The use of beets (*Beta vulgaris*) and molasses (*Saccharum officinarum*) as an herbal therapy by Dominican healers in New York City." *Journal of Ethnopharmacology* 92: 337–339.

King, S.R. 1996. "Conservation and tropical medicine plant research." In *Medicinal resources of the tropical forest: Biodiversity and its importance to human health,* ed. M.J. Balick and S.A. Laird. New York: Columbia University Press, 63–74.

Lee, R.A., M.J. Balick, D.L. Ling, F. Sohl, B.J. Brosi, and W. Raynor. 2001. "Cultural dynamism and change in Micronesia." *Economic Botany* 55: 9–13.

Milliken, W., and B. Albert. 1996. "The use of medicinal plants by the Yanomami Indians of Brazil." *Economic Botany* 50: 10–25.

Ososki, A.L., P. Lohr, M. Reiff, M.J. Balick, F. Kronenberg, A. Fugh-Berman, and B. O'Connor. 2002. "Ethnobotanical literature survey of medicinal plants in the Dominican Republic used for women's health conditions." *Journal of Ethnopharmacology* 79: 285–298.

Reiff, M., B. O'Connor, F. Kronenberg, M.J. Balick, P. Lohr, M. Roble, A. Fugh-Berman, and K.D. Johnson. 2003. "Ethnomedicine in the urban environment: Dominican healers in New York City." *Human Organization* 62: 12–26.

Setchell, K.D.R. 1998. "Phytoestrogens: The biochemistry, physiology, and implications for human health of soy isoflavones." *American Journal of Clinical Nutrition* 68: 1333S–1346S.

US Census Bureau. 2000. Census. Retrieved 17 January 2006 from http://www.census.gov.

US Census Bureau. 2005. International data base. Retrieved 17 January 2006 from http://www.census.gov/ipc/www/idbnew.html.

Venes, D., and C.L. Thomas, eds. 2001. *Taber's cyclopedic medical dictionary.* Philadelphia: F.A. Davis Company.

Voeks, R.A., and A. Leony. 2004. "Forgetting the forest: Assessing medicinal plant erosion in Eastern Brazil." *Economic Botany* 58: S294–S306.

Chapter 2

Use of Medicinal Plants by Dominican Immigrants in New York City for the Treatment of Common Health Conditions

A Comparative Analysis with Literature Data from the Dominican Republic

Ina Vandebroek, Michael J. Balick, Jolene Yukes,

Levenia Durán, Fredi Kronenberg, Christine Wade,

Andreana L. Ososki, Linda Cushman, Rafael Lantigua,

Miriam Mejía, Lionel Robineau

Introduction

The growth of the Dominican population in the northeastern United States in the past two decades constitutes one of the major immigration waves during the second half of the twentieth century. This movement is equal in magnitude to the massive Puerto Rican migration in the 1950s and 60s (Rivera-Batiz 2002). If current trends continue, the Dominican immigrant population in New York City (NYC) will grow to be larger than that of Puerto Ricans in the next ten years. Fifty-three percent of all Dominicans in the United States live in New York City. Depending on the source, the number of Dominicans in New York City in 2000 was estimated to be 369,200 (NYC Department of City Planning 2004), 407,473 (US Bureau of Census), or 555,000 (Rivera-Batiz 2002). The actual number is likely to be higher given the presence of undocumented immigrants who are not included in the above estimates and whose population is difficult to determine. Dominican immigrants live in all five New York City boroughs, but are primarily concentrated (more than 66 percent) in Manhattan and the Bronx (NYC Department of City Planning 2004).

Washington Heights, located near the northernmost tip of Manhattan, has long been the neighborhood with the largest Dominican community in New York City, together with Inwood (located north of Washington Heights), and Hamilton Heights (adjacent to the South). However, recently the Bronx is taking the lead (unpublished data from the Bureau of Epidemiology Services, NYC Department of Health and Mental Hygiene) in becoming the center of Dominican settlement.

The Dominican community shows a low level of socioeconomic attainment and poor indicators of well-being and health. In 2000, 70 percent of Dominicans living in New York City did not rate themselves as proficient in English, and less than 50 percent graduated from high school. Thirty-one percent were living in poverty, making Dominicans third in poverty measures among twenty NYC immigrant groups, after Mexicans and Bangladeshis (NYC Department of City Planning 2004). Dominicans, like other United States immigrants, disproportionately lack health insurance coverage and receive fewer health services than native-born citizens. Low-income noncitizens are more than twice as likely to be uninsured as compared with low-income citizens (Kaiser Commission on Medicaid and the Uninsured 2003). Dominicans in the United States rank fifth in lack of health insurance coverage (with 36 percent of people uninsured) among sixteen immigrant communities (Carrasquillo et al. 2000). In addition to lack of health insurance coverage, immigrants may encounter limited access to biomedical health care services because of language barriers, citizenship status, and/or the experience of lack of recognition of their cultural beliefs and practices by health care providers (Gomez-Beloz and Chavez 2001; Kaiser Commission on Medicaid and the Uninsured 2003). A health report from the predominantly Dominican neighborhood Washington Heights in Manhattan shows that 26 percent of people do not consider themselves to be in good health, while 17 percent needed care and did not receive it, and 34 percent have no personal doctor. New York City averages for these three health indicators were significantly better (19, 10 and 25 percent, respectively). In addition, the vulnerability of those living in Washington Heights is further demonstrated by the following comparisons (totals for New York City are given in parentheses): 44 (28) percent of adults did not graduate high school, 31 (21) percent live in poverty, and 53 (32) percent feel their neighborhoods are unsafe. These conditions are likely to negatively influence people's ability to increase healthy behaviors and maintain optimal health (Karpati et al. 2003). Hence, there exists an evident need for improved health care for the Dominican community in New York City.

In the United States, *botánicas* and *curanderos* (traditional healers) represent a unique, culturally based health care delivery system that is available outside of community hospitals, clinics, physicians' offices, and pharmacies (Gomez-Beloz and Chavez 2001; Reiff et al. 2003). *Botánicas* are Latino and/or Afro-Caribbean shops and "sites of healing and community support" (Hernández and Jones 2004) that deal with physical, spiritual, and/or religious well-being and often function as herbal pharmacies by selling fresh, dried, and processed herbs grown in the United States, or imported from the Caribbean or elsewhere (Balick et al. 2000). By drawing on religious and healing traditions that recognize the multiple (physical, mental/emotional, spiritual, social, and environmental) dimensions of health and well-being in a holistic fashion, *botánicas* and the traditional healers associated with them are important resources for Latino immigrant health care. In academic publications, *botánicas* have been described in myriad ways: as "herb shops" and folk medical pharmacies (Fisch 1968), "small shops selling *artículos religiosos*" or religious supply stores (Borrello and Mathias 1977), "botanical shops" which function as "community-based establishments" and "outlets for important cultural traditions pertaining to healing and spirituality" (Delgado 1996; Delgado and Santiago 1998), "invisible hospitals" (Jones and Polk 2001), and "ethnic healing-religious stores" (Viladrich 2006). A large number of *botánicas* are situated in Manhattan and the Bronx, two boroughs of New York City with high populations of Latinos, including Dominicans. During a two-day systematic street survey through two ZIP-codes in the Bronx (representing 8 percent of all Bronx ZIP-codes), we mapped sixteen *botánicas* (unpubl. results). Therefore, we expect to map many more botánicas when we expand the survey. In addition, seventy-five *botánicas* were identified in the Bronx based on a review of local, Spanish-language telephone books, newspaper advertisements, and Internet searches (Yukes unpubl. results). Since most *botánicas* do not openly advertise themselves in print or online, the actual number of *botánicas* in this borough is probably considerably higher.

Botánicas offer culturally appropriate health care to their customers (Gomez-Beloz and Chavez 2001), and an herbalist, healer, or spiritual counselor (who may be called a *curandero/a, Espiritista, Santero/a,* or simply *Señor/Señora* or *Don/Doña*) can frequently be consulted on the spot. *Botánica* staff usually share the same cultural background (i.e., of Latino/a or Caribbean descent but not necessarily of the same nationality) as their clientele. Hence, explanations of the cause of illness and treatment regimes are generally understood by customers-patients, and herbal remedies may be familiar to those seeking them since they are also used in the immigrant's home country. Moreover, medicinal plants can be purchased in small quantities at relatively low cost, which allows for more flexibility in spending as compared to the one-time purchase of expensive whole packages of over-the-counter or prescribed pharmaceuticals. The particular modalities offered by *botánicas* and the reciprocal dialogue on health, illness, and treatment between customer (patient) and shopkeeper (alternative health care provider) may be convenient, culturally familiar, and cost-effective for many Dominicans.

Also, by providing a space for social interaction, cultural expression, and spiritual practices associated with Afro-Caribbean and Latin American religious traditions, *botánicas* are a valuable community resource.

Medicinal plants—or *plantas de la tierra*, *matas*, or *hojas*, as Dominicans call these herbal remedies—are an important aspect of Dominican culture (Robineau 1986; Brendbekken 1998). *Botánicas* exist in the Dominican Republic (DR) as well, and medicinal plants are also easily available at open air markets, local grocery stores, and even at supermarkets from international chains, some of which sell locally made bottles of medicinal roots (*botellas*) and packets of dried herbs for the promotion of health and well-being. Most Dominican homes in the Dominican Republic have patio gardens or agricultural land (*conucos*) that are another source for obtaining *matas*. It has been shown that after immigration to the United States, Latinos in general—and Dominicans in particular—continue to use medicinal plants for health care (Allen et al. 2000; Balick et al. 2000; Ososki et al. 2002; Mikhail et al. 2004).

In spite of the popularity of medicinal plant use, relatively few systematic ethnobotanical inventories have been compiled so far that focus specifically on Dominican (medicinal) plant use, either in the United States or in the Dominican Republic (Robineau 1986; Polanco et al. 1998; Balick et al. 2000; Peguero et al. 2001; Peguero 2002). An exception is the extensive ethnopharmacological survey conducted by TRAMIL (Traditional Medicine in the Islands) in twenty-three Caribbean countries including several communities in the Dominican Republic. However, although the survey records all the medicinal uses that are reported by participants for treating ten health conditions, the TRAMIL publications (Germosén-Robineau 1997; 2005) only include information on plant species and medicinal uses that were reported in at least 20 percent of the interviews at a given study site. This strategy was chosen as a function of the applied community health objectives of TRAMIL.

The ethnobotanical study on which we report in this chapter is part of an interdisciplinary project entitled "Dominican Herbal Medicine: Plants Used for Inflammation," conducted by researchers at the Institute of Economic Botany at the New York Botanical Garden, and funded by the National Institutes of Health, National Center for Complementary and Alternative Medicine (NIH-NCCAM). The study team consists of a consortium of seven institutes from the United States and Dominican Republic that bridges several disciplines, including botanical gardens (The New York Botanical Garden; Jardín Botánico Nacional Dr. Rafael Ma. Moscoso de Santo Domingo), Dominican community based organizations and networks (Alianza Dominicana; TRAMIL), research centers and universities (National Center for Natural Products research, University of Mississippi; Richard and Hinda Rosenthal Center for Complementary and Alternative Medicine, Columbia University), and clinics (Associates in Internal Medicine, New York Presbyterian Hospital). One of the main objectives of the project is to create an inventory of medicinal plants known and used by Dominicans in New York City and to compare medicinal plant knowledge between New York City and the Dominican Republic.

Here, we report on the results of ethnobotanical fieldwork with 175 Dominican immigrants in New York City. We analyzed a subset of our ethnobotanical database by extracting information on those medicinal plants which participants report to (have) use(d) "only in New York City" (and hence excluding applications reported as "used only in the Dominican Republic" or "used in both New York City and the Dominican Republic"). Subsequently, the medicinal uses of the most frequently mentioned plants were compared with published information on their applications in the Dominican Republic.

Material and Methods

The research and study instruments (questionnaires, flyers, and oral consent forms) were approved by the Institutional Review Board of the City University of New York (IRB # 04-06-0599). Interviews were conducted with 175 Dominican participants, age eighteen or older, in Manhattan and the Bronx between April 2005 and February 2006. Before the start of the survey, a questionnaire consisting of eighty-four structured and semi-structured questions was designed, pilot tested with nine participants, and reviewed several times by different members of the project consortium. The questionnaire covered the following aspects of Dominican medicinal plant use and personal background information: (1) history of a participant's plant use in the Dominican Republic; (2) participant's present plant use in New York City; (3) general use-preference for medicinal plants or pharmaceuticals, and reason(s) for this preference; (4) sociodemographic profile, including age, education, geographical area of origin in the Dominican Republic, household size, income, and type of health insurance; (5) acculturation rating; (6) medicinal plant knowledge for thirty common health conditions; (7) knowledge of potentially harmful medicinal plants; (8) source of plant knowledge; (9) beliefs about illnesses that should be treated exclusively with pharmaceuticals or by a biomedical health care provider (9a), versus beliefs about illnesses that should be treated exclusively with medicinal plants or by a traditional healer (9b).

Participants were questioned about medicinal plant knowledge related to thirty common health conditions and notes were taken about: (1) the local and alternative name(s) for each medicinal plant they mentioned; (2) whether they ever experienced the illness for which a medicinal plant was mentioned; (3) whether they had treated the illness with: (3a) medicinal plants, (3b) pharmaceuticals, or (3c) both medicinal plants and pharmaceuticals; (4) in which country they had used this medicinal plant, with choice of the following options: (4a) in New York City only, (4b) in the Dominican Republic only, (4c) in both New York City and the Dominican Republic. An overview of the list of health conditions included in the survey is provided in table 2.1.

Participants were recruited and interviewed through snowball sampling in New York City and at two specific locations in Washington Heights, New York City: (1) Alianza Dominicana, a Dominican community-based organization, and (2) the Associates in Internal Medicine Clinic (AIM), New York-Presbyterian Hospital. Participants contacted through referral were usually interviewed at home and on a few occasions at the Institute of Economic Botany of the New York Botanical Garden, while those recruited at Alianza Dominicana and the AIM clinic were interviewed on the premises in a private location. Prior informed oral consent was obtained from all participants, and permission was asked to tape record each interview. To ensure confidentiality, an ID-number was assigned to each participant to code their interview materials. On average, interviews lasted between one and two hours. At the end of the interview, participants received a compensation of $20 (US). In total, 175 Dominicans were interviewed, 111 women and 64 men. Data from the pilot interviews (5 women and 4 men) were excluded from data analysis since the questionnaire was substantially revised after pilot testing.

Interview data were entered in a Microsoft/Access database that includes three linked tables: (1) table with participant data (sociodemographic variables); (2) table with illness data (name of illness, use of medicinal plants or pharmaceuticals for treatment, use according to country, and mode of administration of the plant remedy); and (3) table with plant data (including local name, alternative name, whether used individually or in a mixture, and plant part used). For the purpose of data analysis, plant use data reported for "New York City only" were extracted, thus omitting plant use data for herbal remedies that were used "only in the Dominican Republic" and those that were used "in both New York City and the Dominican Republic".

The "NYC use only" data subset was used to calculate the citation frequency of plant species and an Index of Agreement on Remedies (IAR), based on the following formula: IAR= $(n_a-n_r)/(n_a-1)$, with n_a being the citation frequency of the ailment and n_r being the number of different plant remedies cited to treat that ailment (Phillips 1996). IAR-values fluctuate between "0" and "1," with "0" representing no consensus at all and "1" total consensus between participants. The limitation of this index is that it becomes less useful for health conditions that are mentioned infrequently; therefore, we limited calculation of IAR-values to health conditions mentioned at least ten times.

Three methods were used to cross reference local (Spanish) and scientific names of plants. The first method consisted of collecting voucher specimens together with Dominican participants in green spaces in New York City. This is the voucher collection of the first author (IV). The second sampling method involved collecting samples of dried plants or plant parts purchased from *botánicas,* defined here as the reference collection (Ref-IV); these samples were often sold as small pieces or fragments of sterile parts (not flowering and/or fruiting), and typically sold in bags or in bulk. A third method consisted of using a plant portfolio displaying color photos of commonly used plants with numbered pictures for botanical identification purposes (Pic-Dom). The plant portfolio is particularly useful for verification of food plants. Since food plants are well known from a botanical perspective, no voucher collection was made from them. Photographs included in this portfolio were taken in New York City during visits to *botánicas* and Latino grocery stores (*bodegas*), and in the Dominican Republic in markets and in the field. Researchers were accompanied by knowledgeable Dominican participants who confirmed the local Spanish names of the plants. Plant images were provided by Irina Adam, Michael Balick, Flor Henderson, Andreana Ososki, and Ina Vandebroek.

Results and Discussion

Ninety-seven of 175 Dominican respondents mentioned medicinal plant remedies which they use(d) "only in New York City" (thereby excluding plants used "only in the Dominican Republic" or "in both New York City and Dominican Republic"). In total, there were 596 responses of herbal remedies to treat specific health conditions, and a total of 762 individual plant use reports, indicating that medicinal plants are often used in mixtures. Forty-one plants account for 74 percent of all the responses, and twenty-six plants are mentioned at least 10 times. Table 2.1 lists the citation frequency (of plant remedies) and IAR-values for the health conditions studied, as well as the number of different plants that were reported for treatment of a particular health condition. IAR-values fluctuate between "0" and "1," with "0" representing no consensus, "0.5" average consensus, and "1" total consensus.

Participants most frequently cited medicinal plants (fifty citations or more) for the following ailments (in descending order): flu and common cold, elevated levels of cholesterol, diabetes, asthma, bronchitis and cough. Consensus about plant remedies was high for flu and common cold, and for cholesterol (IAR-value of 0.73 and 0.76, respectively), indicating that participants know very well how to treat these conditions with medicinal plants and frequently cite the same plant remedies. The same holds true for asthma (IAR=0.62). On the other hand, consensus is borderline for bronchitis and cough (IAR=0.51), and low for diabetes (IAR=0.46). This shows that, in general, participants use a high diversity of plant remedies in New York City to treat diabetes, and that particularly popular and hence widely known plant remedies may not exist for this health condition among the study group.

The total number of 762 individual plant use reports corresponded with 125 different medicinal plants. Hence, the same plants are cited repeatedly for similar or different ailments, and some plants are cited more frequently than others. Sixty-one percent of these were food plants that are used as medicine, meaning

Table 2.1. Index of Agreement on Remedies (IAR) Values for Common
Health Conditions

Health condition/Ailment (*Spanish name*)	Response frequency in interviews (n=97)*	Number of plant remedies cited	IAR-value**
flu and common cold (*gripe y resfriado/catarro*)	107	30	0.73
high cholesterol (*colesterol alto*)	79	20	0.76
diabetes (*diabetes, azúcar*)	62	34	0.46
asthma/chest congestion (*asma/pecho apretado*)	59	23	0.62
bronchitis and cough (*bronquitis y tos*)	52	26	0.51
hypertension (*presión alta*)	48	23	0.53
sinusitis (*sinusitis*)	29	16	0.46
kidney problems and stones (*riñones y cálculos/botar piedras*)	28	14	0.52
rheumatism (*artritis/reumatismo*)	28	17	0.41
diarrhea (*diarrea*)	25	13	0.50
menstrual pain (*dolor menstrual*)	15	8	0.50
fungal skin infections (*hongos de la piel, mazamorra, paños, manchas blancas en la piel*)	15	11	0.29
burns (*quemaduras*)	14	6	0.62
labor pain and puerperium (*dolor del parto, posparto, entuerto*)	13	7	0.50

vaginal discharge (*flujo vaginal, flujo blanco*)	11	9	0.20
hypotension (*presión baja*)	8	4	—
infertility (*infertilidad, cuando alguien no puede tener bebes y quiere tenerlos*)	6	6	—
wounds (*heridas*)	5	3	—
back pain (*dolor de espalda*)	4	3	—
impotence (*impotencia*)	4	4	—
furuncles (*nacíos*)	3	3	—
sprains (*torcedura de huesos, huesos torcidos, esguince, zafadura de mano, pie, tobillo, dedos*)	3	2	—
trauma (*traumatismo de golpe, chichón, moratón, para debaratar golpes*	3	2	—
herpes zoster (*culebrilla*)	2	2	—
birth control (*contracepción, para no tener más niños*)	2	2	—

*The response frequency refers to the total number of times a plant remedy (either a plant mixture or a single plant) was reported for treating a specific health condition in New York City by the sample population. Out of 175 interviewees, 97 mentioned medicinal plants that were used "only in New York City" for treatment of health problems.

**IAR-values were calculated for health conditions mentioned at least ten times, based on the following formula: $(n_a-n_r)/(n_a-1)$, with n_a representing the citation frequency of an ailment and n_r being the number of different plant remedies cited to treat that ailment. IAR-values fluctuate between "0" and "1," with "0" representing no consensus at all and "1" total consensus.

that they have a primary culinary application (as fruits, vegetables, staple foods, or condiments) and are also used by Dominicans for medicinal purposes. Table 2.2 lists twenty-six plants used in New York City that were cited at least ten times. Foods constituted the majority of the most frequently cited plants (twenty-two of twenty-six, or 85 percent).

According to table 2.2, *Aloe vera* (L.) Burm.f., commonly known by Dominicans as *sábila* (English common name: aloe), is the most popular and most versatile species. It is used in New York City to treat twenty-two different ailments, and—based on the citation frequency for individual ailments—its major uses include diabetes, obesity, and cancer. The plant part that is used is the inner leaf gel (called *cristál*) that is obtained after peeling the leaves. Several of the uses recorded for *sábila* in New York City are corroborated by literature from the Dominican Republic or the Caribbean, including its application for flu and common cold, asthma, wounds and other skin conditions, hair care, and constipation (see table 2.2 for references). However, other applications for this species according to the published literature were not recorded in the data subset of New York City. These include its use for menstrual disorders, liver ailments, inflammation, burns, and as a vermifuge. Although menstrual disorders were not mentioned specifically in the present study for the NYC data subset, other uses reported for *sábila* in New York City that are related to the genito-urinary system include vaginal discharge, cleansing of the reproductive organs, and infertility. Balick et al. (2000) report on the use of *Aloe vera* by Latino healers in New York City for treatment of uterine fibroids (benign connective tissue tumors). NYC participants cited the raw tubercles of *papa* (potato, *Solanum tuberosum* L.), and to a lesser extent the fried seeds of *bija* (lipstick tree, *Bixa orellana* L.), rather than *sábila* for burns. No NYC uses as a vermifuge were reported for any plant species in the present study, probably because intestinal parasites are common in the tropics but not in a temperate urban setting.

A comparison of plants reported as used "in New York City only" and literature data from the Dominican Republic and the Caribbean (table 2.2) shows that the literature confirms most of the major medicinal uses for different species in New York City. Hence, the use of any of the following plants to treat flu and/or common cold is similar between New York City and the Dominican Republic: *limón* (lime and lemon, *Citrus aurantifolia* (Christm.) Swingle and *Citrus limon* (L.) Burm.f., respectively) (juice, 19 citations); *cebolla* (onion, *Allium cepa* L.) and *cebollín* (shallot, *Allium cepa* var. *aggregatum* G. Don) (juice, 16 citations); *manzana* (apple; *Malus* spp.) (tea from fruit, 11 citations); *canela* (cinnamon, *Cinnamomum verum* Presl) (tea from bark, 11 citations); *higuereta* (castor bean, *Ricinus communis* L.) (oil from seeds, 9 citations); *berro* (watercress, *Nasturtium officinale* R. Br.) (juice, 6 citations); *jengibre* (ginger, *Zingiber officinale* Roscoe) (tea from root, 6 citations); *naranja agria* (bitter orange, *Citrus aurantium* L.) (tea from leaves, 6 citations); *sábila* (juice, 4 citations).

Table 2.2. Medicinal Plants Used by Dominican Immigrants in New York City That Were Cited at Least Ten Times During the Survey. Plants are ranked according to decreasing cumulative citation frequency (cit. freq.) and are presented as follows: local name {scientific name, family} [reference number**]

Cit. freq.	Medicinal use	Alone (A) or Mixed with other plants (M)	Administration	Use in the Dominican Republic and the Caribbean* according to literature data
56 citations: *sábila* {*Aloe vera* (L.) Burm.f., Asphodelaceae}[Pic-Dom-130, Pic-Dom-133, Ref-IV-20]				
6	diabetes	A, M	oral	
5	obesity	A, M	oral	
5	cancer, incl. leukemia (prevention and treatment)	A, M	oral	
4	high cholesterol	A, M	oral	
4	flu and common cold	A, M	oral	flu and common cold[2,6]
3	asthma	M	oral	asthma[2]
3	bronchitis and cough	A, M	oral	
3	skin fungi, incl. *Pityriasis versicolor* (*paños*)	A	oral, topical	skin patches[2]
3	wounds, cuts and scratches	A	oral, topical	cuts and scratches[2], anti-hemorrhagic (styptic)[1]
3	skin and hair care	A	oral, topical	baldness[2]
3	vaginal discharge	A	oral, "ovule"	
2	constipation	A, M	oral	digestive, constipation[1]
2	trauma, to increase muscle strength	M	oral, topical	
2	sinusitis	A, M	oral	
1	arthritis	A	oral	
1	furuncles (*nacíos*)	A	topical	
1	insect bites, swelling of the legs	A	oral	
1	toothache	A	topical	
1	ophthalmic	A	topical	
1	infertility	A	oral	
1	cleanse the reproductive organs	M	oral	
1	AIDS	A	oral	
				menstrual disorders[1,5] liver ailments[1] inflammation and burns[1] purgative, vermifuge[1]

(continued)

Table 2.2. Continued

47 citations: *ajo* {*Allium sativum* L., Alliaceae} [pic-dom-8]				
21	hypertension	A	oral	hypertension[1,4,7]
4	stomachache, ulcers	M	oral	flatulence and stomach problems[2,6,7]
3	hypotension	A	oral	
3	high cholesterol	A, M	oral	
3	fungal skin infections	A	topical	skin problems, including candidiasis and rash[2]
2	rheumatism	A, M	oral, topical	rheumatism[1]
2	asthma	M	oral	
2	obesity	A, M	oral	
2	bronchitis and cough	A, M	oral	
1	insomnia	M	oral	
1	diarrhea	A	oral	
1	flu and common cold	M	oral	febrifuge[2,4]
1	depression	M	oral	
1	uterus (unspecified)	M	oral	uterus ("raises the uterus")[5] vermifuge[1,2,4] toothache and earache[2,4] "bad blood"[2]

43 citations: *cebolla* {*Allium cepa* L., Alliaceae} [Pic-Dom-32] and *cebollín* {*Allium cepa* var. *aggregatum* G. Don, Alliaceae} [Pic-Dom-36]				
16	flu and common cold	A, M	oral	common cold[1]
11	asthma, chest (unspec.)	A, M	oral	"chest congestion" (*pecho apretado*)[7]
10	bronchitis and cough	A, M	oral	bronchitis, respiratory infections[1,2,8]
2	sinusitis	M	oral	
1	rheumatism	M	oral	
1	acne	A	topical	
1	ulcers (unspec.)	A	oral	
1	obesity	M	oral	
				edema[1] kidney problems[7] oral candidiasis[2] infertility[5]

40 citations: *limón/limon agrio* {*Citrus aurantifolia* (Christm.) Swingle, Rutaceae} [Pic-Dom-78] and {*Citrus limon* (L.) Burm.f., Rutaceae} [Pic-Dom-75]				
19	flu, fever and common cold	A, M	oral, internal	flu and common cold[1,2,3,6,7] febrifuge[1,2,3,6]
4	cough, chest (unspec.)	A, M	oral	cough[2]
4	high cholesterol	M	oral	

3	asthma	M	oral	
3	rheumatism	A, M	oral	rheumatism[9]
3	diarrhea	A, M	oral	diarrhea[1,2,3,7,9], dysentery[1]
2	sinusitis, headache	A	oral, internal	headache[2]
2	trauma, bruises	A	topical	
2	diabetes	M	oral	
1	vaginal discharge	A	oral	
1	herpes zoster (culebrilla)	A	ritual	
1	kidney problems	M	oral	
1	fungal skin infections	A	topical	fungal skin infections (mazamorra)[7]
1	diuretic	M	oral	
				ophthalmic[1,2]
				abundant menstruation[5]
				gonorrhea[1]
				earache[2]
				scurvy[1]
				flatulence[9]
				inflammation[9]
				antidote for Euphorbiaceae poisoning[1]

30 citations: *avena* {*Avena sativa* L., Poaceae}[Pic-Dom-25]

26	high cholesterol	A, M	oral	
1	diabetes	A	oral	
1	obesity	M	oral	
1	hypertension	A	oral	
				spasmolytic[1]
				CNS tonic[1]

29 citations: *manzanilla* {*Matricaria recutita* L., Asteraceae}[Pic-Dom-92, Ref-IV-27]

9	menstrual pain, labor pain, and puerperium	A, M	oral	menstrual pain and puerperium[2,4], uterine colic, menstrual disorders[5]
6	insomnia and CNS tranquilizer	M	oral	nervousness[5]
3	stomachache and flatulence	A, M	oral	stomachache, diarrhea, spasmolytic and carminative[1,2,4]
3	bronchitis and cough	M	oral	
2	asthma	M	oral	
1	blood circulation	A	oral	
1	infertility	A	oral	
1	cysts	M	oral	tumors[5]
1	sinusitis	A	internal	
1	depression	M	oral	

(continued)

Table 2.2. Continued

1	flu and common cold	M	oral	febrifuge (*Chrysanthemum parthenium*)[1] wound healing[1]

27 citations: *berro* {*Nasturtium officinale* R. Br., Brassicaceae}[Pic-Dom-23]

12	asthma, chest (unspec.), cleanse lungs	A, M	oral	expectorant[1]
9	bronchitis and cough	A, M	oral	bronchitis[1]
6	flu and common cold	A, M	oral	flu[6]
				diuretic[1] tuberculosis[1] diabetes[1] aphrodisiac[1] scurvy[1]

24 citations: *jengibre* {*Zingiber officinale* Roscoe, Zingiberaceae}

6	flu and common cold	M	oral	flu and common cold[1,2,3,4,6], febrifuge[2,4,6]
4	labor pain and puerperium	A	oral	
4	rheumatism	A, M	oral, topical	rheumatism[1,7]
2	sinusitis	A, M	oral, topical	
2	sprains and back pain	M	oral, topical	
1	obesity	M	oral	
1	constipation	M	oral	
1	skin fungi	A	topical	
1	cough	A	oral	cough[2,4], respiratory infections[1]
1	diabetes	M	oral	
1	hypothermia	A	oral	hypothermia[1] tonic[1] indigestion, carminative, stomachache[1,2,4,6, 9] toothache[1] aphrodisiac[1] asthma[2] diarrhea[2,4] vomiting[2,4] regulates menstruation[5] facial paralysis[1]

23 citations: *higuereta* {*Ricinus communis* L., Euphorbiaceae}[Pic-Dom-68]

9	flu and common cold	A, M	oral	flu and common cold[1,6]
5	asthma	A, M	oral	asthma (*pecho apretado*)[2,4,7,8, 9]
4	sinusitis	M	topical	headache[1,2,4]
3	bronchitis and cough	A, M	oral	chest (unspec.)[1], respiratory infections[2,4]
1	rheumatism	A	topical	rheumatism[2,4]
1	AIDS	M	oral	
				toothache, earache[2,4,7,8, 9]
				constipation[2,4]
				burns[2,4]
				trauma[2,4]
				sprains[2,4]
				tumors[5]
				purgative, vermifuge[1]
				wounds, abscesses (veterinary use)[1]

22 citations: *canela* {*Cinnamomum verum* Presl, Lauraceae}[Pic-Dom-31]

11	flu and common cold	M	oral	febrifuge[7]
3	hypotension	A, M	oral	
2	asthma, chest (unspec.)	M	oral	
2	diarrhea, intestinal problems	M	oral	diarrhea[2,4]
2	bronchitis and cough	M	oral	
1	menstrual pain	A	oral	
1	hair loss	M	topical	
				vomiting[2,4,7, 9]
				headache[7]

19 citations: *apio*/celery {*Apium graveolens* L., Apiaceae}[Pic-Dom-18]

6	high cholesterol	M	oral	
3	bronchitis and cough	A, M	oral	hoarseness[1]
3	hypertension	M	oral	
2	kidney problems	M	oral	edema (diuretic)[1]
2	diabetes	M	oral	
2	obesity	M	oral	
1	anemia	M	oral	
				rheumatism[1]
				scurvy[1]
				carminative[1]
				bruises[1]
				ulcers[1]

(continued)

Table 2.2. Continued

18 citations: *piña* {*Ananas comosus* (L.) Merr., Bromeliaceae}[Pic-Dom-114]				
7	high cholesterol	A, M	oral	
3	obesity	M	oral	
2	rheumatism	A, M	oral	
2	diabetes	M	oral	
2	hypertension	M	oral	
1	diuretic	M	oral	diuretic[1,6]
1	constipation	M	oral	
				parasites, vermifuge[1]
				digestive[1]
				ulcers[1]
				antidote for poisoning[1]
				regulates menstruation[1,5]
				cough[6]

17 citations: *rábano* {*Raphanus sativus* L., Brassicaceae}[Pic-Dom-118]				
7	asthma, chest, lungs (unspec.)	M	oral	
7	bronchitis and cough	M	oral	respiratory infections[1]
3	flu and common cold	M	oral	
				diuretic[1]
				gallstones[1]

15 citations: *naranja agria* {*Citrus aurantium* L., Rutaceae}[Pic-Dom-101, Pic-Dom-104, Pic-Dom-105]				
6	flu	M	oral	flu, febrifuge[1,2,3,6,7,9]
5	sinusitis, headache	A, M	oral, topical	headache[2,7,9]
2	hypertension	A, M	oral	hypertension[7,9]
1	obesity	M	oral	
1	high cholesterol	M	oral	
				laxative, purgative[1]
				antiseptic[1]
				hemostatic[1]
				colic, flatulence[2], stomach problems[7], vomiting[9]
				opthalmic[2]
				diarrhea[2]
				intestinal parasites[2]
				cough[2]
				chest congestion (*pecho apretado*)[6]
				sweating[5]
				burns[7,9]

15 citations: *tilo* {*Tilia* spp., Malvaceae}[Ref-IV-26]			
5	CNS tranquilizer	A, M	oral
5	insomnia	A, M	oral
2	flu and common cold	M	oral
1	intestinal problems	M	oral
1	depression	M	oral
1	menstrual pain	M	oral

15 citations: *manzana* {*Malus* spp., Rosaceae} [Pic-Dom-95]				
11	flu and common cold	A, M	oral	common cold[1]
2	diarrhea	A, M	oral	
1	cough	M	oral	cough[1]
1	depression	M	oral	
				mild laxative[1]

14 citations: *pepino* {*Cucumis sativus* L., Cucurbitaceae}[Pic-Dom-116]			
7	high cholesterol	A, M	oral
3	hypertension	A, M	oral
2	obesity	M	oral
2	diabetes	A, M	oral

14 citations: *anís de estrella* {*Illicium verum* Hook. f., Illiciaceae}[Pic-Dom-17]			
3	flu and common cold	M	oral
3	menstrual pain	M	oral
2	abdomen, intestines (unspec.)	M	oral
2	vaginal discharge	M	oral
1	insomnia	M	oral
1	CNS tranquilizer	M	oral
1	depression	M	oral
1	cysts of the reproductive organs	M	oral

13 citations: *bija* {*Bixa orellana* L., Bixaceae}[Pic-Dom-20, Ref-IV-15]				
3	anemia	A, M	oral	"weakness"[7,8, 9]
3	diabetes	A, M	oral	
3	burns	A, M	topical	burns[2,4,6,7,8, 9]
2	vaginal discharge	A, M	oral	vaginal infections[1]
1	leukemia	M	oral	
1	ulcers (unspec.)	A	oral	
				antidote for poisoning[1]
				trauma[6,7,8, 9]
				flu[6], febrifuge[1]
				insect repellant[1]

(continued)

Table 2.2. Continued

13 citations: *perejil* {*Petroselinum crispum* (Mill.) Nyman ex. A.W. Hill, Apiaceae}[Pic-Dom-113]

4	high cholesterol	A, M	oral	
3	rheumatism	A, M	oral, topical	
2	regulates blood pressure, circulation	A	oral	
1	diabetes	M	oral	
1	obesity	A	oral	
1	cough	M	oral	
1	diuretic	M	oral	diuretic[1] carminative[1] febrifuge[1] contusions[1] insect bites and stings[1] emmenagogue, regulates menstruation[1,5] infertility[5] sweating[5] kidney ailments[1]

12 citations: *malagueta* {*Pimenta dioica* (L.) Merr., Myrtaceae}[Pic-Dom-97, Ref-IV-25]

6	diabetes	A, M	oral	
4	diarrhea and vomiting	A, M	oral	vomiting[2,4,7,8, 9]
1	labor pain and puerperium	M	oral	
1	chest (unspec.)	M	oral	

11 citations: *zanahoria* {*Daucus carota* L., Apiaceae}[Pic-Dom-140]

3	high cholesterol	A, M	oral	
1	impotence	M	oral	
1	vision	A	oral	
1	flu	M	oral	
1	diabetes	M	oral	
1	immunostimulant, "good for blood"	M	oral	tonic[1]
1	diarrhea	M	oral	diarrhea, intestinal ailments[1]
1	fungal skin infections (*paños*)	M	oral	
1	hypertension	M	oral	
				stomach and liver ailments[1] regulates menstruation[5] tumors[5]

10 citations: *yerba buena, hierba buena* {*Mentha* spp., Lamiaceae}[Pic-Dom-70]				
3	stomach ailments, incl. ulcers and gastritis	A	oral	stomach ailments, indigestion, vomiting, colic[1,4,7]
2	hair care	M	topical	
1	diabetes	M	oral	
1	furuncles (*nacíos*)	A	topical	
1	burns	A	topical	
1	hypertension	A	oral	hypertension[8, 9]
1	menstrual pain	A	oral	
				cough[1]
				flu[3,4,6], febrifuge[3,6]
				diarrhea[4]

10 citations: *maiz* {*Zea mays* L., Poaceae}[Pic-Dom-87]				
7	kidney problems, incl. stones	A, M	oral	kidney problems, incl. stones[1,2,3,4,6,7,8, 9]
2	vaginal discharge	A, M	oral	
1	urinary infection	M	oral	
				diuretic[1]
				heart ailments[1]
				rheumatism[1]
				prostatitis[1]
				liver ailments[1]
				inflammation[2,4]
				edema[2,4]
				regulates menstruation[5]

10 citations: *papa* {*Solanum tuberosum* L., Solanaceae}[Pic-Dom-109]				
6	burns	A, M	topical	burns[7]
1	ulcers (unspec.)	A	oral	gastroduodenal ulcers[2]
1	cough	M	unspec.	
1	pain (unspec.)	A	topical	
1	sinusitis	M	topical	headache[2,4]
				baldness[2]
				hematoma[2]
				flu[6]

*The following countries in the Caribbean are included: Dominican Republic, Haiti, Antigua, Barbados, Belize, Costa Rica, Cuba, Dominica, Granada, Guadalupe, Guatemala, Honduras, Martinique, Nicaragua, Panama, Puerto Rico, and Caribbean regions of Mexico, Columbia, and Venezuela

**The reference number refers to the Dominican plant portfolio collection (Pic-Dom) and/or the bagged samples reference collection from Ina Vandebroek (Ref-IV) (see Material and Methods for more details). Plant family names are in agreement with Haston et al. (2007). [1]Liogier (2000); [2]Germosén-Robineau (2005); [3]Peguero et al. (2001); [4]Germosén-Robineau (1997); [5]Ososki et al. (2002); [6]Peguero (2002); [7]Robineau (1986); [8]Germosén-Robineau (1995); [9]Germosén-Robineau (1991)

In both New York City and the Dominican Republic, several of the species recorded for treatment of flu and common cold are also used for respiratory conditions, including watercress (asthma, 12 citations; bronchitis and cough, 9 citations); *cebolla* and *cebollín* (asthma, 11 citations; bronchitis and cough, 10 citations); *rábano* (radish, *Raphanus sativus* L.) (asthma, 7 citations; bronchitis and cough, 7 citations); *higuereta* (asthma, 5 citations; sinusitis, 4 citations; bronchitis and cough, 3 citations); bitter orange (sinusitis, 5 citations); lime and lemon (cough, 4 citations); and *sábila* (asthma, 3 citations).

Apart from these two large clusters of health conditions (common cold/flu and disorders of the respiratory system), there are other major uses that overlap between those reported in New York City and the Dominican Republic. These include: (a) oral administration of *dientes de ajo* (garlic cloves, *Allium sativum* L.) to treat hypertension (21 citations) and stomach problems (4 citations); (b) tea of *manzanilla* (chamomile, *Matricaria recutita* L.) to calm menstrual and labor pain (9 citations) and insomnia/nervousness (6 citations); (c) tea of *barba de maíz* (corn silk, *Zea mays* L.) for kidney problems and kidney stones (7 citations); (d) application of *papa* (potato, *Solanum tuberosum* L.) to soothe burns (6 citations); (e) application of ginger root for rheumatism (4 citations); and (f) tea of *malagueta* (allspice, *Pimenta dioica* (L.) Merr.) for vomiting (4 citations).

As table 2.2 shows, minor medicinal uses that were mentioned only once or twice in the dataset of plants used "in New York City only" are generally not supported by literature from the Dominican Republic. This illustrates the usefulness of conducting quantitative surveys with a large group of participants and querying the sample population systematically about a given set of health conditions. This methodological approach allows for: (a) identification of those ailments for which people know many herbal remedies; and (b) evaluation of the popularity and specificity of each herbal remedy as expressed by its citation frequency for different ailments. In the present study, herbal remedies were most frequently cited for flu and common cold (107 use reports, see table 2.1). These 107 use reports involved thirty different plants. Of these, nine plants were cited five times or more, and eight of them are listed in table 2.2. Moreover, a comparison with literature data from the Dominican Republic shows that for each of these eight plants, their use in New York City as a remedy for treating flu and common cold is in agreement with their use in the Dominican Republic. Hence, after migration to New York City, Dominicans continue to make use of the same remedies they previously used in the Dominican Republic for treating flu and common cold, and this knowledge seems to be generalized among immigrants. A similar conclusion can be drawn for respiratory conditions. Asthma and bronchitis/ cough respectively rank fourth and fifth according to frequency of mention, and the pooled number of different medicinal plants reported for both health conditions is thirty-four. Seven plants from this group were cited more than five times (aloe, radish, watercress, onion, lemon and lime, shallot and castor bean) and their use for respiratory conditions is similar to that in the Dominican Republic and/or the Caribbean.

Medicinal plant applications recorded in New York City that were not corroborated by literature data from the Dominican Republic are major uses for elevated levels of cholesterol by oral administration of the following plants: *avena* (oats, *Avena sativa* L., 30 citations); *piña* (pineapple, *Ananas comosus* (L.) Merr., 7 citations); *pepino* (cucumber, *Cucumis sativus* L., 7 citations), *apio* (celery, *Apium graveolens* L., 6 citations); *limón* (lemon and lime, 4 citations); *sábila* (aloe, 4 citations) *perejil* (parsley, *Petroselinum crispum* (Mill.) Nyman ex. A.W. Hill, 4 citations); and *zanahoria* (carrot, *Daucus carota* L., 3 citations). Neither is the use of *sábila* to treat diabetes (6 citations), cancer (5 citations), and obesity (5 citations) confirmed by DR literature data. A similar observation can be made for other plant species that are used in New York City to treat hypotension, hypertension, obesity, and insomnia, such as: *tilo* (linden, *Tilia* spp., a tea of linden flowers and leaves serves as a tranquilizer and for insomnia, 10 citations); cinnamon (tea for hypotension, 3 citations); pineapple (supernatant from the fruit skin soaked in water taken internally for obesity, 3 citations); cucumber (juice for hypertension, 3 citations). All these plants, except *sábila* and *tilo,* are food plants. It is likely that Dominicans in New York City are more frequently diagnosed with these health conditions than their peers in the Dominican Republic because of a more generalized use of screening facilities for these "urban lifestyle illnesses" in New York City as compared to the Dominican Republic, and/or a higher prevalence rate of these illnesses in New York City. It is known that cities are both the source of serious threats to the health of the public and the source of many public health innovations (Galea et al. 2005). Hence, these plant uses may well represent "newly acquired" applications after immigration to New York City.

Some Caribbean reference books and more anecdotal publications were not included in the literature review (e.g., Cordero 1978; Núñez Meléndez 1982; Liogier 1990; Estevez and Baez 1992; Longuefosse 1995). The reasons for excluding a literature source were: (1) it was not clear whether the publication was based on a voucher collection, which is important for the correspondence between local and scientific plant names (Cordero 1978; Estevez and Baez 1992); or (2) the precise (country) origin of specific ethnobotanical information provided in the literature source was unclear, or the information presented was based on a compilation from other literature sources, sometimes outside the Caribbean area (Cordero 1978; Núñez Meléndez 1982; Liogier, 1990; Estevez and Baez 1992); or (3) the Dominican Republic is not the principal Caribbean study site (Núñez Meléndez 1982; Longuefosse 1995). Although the Dominican ethnobotanical literature references cited here are not exhaustive, they do include the primary available publications that are specific to the Dominican Republic. The literature data presented in table 2.2 provide a suitable comparison with our NYC data subset since it was derived from the most important, systematic, eth-

nobotanical studies based on good botanical practices conducted in the Dominican Republic to date (Balick 1999).

Conclusion

A comparison of medicinal plant uses reported by Dominican immigrants in
New York City for the treatment of thirty common health conditions with a review of the Dominican ethnobotanical literature has resulted in the documentation of similarities and differences in the medicinal applications of herbal
remedies between host and home country. The use of specific plant species to
treat flu/common cold and respiratory disorders in New York City is in agreement with their use in the Dominican Republic. Moreover, knowledge about
herbal remedies for these two groups of health conditions is generalized among
respondents as evidenced by the high citation frequency of individual plant
species and the high index of agreement on remedies for these illnesses. On the
other hand, plant uses reported for elevated levels of cholesterol were not confirmed by the DR literature and hence may be specific to New York City. The
incorporation and disappearance of medicinal plants in healing pharmacopoeias
is a continuous process that responds to both external and internal challenges.
These include limited availability of herbal remedies, prevalence of diseases, and
the loss of cultural practices and plant knowledge. Nearly all the plants used to
treat cholesterol are food plants, which are readily available in New York City.
Hence, it is likely that the use of these plants has evolved in response to immigrants
having to deal with a new lifestyle in a different environment, where they might
be diagnosed with diseases previously unfamiliar to them, while at the same time
they are confronted with a limited number of herbal healing resources. All the
plants listed in the present study are currently available in the Dominican Republic as well as in New York City, but many may not always have been easily
accessible in the past. For example, apples have only become available on a year
round basis in markets and supermarkets in Santo Domingo since about twenty
years ago, and even more recently in rural areas in the Dominican Republic
(Peguero, pers. comm.). Further research is needed to determine whether these
plants are also currently used in the Dominican Republic to treat cholesterol,
hypertension, insomnia, and diabetes. Presently, we are conducting comparative
research in the Dominican Republic using the same questionnaire that was administered in New York City with minor modifications. This will allow us to make
a direct transnational comparison of frequently treated illnesses and herbal remedies.

Acknowledgements

The authors wish to thank the Dominican community in New York City and our study participants in particular for their friendship, their generosity in sharing plant knowledge, their

patience in collaborating until the end of interviews which often took as long as two hours to complete, and their help with referring us to other participants. This project was supported by a grant from the National Institutes of Health/National Center for Complementary and Alternative Medicine (NIH-NCCAM) (PI: Dr. Michael J. Balick, Grant #5 R21 AT 001889-02).

References

Allen, R., L.F. Cushman, S. Morris, J. Feldman, C. Wade, D. McMahon, M. Moses, and F. Kronenberg. 2000. "Use of complementary and alternative medicine among Dominican emergency department patients." *American Journal of Emergency Medicine* 18: 51–54.

Balick, M.J. 1999. "Good botanical practices." In *Botanical medicine: Efficacy, quality, assurance, and regulation,* ed. D. Eskinazi, M. Blumenthal, F. Farnsworth, and C. Riggins. New York: Mary Ann Liebert, 121–125.

Balick, M.J., F. Kronenberg, A.L. Ososki, M. Reiff, A. Fugh-Berman, B. O'Connor, M. Roble, P. Lohr, and D. Atha. 2000. "Medicinal plants used by Latino healers for women's health conditions in New York City." *Economic Botany* 54: 344–357.

Borrello, M.A., and E. Mathias. 1977. "Botánicas: Puerto Rican folk pharmacies." *Natural History* 86: 65–73.

Brendbekken, M. 1998. *Hablando con la mata.* Santo Domingo, Dominican Republic: Instituto de Medicina Dominicana.

Carrasquillo, O., A. Carrasquillo, and S. Shea. 2000. "Health insurance coverage of immigrants living in the United States: Differences by citizenship status and country of origin." *American Journal of Public Health* 90: 917–923.

Cordero, A.B. 1978. *Manual de medicina domestica (plantas medicinales Dominicanas).* Santo Domingo, Dominican Republic: Editora Taller.

Delgado, M. 1996. "Puerto Rican elders and botanical shops: A community resource or liability?" *Social Work in Health Care* 23: 67–81.

Delgado, M., and J. Santiago. 1998 "HIV/AIDS in a Puerto Rican/Dominican community: A collaborative project with a botanical shop." *Social Work* 43: 183–186.

Estevez, A., and F. Baez. 1992. *Plantas curativas: Usos populares y científicos.* Santo Domingo, Dominican Republic: Talleres de Servicios Gráficos Oriental.

Fisch, S. 1968. "Botánicas and spiritualism in a metropolis." *The Milbank Memorial Fund Quarterly* 46: 377–388.

Galea, S., N. Freudenberg, and D. Vlahov. 2005. "Cities and population health." *Social Science & Medicine* 60: 1017–1033.

Germosén-Robineau, L., ed. 1991. *Towards a Caribbean pharmacopoeia. TRAMIL 4 Edition.* Santo Domingo, Dominican Republic: enda-caribe, UNAH.

———. 1995. *Hacía una farmacopea Caribeña. Investigación científica y uso popular de plantas medicinales en el Caribe. Edición TRAMIL 7.* Santo Domingo, Dominican Republic: enda-caribe, UAG & Universidad de Antioquía.

———. 1997. *Farmacopea Caribeña (primera edición).* Fort-de-France, Martinique: Ediciones Emile Désormeaux.

———. 2005. *Farmacopea vegetal Caribeña. Edición Especial Dominicana.* León, Nicaragua: Editorial Universitaria UNAN-León.

Gomez-Beloz, A., and N. Chavez. 2001. "The botánica as a culturally appropriate health care option for Latinos." *The Journal of Alternative and Complementary Medicine* 7: 537–546.

Haston, E., J.E. Richardson, P.F. Stevens, M.W. Chase, and D.J. Harris. 2007. A linear sequence of Angiosperm Phylogeny Group II families. *Taxon* 56: 7–12.

Hernández, C.J., and M.O. Jones. 2004. "Botánicas: Sites of healing and community support." In *Botánica Los Angeles: Latino popular religious art in the City of Angels,* ed. P.A. Polk. Los Angeles: UCLA Fowler Museum of Cultural History, 46–55.

Jones, M.O., and P.A. Polk. 2001. "Invisible hospitals: Botánicas in ethnic health care." In *Healing logics: Culture and medicine in modern health belief systems,* ed. E. Brady. Logan: Utah State University Press, 39–87.

Kaiser Commission on Medicaid and the Uninsured. 2003. "Immigrants' health care coverage and access." Retrieved 26 June 2006 from http://www.kff.org/uninsured/upload/Immigrants-Health-Care-Coverage-and-Access-fact-sheet.pdf

Karpati, A., X. Lu, F. Mostashari, L. Thorpe, and T.R. Frieden. 2003. "The health of Inwood and Washington Heights." *NYC community health profiles* 1 (24): 1–12.

Liogier, A.H. 1990. *Plantas medicinales de Puerto Rico y del Caribe.* San Juan, Puerto Rico: Iberoamericana de Ediciones.

———. 2000. *Diccionario botánico de nombres vulgares de la Española.* Santo Domingo, Dominican Republic: Jardín Botánico Nacional Dr. Rafael Ma. Moscoso.

Longuefosse, J.L. 1995. *100 plantes médicinales de la Caribe.* Trinité, Martinique: Gondwana Editions.

Mikhail, N., S. Wali, and I. Ziment. 2004. "Use of alternative medicine among Hispanics." *The Journal of Alternative and Complementary Medicine* 10: 851–859.

New York City Department of City Planning. 2004. "The newest New Yorkers." Briefing booklet. "Immigrant New York in the new millennium." Retrieved 18 October 2005 from http://www.fahsi.org/2000%20Immigration%20Booklet.pdf

Núñez Meléndez, E. 1982. *Plantas medicinales de Puerto Rico.* Río Piedras, Puerto Rico: Editorial de la Universidad de Puerto Rico.

Ososki, A.L., P. Lohr, M. Reiff, M.J. Balick, F. Kronenberg, A. Fugh-Berman, and B. O'Connor. 2002. "Ethnobotanical literature survey of medicinal plants in the Dominican Republic used for women's health conditions." *Journal of Ethnopharmacology* 79: 285–298.

Peguero, B. 2002. "Estudio etnobotánico de las comunidades ubicadas dentro y en la periferia del Parque Nacional Juan Bautista Pérez Rancier." In *Evaluación ecológica integradra del Parque Nacional Juan B. Pérez Rancier (Valle Nuevo),* ed. F. Núñez. Santo Domingo, Dominican Republic: Secretária de Estado de Medio Ambiente y recursos / Fundación Moscoso Puello, 57–79.

Peguero, B., F. Jiménez, and A. Veloz. 2001. "Estudio etnobotánico en El Cachote, Provincia Barahona, Republica Dominicana." *Moscosoa* 12: 79–104.

Phillips, O.L. 1996. "Some quantitative methods for analyzing ethnobotanical knowledge." In *Selected guidelines for ethnobotanical research: A field manual,* ed. M.N. Alexiades. New York: The New York Botanical Garden, 171–197.

Polanco, D., B. Peguero, and F. Jiménez. 1998. "Estudio etnobotánico en siete comunidades rurales de Bayaguana, República Dominicana." *Moscosoa* 10: 86–113.

Reiff, M., B. O'Connor, F. Kronenberg, M.J. Balick, P. Lohr, M. Roble, A. Fugh-Berman, and K.D. Johnson. 2003. "Ethnomedicine in the urban environment: Dominican healers in New York City." *Human Organization* 62: 12–26.

Rivera-Batiz, F.L. 2002. "The socioeconomic status of Hispanic New Yorkers: Current trends and future prospects." Pew Hispanic Center Study Report. Washington, DC: Pew Hispanic Center.

Robineau, L. 1986. "Encuesta sobre medicina tradicional popular en una zona rural y una zona urbana marginal de la República Dominicana." *Moscosoa* 4: 226–265.

Viladrich, A. 2006. "*Botánicas* in America's backyard: Uncovering the world of Latino healers' herb-healing practices in New York City". *Human Organization* 65: 407–419.

Chapter 3

Between Bellyaches and Lucky Charms

Revealing Latinos' Plant-Healing Knowledge and Practices in New York City

Anahí Viladrich

Introduction

This chapter provides a contribution to the fields of urban ethnomedicine and medical anthropology by examining the reliance on alternative forms of healing among Latino immigrants, via the provision of and access to traditional plants and herbs in New York City. In particular, this chapter explores the migratory careers of Latino healers in New York City, and the ways in which they adapt their prescription of herbs and plants both to what is affordable and available to their Latino clientele.

Two bodies of research inform the focus of this chapter: (1) classic studies on *Santería*, and; (2) recent work on the role of *botánicas* (ethnic healing stores) that serve the Latino and Caribbean population in the United States. Seminal studies on *Santería* and other spiritual faiths, including *Espiritismo* (Spiritism), have built the foundations for our understanding of the impact of collective religious practices as resistance and accommodation against oppression (Baer 1984; Singer and Baer 1995), as well as on the formation of alternative mental health systems (Koss-Chioino 1975; Garrison 1977; Sandoval 1977; Koss-Chioino 1992) that serve as multifunctional institutions in building community resources (Harwood 1977a; 1977b). Both *Santería* and *Espiritismo* rest on the belief of divine beings interacting with those alive, as well as on the veneration of the spirits of the dead. Studies on African-related religious systems (e.g., Voodoo, *Santería*, the Orisha religion) have uncovered the role and strength of religious healing in the construction of diaspora communities (Cabrera 1971; Sandoval 1977; González-

Whippler 1989; McCarthy Brown 1991; Gregory 1999; Moreno Vega 2000; Fernández Olmos and Paravisini-Gebert 2003; De la Torre 2004). These faiths are also understood as adaptive coping systems that help Caribbean and Latino immigrants during times of stress (Paulino 1996; Viladrich 2006a).

Santería and other Yoruba and Congolese traditions rely on divination (as the main consultation vehicle between divinities and their followers), animal sacrifice to honor the pantheon's divinities, spirit possession, and secrecy. *Santería* also underscores functional aspects by providing conflict resolution skills to their followers, social stability (Martinez and Wetli 1982), and stress-reduction guidelines (Perez y Mena 1977).

Botánicas, which refer literally to botany, are local dispensaries that offer healing, religious, and spiritual products and services to a diverse clientele comprised mostly of immigrants from Latin American and Caribbean countries, including those of Cuban and Haitian descent.[1] Recent studies have focused on the contributions of *botánicas* to ethnobotany via the selling of plants and herbal infusions for the treatment of specific health problems, including the processes through which healers diagnose and treat women's health conditions (Balick et al. 1999; Ososki et al. 2002; Reiff et al. 2003). Researchers have also noted their role as a viable substitute of health care, given the financial and cultural barriers to Western medicine (Gomez-Beloz and Chavez 2001; Balick and Lee 2001). Today's *botánicas* offer a much more diversified set of healing and religious practices, including Afro-Caribbean religious beliefs and traditions such as *Santería,* Palo Monte, Voodoo and Spiritism (Fernandez Olmos and Paravisini-Gebert 2003; Polk 2004).

Botánicas represent quintessential cultural mélanges where magic, religious syncretism, and herbal recipes are displayed in convivial liaisons with African, Asian, and Latin American spiritual traditions (Romberg 2003). Defined as "invisible hospitals" and as ethnic healing-religious enterprises (Jones et al. 2001), *botánicas* have evolved along with the changes that have taken place in the urban fabric. In recent years, they have attracted an international clientele searching either for traces of the Americas' indigenous cultural forms or for innovative versions of New Age healing therapies (Pagel 2004; Polk 2004). If, on the one hand, the increasing publicity and attention of *botánicas* has brought legitimacy to practices long degraded as "uncultivated," they have also contributed to the emergence of innovative lucrative forms of transnational healing practices (Romberg 2003). In this regard, critical voices point out that *botánicas* have been swept away from their authentic ethnic and cultural roots, at the expense of turning into tools for profit by petty capitalist enterprise, delivering venal entertainment treats to buyers of novel esoteric items (Jacobs 2001; Long 2001).

In spite of the valuable studies summarized above, there is still a paucity of research on the overlapping fields of religious beliefs and healing practices among immigrants, particularly when it comes to the use of herbs, fruits, and plants.

Secrecy and the fear of being held accountable for practicing *brujería* (witchcraft) and illegal medicine has kept skeptical practitioners away from revealing the meaning and depth of their practices to outsiders (Mautino 1999). In addition, oral tradition and rules of initiation involved in religious-healing belief systems prevent folk healers from sharing their knowledge with non-practitioners (Espín 2003). Questions that remain to be fully answered relate to the combination of herbs according to diverse healing and religious traditions (e.g., *Santería* and Spiritism) vis-à-vis their organic and chemical properties.

In the following pages, this article will examine the role of *botánicas* (ethnic healing-religious stores) as main outlets for the supply of herbal products and for the provision of informal therapy to a pan-ethnic Latino population in New York City. *Botánicas* in New York City have become a conspicuous door to a disguised market of folk healing products that caters to a steady and ever growing Latino immigrant population in the city. Salespeople, healers, and their clients are not only exposed to a rich diversity of items offered in botánicas, but they are also influenced by a multinational milieu that welcomes assorted belief systems and practices.

By relying on a multimethod ethnographic research design, this chapter examines the careers of healing practitioners in the context of their migratory experiences, as well as their use of plants and herbs for the treatment of diverse ailments vis-à-vis their magical and curative properties. Healers deal with a variety of issues that Latino immigrants face in urban milieus, along a range that takes account of diverse hurdles, including their undocumented status, low-paid jobs, family separation, and major life threatening challenges (e.g, AIDS, cancer). The term *immigrant continuum* is used in this article to refer to the complex set of problems brought by Latino immigrants to a consultation, ranging from stress-related syndromes due to family crises and separation, to the management of chronic and acute life conditions (e.g., arthritis, AIDS).[2] In addition, this article examines the role of *botánicas* in supporting religious networks instituted on the basis of the practice of *Santería,* which find in *botánicas* both a symbolic and physical space for trade and exchange among Latino immigrants in the city.

Uncovering Latinos' Urban Healing Fabric: Research Design and Methods

Between the spring of 2004 and the fall of 2005, a team of ethnographers, sponsored by the Immigration and Health Initiative at Hunter College, was devoted to the identification and mapping of botánicas throughout the five New York City boroughs. The first stage of the project focused on finding botánicas that had healers working on their premises. Researchers were teamed up in groups of two or more for the purpose of visiting specific areas selected for recruitment and follow up. Ph.D. graduates and students, as well as master students of urban health and anthropology, joined the team at different times. Information was collected via a variety of means, including phone listings (e.g., yahoo, yellow pages) that were combined with "on site" visits to neighborhoods presenting the highest concentrations of Latinos.

Ethnographers were careful to develop a rapport with potential key informants which often led to in-depth interviews with qualified practitioners. Criteria for sample selection included being an immigrant from Latin American (first or second generation), and providing consultations for health-related conditions either on a part-time or a full-time basis in exchange for either money or bartering (e.g., offering healing practices for either products or services). Nevertheless, healing exchanges were based on money transactions in all the cases studied. Once eligible informants agreed to participate in the study, a two-phase tape-recorded interview took place in locations and times convenient for them.

Several consent forms were prepared and approved following the regulations of the Hunter College Institutional Review Board aimed at protecting the rights of human subjects. The first part of the interview addressed the healers' migratory history, their methods of diagnosis and treatment, as well as their healing philosophy. The second part focused on the characteristics of their Latino clientele, including their reasons for consultation, particularly with regard to the conditions most frequently treated, symptoms, and prognosis. A qualitative software package (Atlas 5.0) was used for the analysis of all collected data, including field notes, interview schedules, and follow-ups. SPSS and Excel were used for the quantitative measures and graphing of sociodemographic data.

The total sample consisted of 56 healers, 26 males and 30 females. Ethnographers assessed the healers' practice by first asking participants to define themselves, information that was later complemented by field notes as well as by additional data drawn from the in-depth interviews. Nineteen participants defined themselves as *Espiritistas* (Spiritists), 11 as *Santeros* (*Santería* practitioners) and 4 shared these two belief systems. The remaining group of 22 healers followed diverse therapeutic disciplines, considering themselves among other things as herb specialists and clairvoyants. In terms of nationality, 23 healers were from the Dominican Republic, 11 from Puerto Rico, 5 from Cuba, 3 from Colombia, and 7 were born in the United States of Latin American parents (Puerto Rican, Salvadorean, Cuban, and Mexican). One healer was born in Trinidad, although he identified himself as Latino since he had Hispanic ancestry. The remaining group of 6 practitioners comprised nationals from Central and South America, including healers from Guatemala, Peru, Venezuela, Ecuador, Uruguay, and Argentina.[3]

The collection of healers' migratory paths mostly addressed the process that led them to leave their countries of origin for the purpose of settling in the United States. Information on the origin of their customers was assessed via participation observation (recorded in field notes) and conversations held with clients of *botánicas.* Although most healers stated the "international character" of their clientele, customers specifically of Latino origin are predominant in certain neighborhoods.[4] For example, immigrants from the Dominican Republic are more often regulars of *botánicas* located in the Washington Heights area, and Puerto Ricans are more numerous in El Barrio neighborhood located on Manhattan's East Side.

Researchers visited community-based organizations as well as ethnic businesses such as *bodegas,* which are popular grocery stores that provide accessible information to Latinos on a variety of issues. In addition, ethnographers participated in *Santería* rituals (e.g., tambores/drumming and *toques de Santos*) and private ceremonies. In search for practitioners working at home, ethnographers followed up on leads provided by salespeople and other key informants at *botánicas.* This strategy was crucial from a research design perspective, as it allowed us to find a hard-to-recruit population that would probably not have been reached otherwise, as healers working from home are not as visible as their peers practicing on the botánicas' premises.

The Healing Trade: Migration and Transnational Flows

According to most interviewees, herbs and plants sold at the botánicas come from many places, including the tri-state New York City area, Miami, Puerto Rico, and the Dominican Republic, and increasingly from Mexico, Central America, and Brazil. A couple of informants mentioned Miami as the easiest point of access for distributors who receive their shipments in New York City and then sell them to the botánicas, as it is apparently easier to clear customs in Miami than in the JFK airport. In recent years, the international trade of healing products (including herbs and plants) has been encouraged by a widespread access to telecommunications and Internet services, as well as by the promotional costs of airfares, particularly during low seasons. For example, we encountered the case of the owner of one botánica whose father handles the plants and herbs at the source, sending them to her from the Dominican Republic.

In addition, healers express a preference for certain herbs coming from specific countries of origin (or from a specific place within a country) because of their potency, quality, and availability. Fresh herbs grown in the Caribbean and in other subtropical regions are expected to have a special kind of energy (*ashé*), which is supposedly missing among those grown in more artificial environments and sold in dried and mixed forms. Some healers may rely on a particular herb, but doubts about the place of origin may discourage its use. In some cases, practitioners become very punctilious about the sources of plants as they are concerned about non-organically grown herbs, which may have been exposed to dangerous chemicals that affect their quality and safety. Interviewee 100, a respected *babalao* (*Santería* priest) born in Puerto Rico, tells us:

"There is a problem in the United States, at least in my spiritual field, we cannot use natural medicines because of the contamination that has damaged the plants a lot; the plants here in the United States are contaminated and do not have the same effect that the ones in Central America, Peru, Costa Rica [have]. . . . If you think where the natural medicine comes from, it never comes from here because they [plants] are contaminated, and they made them out of chemicals and not in a natural form. . . . That is why in countries like Peru, Mexico, Ecuador, they have the plants that really help, that is why they come from there."

Most of our informants, who learned about their trade in their countries of origin, did not recall having *botánica*-like stores as singular places for healing exchanges when growing up. Plants and herbs, far from constituting esoteric entities, were part of an indigenous pharmacopoeia easily accessible in the open *monte* (forest, mountain), street fairs, and local stores (Cabrera 1971; Viladrich 2006a). For most healers who grew up abroad, their own migratory journey to the United States provided them with new venues to deal with herbs and plants that had been familiar staples in their upbringing. Although most study participants stated their preference for fresh herbs, they were aware of the difficulties in finding their preferred specimens in New York City due to the harsh climate, lack of space, and pressures to keep the equation cost-to-quality within an acceptable range. When recalling their healing careers in both countries, healers often made comparisons between what they could get and use at "home" (their countries of origin) and in New York City. For example, an *Espiritista* from the Dominican Republic (interviewee 313, male) emphatically argued:

"Greens [herbs] come with more [energy] when they are greener, they are like leaves, they have flowers. . . . When they are dry, they have already been gathered, they don't have the same spirit, so it is not the same, they don't have the same strength. . . . In Santo Domingo, there are leaves that you find right there in your neighbor's backyard."

While healers working at *botánicas* are more eager to support the quality and price of the products sold on their premises, those working at home are adamant about the better quality of herbs and plants found in their countries of origin and even in other outlets, such as flower shops. Indeed, some prefer to buy plants in markets and even go hunting for some wild species (e.g., plantain) in empty lots and public parks. Both groups of healers, however, complain about the high prices of goods in New York City due to the costs of transportation, housing, and rent that make their healing products in general, and herbs in particular, more dependent on the monetary trade.

During fieldwork, team members were able to witness the selling of herbs in *botánicas* and the informal consultations that typically take place between employees and clients. Healers and their customers learn from each other, particularly since *connoisseurs* (healing practitioners like *Santeros* or *herberos*) are among the regulars who visit *botánicas* either to get specific products (e.g., herbs, oils, candles), or just to chat and gossip about their competitors' trade. For example, interviewee 202, a multidisciplinary healer (a male *Santero, Espiritista,* and practitioner of Palo Mayombe) from Puerto Rico told us:

> "Here, there are people who come with the same faculties that I have, the same like me, and they come, we talk, and they tell me about their stories and I tell them mine. There is a communication, and understanding and respect about this. We help each other. . . They talk and you tell them what you think, and that can help them somehow."

Contrary to the narrower webs of apprenticeship in which, for the most part, healers in this study had been involved with in their countries of origins, once in New York City they got exposed to a global exchange of information drawn from different disciplines and belief systems. As other authors note, in New York, Cuban patients are treated by Puerto Rican *curanderos* (traditional healers) and Spiritists as well as by *Santeros* (Garrison 1977; Harwood 1977a; Laguerre 1987; Brandon 1991). This diversity makes healers knowledgeable of the religious and cultural beliefs of their diverse clientele (Viladrich 2006b). According to interviewee 207, a male *Santero:*

> "The Chinese people have the same problems as Africans. Africans have the same problems as Puerto Ricans. They either have a bad neighbor that did something to them, they found stuff on their door. . . . You name it. I had them all. I just work with them according to their religion, to help them out. For example, if they believe in, for example, Buddha, I know how to work with that. . . . Cause I can give you plants, I can give you this, none of this [is] gonna work if you don't pray, because it's all belief."

The more a *botánica's* owner or healer knows about different religious practices and belief systems, the more diversified his or her clientele will be, and the greater the likelihood that the *botánica* will have a brisk (and profitable) practice. Rather that being single-minded specialists, successful healers learn from New York City's multicultural diversity to treat clients not only according to their specific needs and traditions, but also to what is affordable and accessible to them. Single religious-healing disciplines that may be the predominant faith in the practitioners' countries of origin achieve a new dimension in New York City as they are combined with different, though related, practices. For example, interviewee 102 (a young male *Santero*) had received an early training in Spiritism while growing up in Puerto Rico, as he belonged to a strong family network of Spiritists. After migrating to New York City a few years ago, he soon became a regular at one of the *botánicas* located in Bushwick, Brooklyn, whose owner had turned into the spiritual father of many young *Santeros* like him. For this young healer, knowledge of the herbal trade had been accompanied by a progressive involvement with *Santería* in New York City as part of his coming of age to become an entrepreneur in the informal, and ever growing, urban economy of healing in New York City. He told us:

> "When I learned about *Santería*, I began to develop Spiritism much more. I learned about other forms of Spiritism, because in Puerto Rico there is one form that is widely practiced: The White Table [*la Mesa Blanca*]. But I learned about other forms, from the Dominican Republic—the Twenty-one Divisions and *el Sansés*, which are very different things that people relate to one religion. And later *Santería*, but *Santería* is much deeper than Spiritism. . . . I practice the White Table, Spiritism, I practice the Twenty-one divisions, *Santería*. . . . *Santería* is what calls my attention the most, it is what I like the most, what I respect the most. . . . Well, all of these religions have a certain connection because the basis is that all of them come from Africa, from different parts of Africa, but their basis is the same."

Santería's Realm and the Collective Construction of Healing

Santería, a descendant of the Yoruba belief systems brought by enslaved populations from Africa to the Caribbean a few centuries ago, achieved its maximum splendor in Cuba during the eighteenth century. *Santería* is currently known extensively as a folk healing and religious system characterized by the syncretic blend of Yoruba gods (Orishas) and Catholic saints. Faiths such as *Santería* and Spiritism are currently widely practiced in Cuba and Puerto Rico, respectively, countries where they have even received some underground blessing from governmental authorities. As is the case with other religious belief systems that have gone "global," Cuba is still the Mecca where *Santería* practitioners achieve their maximum status as priests, and where fans from all over the world get together to reenact their connection with the truthful meaning of the *Orishas'* (deities) healing and religious powers (Wedel 1999).

Although *Santería* is not the only African belief system prevalent in the Caribbean and in Latin America (there is also Voodoo in Haiti, *Candomblé* in Brazil, *Orisha* in Trinidad), it has become one of the most popular religious healing forms in the United States in recent years. The practice of *Santería* in this country has been scattered throughout Spanish-speaking communities of practitioners, particularly via the reproduction of informal networks. Most *botánicas* in New York City are places *par excellence* where healers and clients are able to find almost any *Santería* product they long for (from images to bead necklaces) and where *Santería* networks are reproduced and enacted. Nevertheless, and despite its increasing acceptance, the ritualistic aspects of *Santería* in New York City still hold a secretive nature, partially due to their practitioners' desire to protect themselves from the negative press, and from the New York City Department of Health's inspections that have often made *botánicas* accountable for the selling and dissemination of poisonous substances, including lead and mercury (Laguerre 1987; Zayas and Ozuah 1996; Stern et al. 2003; Prasad 2004). Despite healers' precautionary actions to keep their religion away from outsiders, they have also become conspicuous sponsors of *Santería* practice throughout a myriad of Latino enclaves in the city.

Santería remains a collective religious form in which priests, practitioners, and followers grow to become involved with subtle social webs that find in *botánicas* ideal connecting focal points (Viladrich 2006a). To a certain extent, the social networks of *Santería* get activated in the sites of *botánicas,* either because Santeros and followers go there looking for *cascarilla* (fine eggshell powder used in healing rituals), rue leaves, and flowered water to prepare baths, or to participate in religious ceremonies taking place in their basements. *Santería* webs play an important role in keeping a disperse community of followers connected to each other, while protecting a hierarchical structure in which religious godparents and their godchildren become members of extended families that are subjected to preestablished commands both of duties and rights.

Although in most cases healers, who are also Santeros, acknowledge having been more open about their practices in their countries of origin, they also admit finding in New York City a broader market for experimentation, dissemination, and profit for their healing and religious activities. Some owners of *botánicas* have turned their basements or side rooms into ritualistic temples where followers and priests get together on a regular basis to pay tribute to particular *Orishas* (*Santería* divinities). This is especially relevant from the point of view of understanding the existence, as well as the reproduction, of religious-healing webs among a disperse Latino community in the city.

The teleological use of herbs—oriented, that is, towards an ultimate end beyond the immediate problem addressed—seems to be common among practitioners of African-religious traditions including *Santería* (or *Regla de Ocha*) and *Palo Monte.*[5] Herbs and plants with names such as *abrecaminos* (open paths), *cundeamor* (love spell), *moriyvivi* (I died and lived), *salpafuera* (get out) are prescribed by healters for baths that combine sweet herbs like mint, basil, and cinnamon for the purpose of attracting good energies and loved ones, as well as for repelling the spiritual damage caused by *trabajos* (evil spells) ordered by their clients' enemies. For example, *Santeros* may provide their customers with *trabajos* to either remove haunting spirits from tormenting them or to bring a derailed lover back (see figure 3.1).

Minerals, stones, plants, fruits, and herbs, which are supposed to have the power of delivering a client's wishes, belong to specific *Orishas* in the *Santería's* pantheon, many of whom, as in the case of the *Orisha Babalu Aye,* god protec-

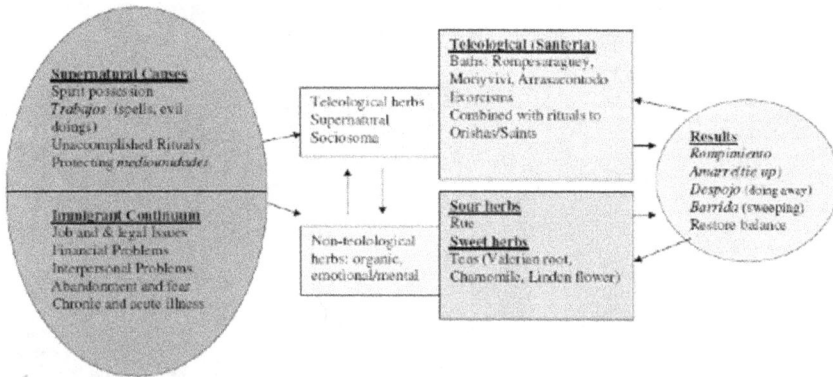

Figure 3.1. *The use of herbs: natural and teleological aims.*
This figure represents two major domains for which herbs and plants are used.
The first one refers to the realm of supernatural causes and effects, and symbolizes
*the spiritual and magical properties (*ashé*) of herbs and plants that impinge upon*
the sufferer's social and physical environment. The second domain, the
immigrant continuum, is related to the organic and chemical properties of plants
(e.g., their calming effect, as in the case of chamomile or valerian root) that
help Latinos deal with their everyday struggles in the United States.

tor of the ill, have Catholic counterparts. This *Orisha* is usually represented by Saint Lazaro, the patron of the poor and the sick, and the one who cures epidemics and infectious diseases (see figure 3.2). Saint Lazaro's baths, easily found at *botánicas,* are popular among those who suffer from leg problems and HIV/AIDS. *Oshún,* another popular *Orisha,* is represented by *la Virgen de la Caridad del Cobre* (the Virgin of the Charity of Copper), a sensuous goddess who embodies the composed energy of sweet rivers and femininity. *Oshún* is able to cure infertility and holds the secrets of witchcraft. *Anis* (star anise) and honey are two of her preferred elements that, given their medicinal and magical properties, are widely used. Interviewee 308 (a male Cuban Santero) tells us about his preference for clove and star anise for special works involving *Oshún's* sexual powers as well as healing:

"The anise, I use it for any [of] *Oshún's* work. Because any work—even anise seed, like in plants or in drinks—any work that you want to do with *Oshún,* you use anise a lot. . . . And also for baths and for its medicinal use, it is also very good. . . . Look, if there is a kid with problems of gases, you give them anise in teas, . . . and it takes all his gases away. It is very good, and it also cures colics in children. When a child is a newborn and has a colic, you give him drops of anise tea, . . . and it takes the gases and the colic away."

Figure 3.2. *San Lazaro, the patron of the sick, embodies the syncretic representation of the* Orisha *Babalu Aye (photo by Maria Gomez)*

The use of herbs in *Santería,* as well as in other Yoruba-related religions, is part of a complex dynamic that envisages Latinos' collective ritualistic participation (Gregory 1999; Wedel 1999). *Omieros,* a mix of herbs that belong to diverse *Orishas* in the *Santería*'s pantheon, are drunk before, during, and after ritual ceremonies, as they are supposed to contain large quantities of *ashé* (divine energy).

Sacrificial knives and sacred stones are also bathed in *omiero*. The *omiero de asiento* is the most important of all, as it is prepared for the initiation into the religion and is supposed to contain the *ashé* of all divinities. Novices bathe and drink in this compound during the seven-day seclusion period of the initiation process. Although according to the *Regla de Ocha's* commands (Ocha's rule) the *omiero de asiento* is expected to contain between one hundred and one hundred and twenty different plants, ecological differences between the United States and the Caribbean has limited access to many of them. Therefore, the use of twenty-one herbs is considered an acceptable number for most *omieros*. Depending on the particular divinity being honored, herbs added to *omieros* may include *meloncillo* (or "coyote gourd"), *kioyo* (known as "basil of big leaves"), *lengua de vaca* (translated into "cow's tongue"), *abrecamino* ("road opener"), and papaya.

Nonetheless, the key to *omieros'* power is not based solely on the particular herbs used, but on the songs and rituals that accompany their preparation. Without the required magic-religious ceremonies, *omieros* will lack the strength to make them work. Most Santeros agree that the fast pace in urban milieus works against some of the traditions followed by the *Santería* ritual. Lack of diversity in terms of herb supply and the busy schedules of followers and priests have contributed to a shortening of many traditional ceremonies. The increasing circulation of the recipes of *omieros* and other beverages made with ritual herbs appalled one particular young Santero, interviewee 102, who explains:

> "And the recipes that they give you are very good, but if you don't have the Saint done [been initiated into *Santería*], how are you planning to prepare it? And even if you have the Saint done, you don't (necessarily) know how to prepare the medicine . . . because all the plants and herbs being used [in *Santería*] have a purpose. . . . And if you are not a Santero [so] you don't understand the purpose of each of them."

For many *Santería* followers, practicing the "religion" in New York City has become an active mechanism against isolation, on the one hand, and a means of creating protective social charms against evil forces and illness on the other. Most of the Santeros we met during fieldwork talked about *Santería* as a "way of life" that tapped into every single aspect of their professional and personal lives. Some healers stated that they had found in *Santería* a response to their needs and a means through which they trained and cured others as well as themselves. In the words of interviewee 250, a Cuban female Santera:

> "Like when I came here [from Cuba to the United States], I got very sick. I would not swallow or drink water, I was very nervous, I was working but I was very nervous. . . . So we looked for a Santero, and the Saint told me that I had to get initiated into the religion with something. . . . I got initiated with a small ceremony, I got slowly into it, and we set up a small *botánica* in Park Avenue, between 110th and 111th street. We started small, and we got bigger until we got where we are now, already thirty years

since then. I used to feel very bad. The doctor was telling me why I wasn't swallowing, why I was so nervous, they prescribed me pills for the nerves, and I was feeling worse and worse. I had something in the stomach that didn't let me swallow. After I got initiated [into *Santería*], I felt much better and I didn't have any problems, and now I am seventy-one years old. . . . I haven't had any illness, I don't have arthritis, I don't have cholesterol, I don't have sugar, I don't have high [blood] pressure, I am not allergic to anything. . . . I have smoked tobacco since I was twelve years old, and it doesn't hurt me."

In sum, the practice of *Santería* in New York City, with *botánicas* as one of its main promotional outlets, has contributed to recreate a community "abroad" of Latino followers who share supportive networks of healing and care. In addition, Yoruba-related practices are rooted in strong healing belief systems that serve well those in need of practical responses to their problems (e.g., from chronic disease to loneliness) without necessarily becoming *Santería* practitioners. Above all, clients of *botánicas* bring new potential apprentices, income, and popularity to *Santería* beliefs and practices, which find in New York City a blossoming social market.

The Healers' Career: Developing the Gifted Call

Despite the specific healing disciplines to which our interviewees ascribe, most of them define themselves as being "generalists" in their trade, able to deal with a vast range of physical and emotional conditions brought by their clients. In some cases, the professionalization of the healers' métier in the United States followed a "calling" they were born with, which often expressed itself via unusual signs in dreams, divinatory predictions, and unexpected changes of mood at some point in their lives. Interviewee 313 tells us how she first got involved in working with plants, as part of her coming of age as a healer:

> "In dreams, plants were talking to me. . . . And one day I had a dream with a sir whose name was Saint Michael, and [he said] that I needed a bath because somebody was doing an evil spell [work] on me. So he told me: "Go and buy *arrasa con todo* [sweep everything] and *epazote* [wormwood]." He told me like five different names and told me what to do. So I went to the *botánica*, and I bought all my leaves. . . . With the faith that I have I boiled everything and I prepared a bath that I poured on myself. . . . Of course, to take away all the bad things."

In other cases, the healers' discovery of their own curative powers became part of their inner apprenticeship that evolved progressively, as in the case of participants who became practitioners of *Santería* and Spiritism in New York City with the guidance of both informal and formal mentors. Certainly, the possibility of "making good money"—whether as spiritual counselors, Santeros, or herbalists—becomes an enticing incentive for many practitioners, who turn either to combining other activities with their healing métier or to becoming full-time practitioners. Healing activities in New York City are a lucrative business for those who are either excluded from the occupations they used to hold in their countries of origin, or whose limited educational credentials will not allow them to make ends meet otherwise. Having a natural (or celestial) "call" for healing, knowing about plants through parents and teachers, or gaining supernatural abilities through spiritual development are three of the main channels through which our interviewees attained tenure in their healing faculties. Despite the specific disciplinary realm healers may subscribe to, the basis for their success as practitioners relies on their reputation as "efficacious" providers, amid a clientele that judges them upon their accomplishments in delivering the outcomes and products they promise. Given the increasingly large number of suppliers of different healing disciplines in New York City (and the high costs of rental space, transportation, and labor), practitioners of folk medicine are eager to provide as many services and products as possible for the purpose of attracting as many customers as they can.

Among our interviewees and key informants, we found a few who had completed professional careers in their countries of origin (e.g., a psychologist, a general doctor, and a lawyer), with titles that had not been upgraded in the United States. For these practitioners, the possibility of working as alternative healers in New York City had somehow become a survival strategy. For example, interviewee 702 (an undocumented female spiritual guide) acknowledged that her new role as a spiritual counselor in New York City had become a sizable form of making a living, something that would have been unthinkable for her in Uruguay, her country of origin:

> "I see here [in New York City] that people. . . I mean, the job market is different, so people have [more] means to get into this, it is different. So . . . let's say, here I do this as a job, which I couldn't do in my country. Let's say if I helped someone there, they would pay me with a dozen of eggs in return! . . . Also this city is wonderful because you see things here, you find things here from every culture, so you learn from the people, it is broader here as an advantage. . . ."

For some study participants, their inability to work officially in their fields of expertise, as in the example above, was somehow compensated for by the possibility of building up a steady clientele within the informal economy of healing, which in most cases is supported by a "rhetoric of efficacy" (Viladrich 2006a). This rhetoric refers to common discursive traits through which healers tend to emphasize the value of their curing powers as compared to their competitors. Given the fact that folk medicine in the United States is not regulated through a formal accreditation process, its practice is mostly supported by assessments of accuracy through informal community networks.

Herbal Baths: Addressing the Immigrant Continuum

"In Cuba, they say: '*The one who doesn't run, flies, and here everybody runs and flies.*'" (interview 300, a female Spiritist from Cuba)

Most *botánicas* keep a separate room, sometimes in the back of the store or in the basement, where a healer working on its premises provides one-on-one consultations with clients who, in exchange for a prearranged fee of usually twenty-five dollars, come in search for solutions to a variety of emotional, experiential, and health ailments. Providers' understanding of physical and emotional disorders is based on three primary sources: (a) supernatural, for ailments caused by witchcraft, envy, and negative energies; (b) natural, determined by biological sources and stressors (e.g., genetic dispositions, organic dysfunctions); and (c) *las preocupaciones,* worries rooted in everyday stressors and major life changes such as divorces, sudden migration, and undocumented status (Viladrich 2005a). During *consultas* (consultations) healers assume the role of counselors, advising their clients about the *causa* (cause, etiology) of their problems either via divination, mediumship, or visual inspection (Koss-Chioino 1975, 1992). This initial part of the consultation is usually followed by a treatment plan that includes *trabajos* (spells) and *limpias* (spiritual cleansings) with plants and herbs.

One of the most striking findings of this study was the assessment of the role of healers in providing time-efficient answers to problems that surpass the field of health, including assisting their clients in regularizing their undocumented status on the basis of combined treatments that consist of baths with herbs and fruits. When it came to discussing *las preocupaciones,* healers adamantly stated the emotional issues faced by Latinos in New York City as linked to job insecurity and debts, family separation, and difficulties in accommodating to the American culture.

Cleansing the client's spiritual energy and "lighting" the social environment are the main purposes for prescribing the preparation and use of baths. By "lighting," healers refer metaphorically to the act of purifying a place from "dark" energies (those from unsteady spirits wandering around or from an enemy's spells) that try to harm their clients' physical and emotional health. As noted by other authors, healers also mentioned the use of eggs and doves to perform *limpias* (cleansing) and *barridas* (which literally means sweeping), for the purpose of removing negative energies from a client's aura and transferring it to an object (Trotter and Chavira 1981; Mulcahy 2005). Cleansing procedures are usually accompanied by praying and offerings to particular saints (e.g., candles, flowers) in order to make sure that wishes will be heard by divine entities, which are usually represented by Catholic saints that embody Yoruba-related gods. The following quote from interviewee 403, a male Santero, who works at a *botánica* with other *Santería* practitioners, conspicuously summarizes these services:

> "We're specialists, we're priests. Our main task is to help people on a daily basis, no matter if they're junkies, no matter if they're old, if they're handicapped, no matter what it is. We help them one way or another, we help them. . . . They come in to get a reading, they have no hope, they have no dreams, they have no direction, so people relate to me more. . . . "Adriano, I'm going to court." So, I prepare candles for them, a bath, a cleaning. They go to court, so clean yourself with this. . . . Because the closer you keep those spirits, the [ones] guiding you, the more of a chance you have."[6]

Not only do some Latino healers combine different sets of disciplines (herb therapy and religious healing) but also insist upon the importance of visiting bio-medical doctors on a regular basis. In addition, rather than "competing" with providers of Western medicine, most folk healers tap into diverse dimensions (e.g., religious and spiritual) that are usually veiled to mainstream scientific knowledge. For example, healers may attempt to alleviate their clients' chronic conditions (e.g., asthma) by prescribing the juice of *Aloe vera* leaves in combination with inviting their clients to honor particular *Orishas* during ritual ceremonies. Also, healers' concerns about being accused of practicing medicine illegally often lead them to insist on the "natural" character of their healing endeavors, while longing for the broader freedom they used to enjoy in their countries of origin. Study participants were eager to point out the secondary pernicious effects of chemically processed substances, as well as the advantages of herbs and plants in counterbalancing the side effects of prescribed drugs. For example, interviewee 251, a female Santera from Cuba, argues:

> "Other than that synthetic shit they pump into us, plants have medicinal properties, like eucalyptus . . . and mint, they uplift you . . . you take a bowl, and you fill it with basil, which in Spanish we call *albahaca*. . . . So you take a bowl and place it into the middle of a room with basil. . . Basil is very good for the nervous system. It gives you a calming effect. . . . Just the aroma, it is like aromatheraphy."

Herbal baths are also commonly used for *despojar* (liberate) and *rompimientos* (breakage) of the dark nodal forces (e.g., usually evil spirits and envy) that keep clients tied either to haunting souls or living enemies. Diagnoses of witchcraft are usually associated to third parties (alive or dead) that interfere with clients' healthy auras (Viladrich 2005b). Herbal baths are undoubtedly the *botánicas'* best sellers as their popularity is somehow based on their accessibility and affordable price and easy preparation, as well as their combination with diverse healing menues (e.g., praying, rituals, and so on). The most common bath preparation consists of boiling herbs, roots, leaves, and fruits in water, followed by cooling the ingredients off, and then splashing the resulting mix onto the body while taking a shower. Baths made with herbs, fruits, and flowers are probably the most common form of healing mixture sold either in prearranged packages or customized by healers on demand. Given the fast pace of life in New York City and the *botánicas'* vast and assorted supply of herbal baths, most healers combine the use of prepackaged herbal compounds with those arranged upon their clients' requests. Baths with flowers, herbs (e.g., basil, parsley, and rue), and fruits and vegetables (e.g., pineapple, lemon, and garlic) are popular cleansing techniques amid the Latino immigrant population in the city. Use of herbs combined with lighting special *velones* (big candles named after particular saints, such as San Gregorio), praying, and the drinking of herbal infusions are typical procedures prescribed for the resolution of many of the everyday hurdles experienced by Latino immigrants and their children. Interviewee 300, a female spiritist from Cuba, referred to the case of a child who had been medically diagnosed with hyperactivity:

"There was this little boy, ten years old, who didn't fit at school. . . . Everybody complained about him, and the little boy what he had, simply, was a high spiritual development, strong, fast, without help from anyone. So I sent him to take a series of spiritual baths, to calm him down. I told the mother to give him baths of lemon verbena, and she gave him baths of verbena for an incredible period of time, along with spiritual praying. Because in this country what they give to these young boys who are like this, what they call hyperactive, it is drugs from medicine. The kid calmed down with baths, praying, and lots of love. The kid didn't need to get Zoloft, which is what they wanted to give him. Who could give Zoloft to a kid? It's unheard of! And I told his mother to put him into a sport. . . . He began going to play ball three times a week, she put him in a reading program. . . . And the kid is now attending the university!"

Herbs are not only physically carried from one place to another, but they also symbolize the spiritual journey that accompanies the immigrant experience. Interviewee 100 provides a paradigmatic example as a "global" healer. Although he is originally from Puerto Rico, he usually alternates his place of residence between New York City and San Juan, along with a traveling schedule that frequently takes him to other Caribbean countries, including Cuba and Haiti. In one opportunity, he was talking about the most serious cases of depression he had found among Latinos after September 11th:

"The immigrant mostly comes [to the United States] looking for a job, to make sure that their [legal] documents won't be asked for, so they won't get caught by the police. There are difficult cases, because it's difficult to work with the law, because the law doesn't have a heart, if it had, it would be easier to work. In these cases, what I do is what I call an exorcism, in which I give them lucky baths for things to go well, for the person to feel secure and confident. It is to give them spiritual strength . . . to get their papers and solve their problems . . . always in the right order. . . . Immigrants in general come with depression because of their papers. . . . And the only thing one can do is to give them some baths so the person can feel secure, the doors can open, people lose their fear and can move ahead in life."

Concluding Remarks

This chapter has addressed the two ends of the immigrant experience: the needs of Latino immigrants who come to the consultation, and the practice of the healers who treat them. In particular, this essay has explored the role of a diverse population of providers along with the role of *botánicas,* which constitute physical and symbolic entry doors to the informal healing economy in New York City. *Botánicas* are mostly retail businesses that often sell herbs among other healing and religious items, frequently imported from Puerto Rico and the Dominican Republic. They are located in street-front stores and are run by small business owners in areas heavily populated by Latinos. Furthermore, they offer cost-efficient and face-to-face services by local experts, sometimes the owner or a Latino employee. In spite of the market surrounding the commercialization of spiritual and religious goods, *botánicas* have remained loyal to their Latino and Caribbean clientele, which finds in their services what other stores do not provide particularly in the form of spiritual guidance, herbs, and other products from Latin America, as well as spiritual counseling.

In a market where *botánicas* play the role as hubs of religious healing, customers and healers get embedded in informal networks in which clients assume the role of mentors and vice versa. As discussed in this chapter, the important function of *botánicas* within the immigrant community exceeds the supplying and selling of herbs and religious objects, since they also play a unique role as informal community centers where neighbors congregate to play dominoes or chess or to follow the afternoon *novela* (Latino soap opera, see figure 3.3). *Botánicas'* networks of care, in the form of *Santería* rituals or *Palero* ceremonies, sustain Latino immigrants in the midst of enduring physical and social stressors (Applewhite 1995).

Figure 3.3. Botánicas *are multifunctional spaces. Here a healer and his friend, a regular at the* botánica, *play chess on a Sunday afternoon*

In New York City, the clientele of healers represents an intercontinental crowd eager to get quick fixes to their pressing health needs, while searching for cost-effective solutions to diverse social and existential ailments. In this context, the use of plants and herbs exceeds their biomedical properties since healers' pharmacopoeias are aimed at helping Latinos deal with the physical, the mental, the emotional, and the spiritual realms. As discussed earlier, plants are combined to provide popular forms of *limpieza* (cleansing) for the purpose of restoring a client's energetic field, amidst a vast range of syndromes that shape the immigrant continuum. Plants achieve a new dimension in a challenging urban milieu, where baths have become the quintessential multilevel therapy for the treatment of a variety of problems, ranging from the amelioration of an unbearable indigestion to repelling evil spells sent to a client by a jealous relative or an evil coworker.

Not surprisingly, in such a multicultural context healers do not restrict their claims of expertise to a single métier but instead ascribe their capabilities to a complex realm of disciplines. In New York City, healers' entrepreneurship has become a profit-making activity somehow deprived of the informal exchange that may have taken place at home. As such, plants are not only vehicles for clients' relief but also concrete means of survival. Most healers are quite aware of their customers' needs as well as of the possibilities and limits of their trade. They may long for the plants they used to gather in the open *monte* (forest) back home, but for a specified sum the transnational herbal economy has made sure that they can get specimens from almost anywhere in Latin American. Therefore, the healing career in New York City challenges single-minded religious belief systems and opens new venues for knowledge and experimentation.

Acknowledgements

I want to express my deep gratitude to the following collaborators during the ethnographic phase of this study: Antonella Fabri (fieldwork coordinator during the first stage of the project), Andrea Feduzi, Jasmine Gartner, Vincent Goldberg, Joel Naatus, Helga Perez, Martha Rodriguez, and Cassandra Torrico. Lori Bukiewicz and Maria Angela Soto provided crucial assistance in the analysis of the research results; Maria Gomez successfully continued their work. Sabina Gritta contributed to the success of this project by coordinating and supervising the transcriptions and interview outcomes. I also appreciate the thoughtful comments provided by an anonymous reviewer that greatly contributed to improving the final product. Ina Vandebroek has been a wonderful editor and colleague, whose support has deeply influenced the direction and contents of my work on ethnobotany. As usual, Stephen Pekar provided considerable time and effort in reading and editing this article. This project has been possible due to a generous grant awarded by the Russo Gift and two PSC-CUNY grants, and the ongoing sponsorship of the Schools of the Health Professions, the School of Health Sciences, and the Urban Public Health Program at Hunter College of the City University of New York.

Notes

1. The origin of *botánicas* in New York City has been traced to the drug dispensaries (*droguerías*) that appeared as early as the 1900s and served Southern blacks and West Indians looking for herbal remedies, lotions, and powders (Long 2001; González-Wippler 1989).
2. Most healers advertise their practices within the realm of spiritual and religious faiths while protecting themselves from being identified as unlawful practitioners of medicine. The attempt to recruit potential interviewees encountered several obstacles during fieldwork, including healers' mistrust and concealment of their healing practice. These findings coincide with other studies that note the difficulties in doing research on this topic among Latino and Chinese immigrant communities (Reiff et al. 2003).
3. The two largest national groups represented in our sample (Dominicans and Puerto Ricans), correspond with the two dominant Latino groups in New York City according to the United States Census 2000.
4. Although the study design focused on healers' performance, and not on their customers' health outcomes, participant observation and in-depth interviews held with Latino healers provided insights on the characteristics of the clientele as well as on the nature of their social interactions.
5. The word *teleology* (from the Greek word *telos,* which means "end," and *logos,* which means "discourse"), has its roots in Greek philosophy. Teleology assumes that there are final ends or causes (as a directive principle) that escape the simple works of nature. The supernatural becomes the realm in which final purposes are disinherited. The concept of teleology is supported by the notion that the universe has an invisible design built upon a chain of events leading to final causes.
6. The *Vela para Todo* (Candle for Everything) can be used for any purpose as it has seven colors, each representing a particular saint.

References

Applewhite, S.L. 1995. "Curanderismo: Demystifying the health beliefs and practices of elderly Mexican Americans." *Health & Social Work* 20: 247–553.

Baer, H.A. 1984. *The Black spiritual movement: A religious response to racism.* Knoxxille: University of Tennessee Press.

Balick, M.J., F. Kronenberg, A.L. Ososki, M. Reiff, A. Fugh-Berman, B. O'connor, M. Roble, P. Lohr, and D. Atha. 1999. "Medicinal plants used by Latino healers for women's health conditions in New York City." *Economic Botany* 54: 344–357.

Balick, M.J., and R. Lee. 2001. "Looking within: Urban ethnomedicine and ethnobotany." *Alternative Therapies* 7: 114–115.

Brandon, G. 1991. "The uses of plants in healing in an Afro-Cuban religion, *Santería.*" *Journal of Black Studies* 22: 55–76.

Cabrera, L. 1971. *El monte.* Miami: Colleccion Chichereku.

De la Torre, M.A. 2004. *Santería: The beliefs and rituals of a growing religion in America.* Grand Rapids: Wm. B. Eerdmans Publishing Company.

Espín, O. 2003. *Latina healers.* Mountain View: Floricanto Press.

Fernández Olmos, M., and L. Paravisini-Gebert. 2003. *Creole religions of the Caribbean: An Introduction from Voodoo and Santería to Obeah and Espiritismo.* New York: New York University Press.

Garrison, V. 1977. "Doctor, Espiritista or psychiatrist? Health seeking behavior in a Puerto Rican neighborhood of New York City." *Medical Anthropology* 1: 65–191.

Gomez-Beloz, A., and N. Chavez. 2001. "The *botánica* as a culturally appropriate health care option for Latinos." *The Journal of Alternative and Complementary* Medicine. 7: 537–546.

González-Wippler, M. 1989. *Santería: The religion. A legacy of faith, rites, and magic.* New York: Harmony Books.

Gregory, S. 1999. *Santería in New York City. A study in cultural resistance.* New York and London: Garland Publishing.

Harwood, A. 1977a. "Puerto Rican Spiritism. Part I—Description and analysis of an alternative psychotherapeutic approach." *Culture, Medicine and Psychiatry.* 1: 69–95.

Harwood, A. 1977b. "Puerto Rican Spiritism. Part 2—An institution with preventive and therapeutic functions in community psychiatry." *Culture, Medicine and Psychiatry* 1: 135–53.

Jacobs, C.F. 2001. "Folk for whom? Tourist guidebooks, local color, and the spiritual churches of New Orleans." *Journal of American Folklore* 114: 309–330.

Jones, M.O., P.A. Polk, Y. Flores-Peña, and R. J. Evanchuk. 2001. "Invisible hospitals: Botánicas in ethnic health care." In *Healing logics. culture and medicine in modern health belief systems,* ed. E.B. Logan. Utah: Utah State University Press, 39–87.

Koss-Chioino, J. 1975. "Therapeutic aspects of Puerto Rican cult practices." *Psychiatry* 38: 160–71.

Koss-Chioino, J. 1992. *Women as healers: Women as patients: Mental health care and traditional healing in Puerto Rico.* Boulder, San Francisco, Oxford: Westview Press.

Laguerre, M. 1987. *Afro-Caribbean folk medicine.* South Hadley: Bergin and Garvey.

Long, C.M. 2001. *Spiritual merchants.* Knoxville: The University of Tennessee Press.

Martinez, R., and C.V. Wetli. 1982. "*Santería:* A magico-religious system of Afro-Cuban origin." *American Journal of Social Psychiatry* 2: 32–38.

Mautino, K.S. 1999. "Faith versus the law: traditional healers and immigration." *Journal of Immigrant Health* 1: 125–131.

McCarthy Brown, K. 1991. *Mama Lola: a Vodou priestess in Brooklyn.* Berkeley: University of California Press.

Moreno Vega, M. 2000. *The altar of my soul. The living traditions of Santería.* New York: Random House.

Mulcahy, J.B. 2005. "The root and the flower." *Journal of American Folklore* 118: 45–53.

Ososki, A.L., P. Lohr, M. Reiff, M.J. Balick, F. Kronenberg, A. Fugh-Berman, and B. O'Connor. 2002. "Ethnobotanical literature survey of medicinal plants in the Dominican Republic used for women's health conditions." *Journal of Ethnopharmacology* 79: 285–298.

Pagel, D. 2004. "Browsing through beliefs; Botánica delves into spirituality, capitalism, Latino street shops and the Gods we trust." *Puerto Rico Herald.* Retrieved June 23 2007 from http://www.puertorico-herald.org/issues/2004/vol8n53/LiveEd/53-Browsing.html

Paulino, A. 1996. Spiritism, Santería, Brujería, and Voodooism. A comparative view of indigenous healing systems. *Journal of Teaching in Social Work* 12: 105–124.

Perez y Mena, A.I., 1977. "Spiritualism as an adaptive mechanism among Puerto Ricans in the United States." *Cornell Journal of Social Relations* 12: 125–136.

Polk, P.A. 2004. *Botánica Los Angeles. Latino popular religious art in the City of Angels.* Los Angeles: UCLA Fowler Museum of Cultural History.

Prasad, V.L. 2004. "Subcutaneous injection of Mercury: 'Warding Off Evil'." *Environmental Health Perspectives* 112: 1326–1328.

Reiff, M., B. O'Connor, F. Kronenberg, M.J. Balick, P. Lohr, M. Roble, A. Fugh-Berman, and K.D. Johnson. 2003. "Ethnomedicine in the urban environment: Dominican healers in New York City." *Human Organization* 62: 12–26.

Romberg, R. 2003. *Witchcraft and welfare: Spiritual capital and the business of magic in modern Puerto Rico.* Austin: University of Texas Press.

Sandoval, M. 1977. "*Santería*: Afrocuban concepts of disease and its treatment in Miami." *Journal of Operational Psychiatry* 8: 52–63.

Singer, M., and H. Baer. 1995. *Critical medical anthropology.* Amityville, NY: Baywood Publishing Company.

Stern, A.H., M. Gochfeld, D. Riley, A. Newby, T. Leal, and G. Garetano. 2003. "Cultural uses of mercury in New Jersey." *Environmental assessment and risk analysis element. Research project summary.* Trenton: Division of Science Research and Technology.

Trotter, R., and J.A. Chavira. 1981. *Curanderismo, Mexican American folk healing.* Athens: University of Georgia Press.

Viladrich, A. 2005a. "Overcoming *las preocupaciones* (worries): Latino immigrants' herb-healing practices in New York City." *Anthropology* 47: 54.

———. 2005b. "Can you read my aura? Latino healers in New York City." *Anthropology News* 46: 56.

———. 2006a. "Botánicas in America's backyard: Uncovering the world of latino immigrants' herb-healing practices." *Human Organization.* 65: 407–419.

———. 2006b. "Beyond the supernatural: Latino healers treating Latino immigrants in New York City." *Journal of Latino-Latin American Studies.* 2: 134–148.

Wedel, J. 1999. *Santería healing.* Gainesville: University of Florida Press.

Zayas, L.H., and P.O. Ozuah. 1996. "Mercury use in Espiritism: A survey of botánicas." *American Journal of Public Health* 86: 111–12.

Chapter 4

The Changing Scene of Health Promotion and Disease Prevention Strategies Due to Migration of Indians from the Asian Subcontinent to the United States

Usha R. Palaniswamy

The Population

Indians in the Asian Indian Subcontinent

The history of the Indian subcontinent is a mix of dynasties, religions, and invaders who conquered and ruled India and imposed their own cultural institutions and religions. This has resulted in a complex cultural blending of the native cultures and religions in modern India. Among the major religious influences in Indian history are Hinduism (2500 B.C.), Buddhism (184 B.C.), Islam (A.D. 1526–1707), and Christianity (A.D. 1858–1947). At the time of independence (1947), what had been the Indian nation was divided into India and Pakistan, with the majority of Hindus establishing residency in India and the Muslims in Pakistan. Later, West Pakistan became an independent nation named Bangladesh. Vegetarianism is most popular among Hindus, who form the majority of the Indian population, while eating meat is acceptable to Christians and Muslims, who constitute 12.1 and 2.3 percent of the total population, respectively. Dietary habits of Indians are also influenced by plant species grown in a particular area, availability, and the practice of Indian Ayurvedic medicine, *Ayurveda* meaning "knowledge of life." More than 80 percent of people in India rely on herbal remedies as a principal means of preventing and curing illnesses. Approximately 1,250 plants are used in Ayurvedic medicine (Chatterjee and Pakrashi 1991).

The earliest migration of people from the Indian subcontinent was to Southeast Asia where small groups of individuals settled, merging with the larger local society. During the nineteenth century and until the end of the British Raj, under the indenture system, the migration of slave labor occurred to Mauritius, British Guyana, the West Indies, Fiji, and East Africa; Indians also migrated to work on tea and rubber plantations in Sri Lanka, Burma, and the British Malay (Malaysia and Singapore). Indians emigrated to the United Kingdom, Australia, and New Zealand after independence in 1947. A large number of Indians emigrated on a contractual basis to the Persian Gulf after the 1970s oil boom in the Middle East. The United States became a favored destination after the change in the US immigration law in 1965. Initially considered as an alien "Land of the Whites" not appropriate for "raising children and the upkeep of traditional Indian culture," the United States became a "land of opportunity" for migration in the mid-1980s. In the present study, an Asian Indian is defined as a US resident or citizen who has ancestry originating in the Indian subregion of South Asia and includes Asian Indians born in India who immigrated to the United States, as well as those with Indian ancestry born in the United States.

Asian Indians in the United States

Since the late 1960s and early 1970s, the Asian American population in the United States has grown from 1.5 million to nearly 12 million in 2000, and is projected to grow to 20 million by 2020. The US Census for 2000 revealed that Asians represent the fastest-growing minority group in all regions of the United States, both from 1980–1990 and 1990–2000. The Asian population is projected to have the greatest gains in the West with an increase of 7 million, and in the Northeast with an increase of 2 million during the thirty years from 1995 to 2025. The number of Asian Indians grew from 15,000 in 1965 to over 1.9 million in 2000, making them the third largest group among Asian Americans, making up about 16.4 percent of the Asian American population, after the Chinese (22.6 percent) and the Filipinos (18.3 percent). In the state of Connecticut, there has been over a 100 percent increase in the Asian Indian population in the past decade. The Asian American population in the United States has also been enhanced in its diversity and is now predominantly foreign-born and spread across several different nationalities, extending to various parts of Asia. Based on the 2000 Census and the immigration trends, it has been projected that by 2050, if not sooner, no single ethnic group will constitute more than 50 percent of the total population in the United States. This presents mixed challenges in all dimensions of life, particularly health care issues that are vital to a healthy community. The health status and health needs of immigrants are poorly understood, and the effect of immigration on health is controversial. The United States have identified wide disparities in health status and included a major emphasis in the Healthy People 2010 initiative on eliminating disparities as one of its two overarching goals.

The rigorous process of immigration selects the best immigrants at all stages of approval, on the bases of education, health, language and job skills, characteristics that can facilitate social and economic integration in a new country. Hence, it is not surprising that immigrants who are accepted into the United States are often healthier than the native-born population when they arrive in the new country. However, immigrants lose their healthy advantage over time due to the acculturation process. A Canadian study documented that immigrants were in better health than their Canadian-born counterparts (immigrants born in Canada). However, immigrants who lived in Canada for more than ten years experienced a similar incidence of chronic conditions and long-term disability as the native Canadian-born population (Chen et al. 1996a; 1996b). Such documentation of the effects of acculturation, that are the result of living a substantial number of years in a new host country, calls for the examination of disparities between health behaviors of Asian Indians living in the United States and those living in India.

Asian Indians in the United States represent a group of more recent immigrants who have not been examined in detail for their health beliefs and practices compared to other Asian Americans who migrated to the United States in larger numbers as early as the 1850s. Research into Asian Indian immigrants in several other Western countries, such as the UK and Canada, has found them to be at greater risk for morbidity and mortality from heart disease and diabetes compared to the native population and other immigrants (Pais et al. 1996; Sheth et al. 1999; Wetter et al. 2001). The researchers also report that these diseases seem to be appearing in these populations at earlier stages and at lower risk levels.

Acculturation provides an important framework for understanding the relationship between migration and changes in health following migration. According to this framework, acculturation influences health status indirectly *via* changes in health behaviors, as well as *via* changes in social support and stress (Zambrana et al. 1997; Hull 1979). Recently, more importance is being given to educating the younger populations about risk factors and encouraging them to follow early detection guidelines and prevention practices. A common problem with designing health education programs for risk factor awareness and early detection methods for minority populations is the continuing immigration of people from Asia and increasing complexity of the existing layers of generations. The arrival of people in the United States every year means that there are people with different levels of lingual and cultural diversity. Thus, numbers of immigrants from Asia with similar physical characteristics are increasing, but the individuals vary in their acculturation levels, educational background, health beliefs, and behaviors. So, in health education programs designed for multicultural populations there is a need to identify and factor in the differences based on the length or duration of residence in the United States.

Asian Indian Perceptions of Health and Related Behaviors

Culture plays an integral part in understanding health and illness. Helman's model (2001) proposes that, whereas in nonindustrial nations, illness and health are explained by supernatural causes—such as God, destiny and *karma*—and by other social causes, in industrial nations they are explained by individual behaviors and natural world environments. In Indian societies, supernatural factors are believed to be fundamental to health and illness, and to be affected largely by one's good and bad deeds in past lives (Dalal 2000). Indian health beliefs have been described as holistic, incorporating physical, psychological, and social factors along with the supernatural. Hindu Indians also believe that ill health is the result of the interaction between the laws of *karma* and God's will.

Other studies report poor adherence to prescribed medications such as oral hypoglycemic agents and other prescribed treatment regimens as a serious problem among Asian-Indians and express the need for developing culturally sensitive ways for health care delivery. Asian-Indian views of Western clinical methods and prescriptions are derived from popular Western views of reported side effects of drugs as well as from what prevails in the Indian subcontinent. Such beliefs lead people to regulate their medications themselves; while this may seem 'rational' to them it can go against medical advice (Lawton et al. 2005). Such practices of abandoning the use of prescribed drugs and turning to self-medication are reported widely in the Indian subcontinent (Miller 1997; Marinker et al. 1997) as well as in immigrant Indian populations residing abroad (Gravitas Research and Strategy Limited 2005).

A number of Asian Indians are not familiar with medical insurance and do not prioritize prevention methods identified as belonging to the Western biomedical model. They believe in preventive herbals, diets, and other alternative medical systems such as Ayurveda and often turn to Western medicine as the last resort.

Health Disparities

Historically, race, genetics, and diseases have been linked closely to one another. However, there is accumulating scientific evidence that in many instances the causes of health disparities have little to do with race and genetics but rather are derived from differences in culture, diet, socioeconomic status and social marginalization, access to health care and education, environmental exposures, discrimination, stress, and other factors. Also, there is increasing evidence indicating that observed genetic variations are due to the specific geographic origins of individuals (ethnic roots) and that this accounts for the disparities in ethnic minority

groups originating from a single geographic region. The proposed biological variations in health and disease based on race are fading rapidly, and efforts in health
care settings are geared toward eliminating race as a factor for discrimination and
toward the reduction of racial, ethnic, and socioeconomic disparities as called for
in Healthy People 2010.

Diabetes

INCIDENCE

Diabetes is considered to be an epidemic in many countries around the world.
In the United States, about 2,200 new cases are diagnosed every day, or 800,000
each year (Clark 1998; Burke et al. 1999). Many studies have reported that Asian
Indians have an unusually high prevalence rate of diabetes (Ramaiya et al. 1990;
Ramachandran et al. 2001; Venkatraman et al. 2004). The changing demographic patterns in the United States are expected to increase the number of people who are at risk of diabetes and who will eventually develop the disease, thus
enhancing the public health challenge in the country. Epidemiological studies
have shown that type-2 diabetes results from an interaction between a genetic
predisposition and lifestyle factors including obesity, sedentary behavior, and an
excess intake of calories and dietary constituents (O'Dea 1992; Zimmet 1992).
    Diabetes is a chronic condition that manifests itself in two types: type-1, occur-
ring mainly in children and adolescents eighteen years of age and younger in
which the body does not produce insulin, and thus insulin administration is required to sustain life; and type-2, occurring usually in adults over thirty years of
age in which the body tissues are not able to use their own limited amount of insulin effectively. All diabetics require self-management training, and treatment of
type-2 diabetes currently consists of physical activity, dietary restrictions, oral medications, and insulin. Diabetes has remained the seventh leading cause of death
over the past decade, and its occurrence has increased steadily in all populations,
especially in ethnic minority groups in the United States (Flegal et al. 1991). Diabetes has been associated with end-stage renal disease and a three- to fourfold
increase in coronary heart disease compared to nondiabetics. It is the leading cause
of amputations and blindness among working adults, and contributes to an impaired quality of life and disability (American Diabetes Association 1996; Center for Disease Control and Prevention 1999). The enormous economic and
social cost of this disease makes a persuasive case for prevention in all possible ways.
Several studies clearly suggest that acquired/environmental factors inherent with
a "Western" lifestyle play a significant role in the development of type-2 diabetes.
There is also growing evidence that some groups of individuals or ethnicities
have a particular "predisposition" to develop type-2 diabetes in the presence of
adverse environmental conditions. Numerous epidemiological studies show that

the prevalence of this disease is significantly higher among ethnic minorities such as Hispanics and Asians as compared to Caucasians, even when members of the different groups are exposed to similar environmental conditions (Hanis et al. 1983; Stern et al. 1984; Ramachandran et al. 1992; Simmons et al. 1992). While an increasing prevalence for diabetes is observed in all ethnic groups, Asian Americans have experienced the highest rate of increase in prevalence between the years 1990 and 1998. The enhanced insulin resistance reported for Asian Indians is a primary metabolic defect that may account for excess morbidity and mortality in this group (Abate and Chandalia 2001). The Asian Indians may be predisposed to diabetes due to insulin resistance, and this may be explained either by an environmental or a genetic factor, or a combination of both.

Changing demographics and ethnic composition, increased diversity of the US population, and the progressive enrichment of the population with ethnic groups at high risk for diabetes, such as Asian Americans, may contribute significantly to the worsening of the diabetes epidemic in the United States. Among Asian Americans, immigrants from India, Pakistan, and Bangladesh (all from the Indian subcontinent) have the highest predisposition to develop diabetes. Epidemiological data of those countries from which Indians have migrated document the excessive predisposition to develop diabetes in this ethnic group. Asian Indians living in rural India have only a 2–3 percent prevalence of diabetes (Ramachandran et al. 1992), while those living in urban areas or who have migrated to developed countries experienced about a fourfold increase in a prevalence of diabetes (McKeigue et al. 1989; Dowse et al. 1990; McKeigue et al. 1992). The increase in the incidence of diabetes is also documented in India, which is attributed to lifestyle changes. The prevalence of diabetes in the adult urban population in India increased from 8.2 percent in 1989 to 11.6 percent in 1996 (Ramachandran et al., 1997). In a series of epidemiological surveys in Southern India, Ramachandran et al. (1992; 1997) reported an increasing prevalence of diabetes in the urban population.

It has been estimated that the prevalence of type-2 diabetes in Asian Indians is about 19 percent compared to the 5 percent observed in Caucasians. Venkataraman et al. (2004) reported that, at 18.3 percent, the overall prevalence of diabetes in Asian Indians was the highest among all the groups studied. South Asian women have also been identified as being at higher risk for gestational diabetes (Berkowitz et al. 1992). Since it is projected that by the year 2020 over 50 percent of the US population will consist of people of non-Caucasian ethnic backgrounds, and since new migration greatly involves subjects of Asian Indian descent, the gaps in the health of minority populations in the United States is certain to worsen. Because of the high morbidity and mortality associated with diabetes, this trend evidently poses a public health threat. Therefore, identification of mechanisms involved in the development of excessive type-2 diabetes in Asian Indians becomes a public health priority.

Early Detection and Preventive Practices for Diabetes

It is possible to delay or prevent type-2 diabetes from developing by learning about the risks identified at the prediabetes stage and taking actions to prevent it, if one has or is at risk for diabetes. Before people develop type-2 diabetes, they almost always contract prediabetes, a state in which blood glucose levels are higher than normal but not yet high enough to be diagnosed as diabetes. The American Diabetes Association (ADA) recommends careful monitoring of blood glucose for identification at an earlier stage and taking positive preventive steps that include dietary and lifestyle modifications.

Common herbs used in daily Indian diets as a part of vegetable dishes and spices (Mathew et al. 2000; Sinha et al. 2003) and identified as having anti-hyperglycemic properties include[1]: bitter melon (*Momordica charantia*; Cucurbitaceae), curry leaf (*Murraya koenigii;* Rutaceae), fenugreek (*Trigonella foenum-graecum;* Fabaceae), coriander (*Coriandrum sativum;* Apiaceae), mint (*Mentha* spp.; Lamiaceae), cumin (*Cuminum cyminum;* Apiaceae); ginger (*Zingiber officinale;* Zingiberaceae), turmeric (*Curcuma longa;* Zingiberaceae), garlic (*Allium sativum;* Alliaceae), drumstick (*Moringa indica;* Moringaceae), okra (*Abelmoschus esculentus;* [*Hibiscus esculentus*], Malvaceae), cucumber (*Cucumis sativus;* Cucurbitaceae), snake gourd (*Trichosanthes anguina;* Cucurbitaceae); radish (*Raphanus sativus;* Brassicaceae), tomato (*Lycopersicon esculentum;* Solanaceae); plantain (*Musa* spp.; Musaceae), chili (*Capsicum annuum*; Solanaceae), mango seed powder (*Mangifera indica;* Anacardiaceae); brinjal[2] (*Solanum melongena*; Solanaceae)

Diabetes Study

STUDY SAMPLE

One hundred and forty subjects participated in this study. These were Asian Indians who had been in the United States (1) less than 5 years; (2) 5 to 10 years; (3) more than 10 years; (4) born and brought up in the United States; and (5) White mainstream. There were twenty-eight subjects in each group (n=28). Data were analyzed using the SAS General Linear Models system. The participants were a mix of women residing in Connecticut, who had been in the United States for varying lengths of time. Some were students on campus who may have lived elsewhere before coming to campus, but when the study was conducted all women were residents of Connecticut.

STUDY TOOL

The questionnaire contained knowledge and prevention guidelines described by the American Diabetes Association. There were forty-nine questions organized into six sections.

DEMOGRAPHIC DATA

Demographic data of subjects include age, marital status, number of children, number of years in the United States, country of origin, source of health care and health information, personal experience with diabetes, and diet changes as well as plant use.

ACCULTURATION LEVEL

This section listed three factors strongly associated with acculturation: proficiency in English, proficiency in the native language, and sociocultural preference (Anderson et al. 1993). The highest and the lowest possible scores were 9 and 3 respectively. Subjects who obtained scores between 3 and 5 were grouped as "traditional" and having a low level of acculturation; scores between 6 and 7 were grouped as "bicultural" and having a medium level of acculturation; and scores from 8 to 9 were grouped as "assimilated," having a high level of acculturation.

KNOWLEDGE OF RISK FACTORS

This section assessed knowledge of diabetes, including risk factors, methods for reducing risk, and guidelines from the American Diabetes Association. It listed twenty-nine risk factors and the highest and lowest possible scores were 89 and 29, respectively. Based on the mean and standard deviation observed in the sample population, scores between 55 and 66 were grouped as having a "low knowledge level", scores between 67 and 76 as having a "medium knowledge level"; and scores higher than 76 as having a "high knowledge level".

BARRIERS TO FOLLOWING PREVENTIVE PRACTICES

Questions in this section addressed perceived barriers to preventive practices associated with diabetes and listed four items. The highest and lowest possible scores were 20 and 4, respectively. Based on the mean and standard deviation observed in the sample population, scores between 14 and 17 were grouped as having "high barriers", scores between 9 and 14 as having "medium barriers", and scores between 4 and 8 as having "low barriers".

CONFIDENCE AND HEALTH MOTIVATION

The questionnaire listed nine items and the highest and lowest possible scores were 45 and 9, respectively. Based on the mean and standard deviation observed in the sample population, scores from 17 to 33 were grouped as having "low confidence and health motivation", scores from 34 to 39 as having "medium confidence and health motivation", and scores from 40 to 45 as having "high confidence and health motivation".

Dietary Modifications after Migration

Questions in this section were to identify dietary changes in the study population after migration. A number of studies have implied changes to "traditional diets" as a significant causal factor for the increase in occurrence of cancers and diabetes. Since time immemorial, diets in India have been associated closely with religious and communal affiliations, and there exists a diversity that is unknown to most other countries, with many dietary patterns stemming from strong cultural and religious teachings. A traditional Indian diet can vary widely depending on the native state, community, and religion from which the subject originates. Since the population in the study originated from all over India, it was necessary to limit the number of spices and vegetables and to consider a cuisine more inclusive of India as a nation. Only major components (rice, dhal[2], yogurt, selected vegetables and spices, and traditional garnish) in the daily diet of Indians were considered. The vegetables and spices enumerated in this list included species identified in studies conducted with Indians living in India (Mathew et al. 2000; Sinha et al. 2003). Dietary traditions were evaluated based on the following factors:

a. inclusion of rice in the daily diet
b. inclusion of dhal[3] in the daily diet (red gram dhal)
c. changes in consumption of beverages such as tea, coffee, cocoa, milk, carbonated drinks and knowledge of implications of change
d. inclusion of Indian spices (chili, ginger, garlic, mustard, turmeric, coriander, cumin)
e. list of vegetables consumed (mint leaves, fenugreek (methi) leaves, ginger, garlic, radish, vegetable drumstick, brinjal, cucumber, okra, bitter gourd, snake gourd, tomato and plantain) and awareness of health benefits of the species listed
f. Garnishing with fresh coriander leaves, curry leaves, and/or mustard seeds
g. Reason(s) for not using the vegetables and spices listed
h. Changes in cooking oils and source of fat used
i. Changes in alcohol consumption
j. Comfort foods and comfort zones at home and in the neighborhood.

The diets were classified as traditional, semi-traditional, semi-Westernized, and Westernized/mainstream based on ingredients and frequency of consumption of rice, dhal, vegetables, spices, traditional garnish, and plain yogurt (table 4.1).

Table 4.1. Classification of Diets in Traditional, Semi-Traditional, Semi-Westernized, and Westernized/Mainstream

Ingredient in diet	Traditional	Semi-traditional	Semi-Westernized	Westernized/Mainstream
Rice	Daily	1–2 times/week	2 times/month	Once a month
Dhal	> 2 times/week	Once a week	< Once a month	Occasionally/Never
Vegetables[a]	4 or more/week	< 4/week	> 4/month	< 4/month
Spices[b]	4 or more/day	< 4/day	4 or more/week	< 4/week
Garnish[c]	Daily	Once a week	< Once a week	Occasionally/Never
Plain Yogurt	Daily	>2 times/week	<2 times/week	Occasionally/Never

[a]mint leaves, fenugreek (methi) leaves, ginger, garlic, radish, vegetable drumstick, brinjal, cucumber, okra, bitter gourd, snake gourd, tomato, and plantain; [b]turmeric powder, chili powder, mango powder, dried coriander seeds/powder, and cumin seeds or powder; [c]curry leaves, coriander leaves, and/or mustard seeds

Results and Discussion

Demographic Data

All subjects were less than 45 years old and had a good health insurance plan at the time of the study. About 18 percent had a family history of diabetes. Major sources of health information for all the subjects were their peer-groups, family physicians, local clinics, news media, and magazines.

Acculturation

Average acculturation scores ranged from medium (6.6) in groups who have been in the United States for less than 5 years, to high (8.3) in those who were born and brought up in the United States. Acculturation increased with the duration of stay (table 4.2), and no group was classified as "traditional" (<5) or "completely assimilated" (9).

Knowledge Level of Diabetes

Knowledge of risk factors ranged from 62.6 to 76.7 in the study population. Subjects who have been in the United States for less than five years and those living in the United States between five and ten years had a lower knowledge level than the other groups. Subjects living in the United States for more than ten years or those born and brought up in the United States showed a similar level

Table 4.2. Characteristic Variables Affecting the Level of Knowledge of Diabetes, and Prevention Practices in Asian Indian and Mainstream White Populations (N=140)

Population Group	LR-US[1]	Diabetes (N=110)			
		Acct[2]	KL[3]	B[4]	C-HM[5]
Asian Indian	< 5 years	6.6 c	62.6 c	12.8 a	24.7 c
Asian Indian	5–10 years	6.8 c	68.5 c	12.7 a	30.4 b
Asian Indian	>10 years	7.2 b	76.1 a	12.4 a	33.5 b
Asian Indian	Born in US	8.3 a	76.7 a	9.8 b	36.3 b
Mainstream White	Born in US	—	71.7 b	9.2 b	42.4 a

[1]Length of residence in the United States, [2]Acculturation, [3]Knowledge level, [4]Barriers to practices for prevention, [5]Confidence and health motivation. Significance of differences is computed between the averages in each column by Least Squares Difference (LSD) at $P < 0.05$. Means followed by same letter are not significantly different.

of knowledge (76.1 and 76.7 respectively), which was higher than for the white mainstream group (71.7) (table 4.2).

Barriers to Practicing Preventive Strategies

Asian Indians in the United States for less than five years (12.8), five to ten years (12.7), and more than ten years (12.4) had greater barriers to practicing the preventive strategies compared to those born and brought up the United States (9.8) and the white mainstream (9.2) (table 4.2).

Confidence and Health Motivation for Preventive Practices

Confidence and health motivation for better health practices increased with the number of years that immigrants had spent in the United States but still remained lower than for the mainstream White group (table 4.2).

Dietary Modifications

While none of the Asian Indian groups showed a completely Westernized/mainstream diet, there were differences in the percentages of subjects consuming traditional, semi-traditional, or semi-Westernized diets (figure 4.1). A large percentage (90) of subjects indicated consumption of a traditional diet for taste and smell, and only 10 percent of the subjects had some knowledge about the health ben-

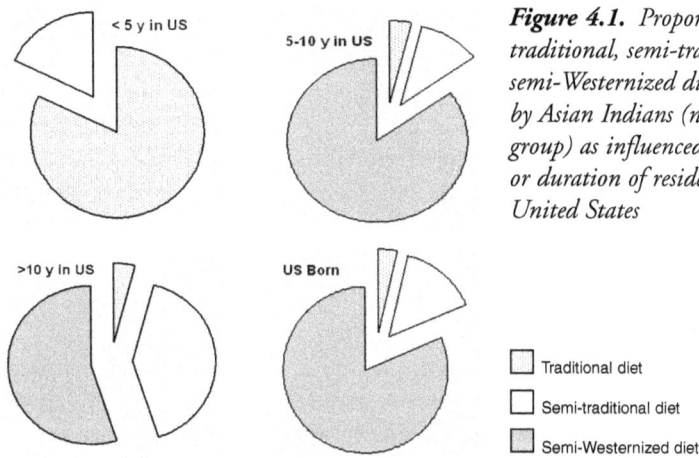

Figure 4.1. *Proportions of traditional, semi-traditional and semi-Westernized diets consumed by Asian Indians (n=22 in each group) as influenced by the length or duration of residence in the United States*

efits of those vegetables and spices listed. While all the Asian Indians recognized all the plant species listed as vegetables and spices, the White mainstream recognized only seven of the eighteen plant species as being edible.

LESS THAN 5 YEARS IN THE UNITED STATES

A large percentage of this group (82) was consuming a traditional diet, while a small percentage ate a semi-traditional diet (18), mostly because they did not consume rice and/or yogurt daily. About 50 percent of the subjects in this group indicated an increased consumption of carbonated beverages, 10 percent indicated a switch from drinking tea as a regular morning beverage to coffee, 5 percent indicated that they were including cocoa-based drinks, 3 percent reported the inclusion of olive oil, and 8 percent indicated that they were starting to drink alcoholic beverages occasionally.

5–10 YEARS IN THE UNITED STATES

A larger percentage of this group (85) indicated that they were following semi-Westernized diets and a smaller percentage consumed semi-traditional diets (11) and traditional diets (4). More subjects fell out of the traditional and semi-traditional diets because of less consumption of rice and less use of yogurt, or because of the use of flavored and sweetened yogurt instead of plain yogurt, which is still the most common form of yogurt consumed daily in India. About 90 percent indicated regular consumption of carbonated beverages, 11 percent said they were drinking tea, 25 percent consumed tea and "other beverages" such as a combination of coffee and cocoa-based drinks. About 8 percent indicated using olive oil; and 18 percent reported that they had started drinking an alcoholic beverage occasionally.

MORE THAN 10 YEARS IN THE UNITED STATES

About 55 percent indicated consuming a semi-Westernized diet, 41 percent a semi-traditional, and 4 percent a traditional diet, respectively. About 92 percent reported regular consumption of carbonated beverages, 91 percent a preference for coffee over tea as a regular morning beverage, 85 percent consumed cocoa-based drinks and coffee only and did not take tea as a hot beverage. About 25 percent indicated using olive oil, and 75 percent mentioned an occasional consumption of alcohol.

BORN AND BROUGHT UP IN THE UNITED STATES

The dietary pattern among this group pointed to 82 percent semi-Westernized, 15 percent semi-traditional and 4 percent traditional Indian diets. About 93 percent indicated regular consumption of carbonated beverages, 11 percent reported a preference for coffee over tea as a regular morning beverage, and all consumed cocoa-based drinks regularly. About 34 percent indicated using olive oil in their cuisine, and 68 percent mentioned occasional consumption of alcoholic beverages.

Comfort Foods and Comfort Zones

Listed comfort foods included pickles, *bhel* (a spicy mix of puffed rice, fresh onions and tomatoes, and a mix of spices); *samosas* (deep-fried snack with stuffed potato and vegetables), and a variety of prepared rice dishes. Comfort zones were cultural centers and common places to hang out on campus or weekend spots in the city, temples, and cultural centers. Some subjects who were in the higher income group (> $60,000), either born in the United States or with a stay of longer than 10 years in the United States, indicated specific comforting plants that they were growing in home gardens. The plants listed included mango, banana, *Bougainvillea,* curry leaf, holy basil, mint, jasmine, hibiscus, eucalyptus, oleander, gardenia, pomegranate, and marigold. The subjects who listed these plants also described the difficulty they have had in bringing them into the United States, either from India when returning from a family visit or from other sources such as specialty nurseries, as well as in maintaining these plants in a temperate climate in Connecticut. Despite the efforts and hard work needed to import these plants—a lack of knowledge about gardening or the basic needs of plants such as placement, temperature, and light requirement, they reported that maintaining a small space in their homes for such use was comforting. Also, they described importing furniture, statues of Hindu Gods, and framed pictures of Indian art and culture for home décor. They reported that this was also comforting to them and a way of teaching their children about their roots and of keeping Indian culture alive for them in the host country.

Correlation Analysis

There was a strong positive correlation between the number of years immigrants had been in the United States, and the knowledge level of diabetes (knowledge of risk factors), and the confidence and health motivation for taking preventive measures against developing diabetes. There was a strong negative correlation between the length of stay in the United States and the barriers to following better health practices (table 4.3). Acculturation was not a significant explanatory factor for health motivation in the Asian Indian groups studied.

Table 4.3. Correlation between Selected Variables Affecting Diabetes Knowledge, and Confidence and Health Motivation to Prevention Practices in Asian Indian and White Mainstream Populations

| | Diabetes (N=110) | | |
	Years in US	Knowledge	Barriers
Knowledge of Risk Factors	0.562***		
Barriers	−0.665***	−0.443**	
Confidence and Health motivation	0.769***	0.386**	−0.357**

*, ** and *** Significant at P < 0.05, P < 0.01 P < 0.001 respectively.

Conclusions

Dietary modifications indicated by the participating Asian Indian groups include the decreased use of plant species listed with anti-hyperglycemic properties, which continue to be used more frequently by Indians living in India on a more regular basis. Thus, immigrants lose the long-term benefits of inclusion of these plant species in their daily diet. Reasons for not using specific plant species include a lack of knowledge of health benefits and not being able to buy them in a grocery store that is close enough to their place of residence. In general, the study population indicated lack of knowledge about the health benefits associated with the listed vegetables and spices. They also indicated no knowledge of threats to health that can arise due to increased coffee intake or carbonated drinks. Subjects who consumed traditional diets indicated regular trips to specific grocery stores in their neighborhood that sell these vegetables. Interestingly, none of the groups studied had completely excluded the traditional Indian diet, and a small proportion of Asian Indians born and brought up in the United States continued to consume a traditional Indian diet. It is encouraging to note that the number of grocery stores listed in the phone directory that sell Indian vegetables in the state of Connecticut has increased from three in 1994 to fourteen in 2005.

The results of this study show that the Asian Indian subjects were not aware of the documented antioxidant and anti-hyperglycemic properties of listed vegetables and spices. Thus, the benefits of traditional diets need to be included in health promotion programs designed for this group. Health promotion is "any combination of health education and related organizational, economic, and environmental support for behavior of individuals, groups, or communities conducive to health" (Green and Kreuter 1999). Interventions based on education enhance and facilitate the voluntary acquisition of specific health-related knowledge, attitudes, and practices associated with achieving health-related behavior changes (Glanz et al. 2002).

In Asian Indians, the Health Belief Model (Becker 1974) can serve as a conceptual framework for the design of health promotion programs. This is particularly useful because it includes all conditions needed for an individual to initiate a health related behavior: (1) awareness of risk factors; (2) perception of susceptibility; (3) perception of effective reduction of incurring the health threat; (4) belief that benefit of preventive action outweighs the consequences of not taking the action; and (5) belief that repeated cuing (prompting or reminding) is necessary for individuals to take action. High motivation exhibited in the groups studied indicates that an educational program model for empowerment may be successful. Freire's Empowerment Theory has been widely used successfully in ethnic minority groups (Wallerstein et al. 1988; 1994).

PROBLEMS WITH THE CURRENT RESEARCH PARADIGM

Although the proportion of incidence of diabetes is higher for Asian Indian as compared to Caucasian females, there is a scarcity of research and specific awareness, early detection and prevention education programs for this minority group. Although some complementary educational programs have been introduced, there has not been sufficient programmatic frequency, intensity, or variety in the educational campaigns offered for minority women. Existing programs pool minority communities such as Asian Americans and do not recognize that within this larger Asian American community there is wide diversity in terms of language, culture, acculturation, genetics, and socioeconomics. This makes it impossible for one generic program to serve such diversity effectively.

Limitations of the Study

The study population represents a small segment of the Asian Indian population, with a high level of educational attainment, high health motivation, medium to high acculturation levels, and good health insurance benefits. Hence the findings from these studies may not apply to all segments of Asian Indian groups in the United States.

Notes

1. It is no longer the trend to cite author names as per the American Society for Horticultural Science, and therefore I did not include author names after the botanical names in this publication. I have verified visually all of the species mentioned in this text.
2. Brinjal is the British English common name of *S. melongena* in India, while eggplant is the American English name
3. Dhal (there are many kinds of dhals used, but most commonly the word "dhal" refers to red gram dhal (*Cajanus cajan,* Fabaceae)

References

Abate, N., and M. Chandalia. 2001. "Ethnicity and type-2 diabetes focus on Asian Indians." *Journal of Diabetes and its Complications* 15: 320–327.

American Diabetes Association, 1996. *Diabetes, 1996.* Vital Statistics. Alexandria, VA: American Diabetes Association.

Anderson, J., M. Moeschberger, M.S.Jr. Chen, P. Kunn, M.E. Wewers, and R. Guthrie. 1993. "An acculturation scale for the Southeast Asians." *Social Psychiatry and Psychiatric Epidemiology* 28: 134–141.

Becker, M.H. 1974. *The health belief model and personal health behavior.* Thorofare, N.J.: C.B. Slack.

Berkowitz, G.S., R.H. Lapinski, R. Wein, and D. Lee. 1992. "Race/ethnicity and other risk factors for gestational diabetes." *American Journal of Epidemiology* 135: 965–973.

Burke, J., K. Williams, S.P. Gaskill, H.P. Hazuda, S.M. Haffner, and M.P. Stern. 1999. "Rapid rise in the incidence of type-2 diabetes from 1987 to 1996. Results from the San Antonio Heart Study." *Archives of Internal Medicine* 159: 1450–1457.

Chatterjee, A., and S.C. Pakrashi. 1991. *The treatise on Indian medicinal plants.* Vol 1. New Delhi: Publication and Information Directorate.

Chen, J., E. Ng, and R. Wilkins. 1996a. "The health of Canada's immigrants in 1994–95." *Health Reports* 7: 33–45.

Chen, J., R. Wilkins, and E. Ng. 1996b. "Health expectancy by immigrant status 1986 and 1991." *Health Reports* 8: 29–37.

Center for Disease Control and Prevention, 1999. National diabetes fact sheet: National estimates and general information on diabetes in the United States. Atlanta: US Department of Health and Human Services (HHS), CDC.

Clarke, C. 1998. "How should we respond to the worldwide diabetes epidemic?" *Diabetes Care* 21: 475–476.

Dalal, A.K. 2000. "Living with a chronic disease: healing and adjustment in Indian society." *Psychology and Developing Societies* 12: 67–81.

Dowse, G.K., H. Gareeboo, P.Z. Zimmet, K.G. Alberti, J. Tuomilehto, D. Fareed, L.G. Brissonnette, and C.F. Finch. 1990. "High prevalence of NIDDM and impaired glucose intolerance in Indian, Creole, and Chinese Mauritians." *Diabetes* 39: 390–396.

Flegal, K., T.M. Ezzati, M.I. Harris, S.G. Haynes, R.Z. Juarez, W.C. Knowler, E.J. Perez-Stable, and M.P. Stern. 1991. "Prevalence of diabetes in Mexican Americans, Cubans, and Puerto Ricans from the Hispanic health and nutrition examination survey, 1982–1984." *Diabetes Care* 14: 628–38.

Glanz, K., B. Rimer, and F.M. Lewis. 2002. *Health behavior and health education. Theory, research and practice.* 3d ed. San Francisco: John Wiley and Sons.

Gravitas Research and Strategy Limited. 2005. Quality of life in New Zealand's largest cities. Residents' survey. Retrieved 22 June 2007 from http://www.manukau.govt.nz/uploaded-Files/manukau.govt.nz/Publications/Surveys/survey_feb_2005.pdf

Green, L.W., and M.W. Kreuter. 1999. *Health promotion and planning. An educational and ecological approach.* 3d ed. Mountain View: McGraw-Hill Humanities/Social Sciences/ Languages. Mayfield Publishing Company.

Hanis, C.L., R.E. Ferrell, S.A. Barton, L. Aguilar, A. Garza-Ibarra, B.R. Tulloch, C.A. Garcia, and W.J. Schull. 1983. "Diabetes among Mexican Americans in Starr County, Texas." *American Journal of Epidemiology* 188: 659–672.

Helman, C.G. 2001. *Culture, health and illness,* London: Arnold.

Hull, D. 1979. "Migration, adaptation and illness. A review." *Social Science and Medicine* 13A: 25–36.

Lawton, J., N. Ahmad, N. Hallowell, L. Hanna, and M. Douglas. 2005. "Perceptions and experiences of taking oral hypoglycaemic agents among people of Pakistani and Indian origin: qualitative study." *British Medical Journal* 330: 1247–1249.

Marinker, M., A. Blenkinsopp, C. Bond, N. Briten, M. Feely, and C. George. 1997. *From compliance to concordance: Achieving shared goals in medicine taking.* London: Royal Pharmaceutical Society of Great Britain.

Mathew, A., P. Gangadharan, C. Varghese, and M.K. Nair. 2000. "Diet and stomach cancer: a case-control study in South India." *European Journal of Cancer Prevention* 9(2): 89–97.

McKeigue, P.M., G.J. Miller, and M.G. Marmot. 1989. "Coronary heart disease in South Asians overseas—a review." *Journal of Clinical Epidemiology* 42: 597–609.

McKeigue, P.M., T. Pierpoint, J.E. Ferrie, and M.G. Marmot. 1992. "Relationship of glucose intolerance and hyperinsulinemia to body fat pattern in South Asians and Europeans." *Diabetologia* 35: 785–791.

Miller, N.H. 1997. "Compliance with treatment regimens in chronic asymptomatic diseases." *American Journal of Medicine* 102: 43–49.

O'Dea, 1992. "Obesity and diabetes in the land of milk and honey." *Diabetes/Metabolism reviews* 8: 373–388.

Pais, P., J. Pogue, H. Gerstein, E. Zachariah, D. Savitha, S. Jayprakash, P.R. Nayak, and S. Yusuf. 1996. "Risk factors for acute myocardial infarction in Indians: A case-control study." *Lancet* 348: 358–363

Ramachandran, A., C. Snehalatha, D. Dharmaraj, and M. Viswanathan. 1992. "Prevalence of glucose intolerance in Asian Indians. Urban-rural difference and significance of upper body adiposity." *Diabetes Care* 15: 1348–1355.

Ramachandran, A., C. Snehalatha, E. Latha, V. Vijay, and M. Viswanathan. 1997. "Rising prevalence of NIDDM in urban population in India." *Diabetologia* 40: 232–237.

Ramachandran, A., C. Snehalatha, A. Kapur, V. Vijay, V. Mohan, A.K. Das, P.V. Rao, C.S. Yajnik, K.M. Prasanna Kumar, and J.D. Nair, Diabetes Epidemiology Study Group in India (DESI). 2001. "High prevalence of diabetes and impaired glucose tolerance in India: National urban diabetes survey." *Diabetologia* 44: 1094–1101.

Ramaiya, K.L., V.R. Kodali, and K.G. Alberti. 1990. "Epidemiology of diabetes in Asians of the Indian subcontinent." *Diabetes/Metabolism Reviews* 6: 125–146.

Sheth, T., C. Nair, M. Nargundkar, S. Anand, and S. Yusuf. 1999. "Cardiovascular and cancer mortality among Canadians of European, South Asian and Chinese origin from 1979 to 1993: An analysis of 1.2 million deaths." *Canadian Medical Association Journal* 161: 132–138.

Simmons, D., D.R. Williams, and M.J. Powell. 1992. "Prevalence of diabetes in different regional and religious South Asian Indian communities in Coventry." *Diabetic Medicine* 9: 428–431.

Sinha, R., D.E. Anderson, S.S. McDonald, and P. Greenwald. 2003. "Cancer risk and diet in India." *Journal of Postgraduate Medicine.* 49: 222–228.

Stern, M.P., M. Rosenthal, S.M. Haffner, H.P. Hazuda, and L.J. Franco. 1984. "Sex difference in the effects of sociocultural status on diabetes and cardiovascular risk factors in Mexican Americans. The San Antonio Heart Study." *American Journal of Epidemiology* 120: 834–851.

US Census 2000: A Report. Special Populations. Washington D.C.: US Census Bureau.

Venkataraman, R., N.C. Nanda, G. Baweja, N. Parikh, and V. Bhatia. 2004. "Prevalence of diabetes mellitus and related conditions in Asian Indians living in the United States." *American Journal of Cardiology* 94: 977–980.

Wallerstein, N., and E. Bernstein. 1988. "Empowerment education: Freire's ideas adapted to health education." *Health Education Quarterly* 15: 379–394.

Wallerstein, N., and V. Sanchez-Merki. 1994. "Freirian praxis in health education: Research results from an adolescent prevention program." *Health Education Research* 9: 105–118.

Wetter, AC., J.P. Goldberg, A.C. King, M. Sigman-Grant, R. Baer, E. Crayton, C. Devine, A. Drewnowski, A. Dunn, G. Johnson, N. Pronk, B. Saelens, D. Snyder, P. Novelli, K. Walsh, and R. Warland. 2001. "How and why do individuals make food and physical activity choices?" *Nutrition Reviews* 59: S11–S20

Zambrana, R.E., S.C. Scrimshaw, N. Collins, and C. Dunkel-Schetter. 1997. "Prenatal health behavior and psychosocial risk factors in pregnant women of Mexican origin: The role of acculturation." *American Journal of Public Health* 87: 1022–1026.

Zimmet, P.Z. 1992. "Challenges in diabetes epidemiology. From West to the rest." *Diabetes Care* 15: 232–252.

Chapter 5

Use of Traditional Herbal Remedies by Thai Immigrant Women in Sweden

Pranee C. Lundberg

Introduction

In Swedish society, which has become multicultural as a result of transnational migrations, different groups—such as Asian, Middle Eastern, African, and European—express their own cultural identities through traditions, values, beliefs, and symbols. Thai immigrant women in Sweden expressed the meaning of health as a state of well-being, absence of illness, and the ability to perform daily role activities and adapt to their new life situation (Lundberg 1999; 2000). The use of herbal remedies is becoming increasingly popular all over the world. Folk medicine involves the use of culturally known herbs and remedies for self-administered treatment of sickness or illness (Hufford 1997; Leininger 1991). To be culturally competent, health care professionals need to have knowledge of ethnomedicine, and of the culture and health practices of their immigrant clients. In particular, knowledge about the use of herbal remedies by Thai immigrant women in Sweden, in the context of their culture, is significant if health care professionals are to provide culturally appropriate and congruent care to this group. Therefore, the purpose of this study was to explore the use of traditional herbal remedies by Thai immigrant women in Sweden.

Theoretical Framework

Leininger's (1991; 1997) theory of cultural care diversity and universality provided the conceptual framework for this study. The central construct of the theory is cultural care in diverse and similar cultural contexts. Leininger observed that care is essential for the growth, well-being, and survival of human beings. Furthermore, factors such as technology, religion, kinship, cultural values, politics, economics, and education influence care and health in different environmental contexts. She also asserts that care is a universal phenomenon with diverse meanings and expressions in different cultures. Two kinds of care are identified, namely, generic and professional. Generic care refers to folk, family, and lay care that is commonly used and relied upon within specific cultural groups (Leininger 1997). In this study, the position was taken that health care professionals who are sensitive to their clients' cultural health care practices can help their clients integrate cultural beliefs, for example, the use of herbal remedies with Western biomedicine in ways that will promote optimal health outcomes for the clients.

Health Beliefs and Health Care Sectors

Health beliefs generally describe beliefs and practices that are held or maintained by individuals from different cultures. Culture and tradition influence everyone's beliefs about health and illness. According to the lay theory model of illness causation (Helman 2000), health beliefs can be divided broadly into three systems, namely biomedical, personalistic, and naturalistic. Illness can be perceived as caused by factors related to the individual (lifestyle, behavior, personality, bacterial invasion, inheritance, physical constitution, and organ function); to nature; to environmental factors; to social relationships; or to the supernatural sphere (influence of fate, spirits, God(s)).

When people become ill or need medical help, they may have several options to seek help depending on where they live, who they are, and the prevailing health care practices in their culture. In one model for health care-seeking behavior (Kleinman 1988), health care may be sought in the popular, folk, or professional sector. The popular sector is the lay, non-professional, non-specialist domain of society, where ill health is first recognized and defined, and health care activities are initiated. In this sector, the main area of health care is that of the family, friends, or relatives. The popular sector usually includes a set of beliefs about maintenance of health and includes beliefs about healthy ways to eat, drink, sleep, dress, work, pray and generally conduct one's life. Most health care in this sector takes place between people already linked to one another by ties of kinship, friendship, neighborhood, or membership of work or religious organizations. Folk health care is provided by non-professional, non-bureaucratic specialists. This sector is composed of a mixture of many different components. Some are closely related to the professional sector, while others are linked to traditional healers, such as monks who are healers in Thailand. The professional sector involves organized, legally sanctioned healing professions and modern—scientifically based—medicine or biomedicine. In most Western societies, this is simply referred to as scientific medicine (Kleinman 1988).

Use of Herbal Remedies

Knowledge and use of herbs and remedies are passed on from one generation to another, primarily through oral history and tradition, usually by mothers, grand-mothers, or others in the community who are identified as healers. The use of herbal medicine in health and illness has been reported frequently in the litera-ture under the topic "alternative medicine." An increasing number of people are turning to complementary and alternative medical therapies to promote health, prevent illness, and treat acute and chronic conditions. Traditional remedies are part of the cultural and religious life of African people. Approximately 80 per-cent of the South African population uses a traditional remedy at some stage of life (Muller 2003). A recent survey reported that 40 percent of Americans had used alternative medicine, including herbs, during the previous year (Astin 1998). Zapata and Shippee-Rice (1999) found three major reasons why Latinos use el-ements of folk healing as a complement to Western physicians and biomedicine: (1) being sick; (2) healing (the process of becoming sound or healthy again) and healers (those who assist in the healing process); and (3) faith and believing. Kraatz (2001) also found that herbal remedies played a part in both health and illness among women in a Brazilian urban slum. The use of herbal remedies in treating common illness in a barrio in Managua, Nicaragua, occurred in 78 per-cent of the surveyed households, and herbal remedies were used for a variety of symptoms (Ailinger et al. 2004).

Traditional East Asian treatments include acupuncture, skin scraping, rub-bing the skin with an object such as a coin, cupping (local suction created by placing a heated cup on the skin to open skin pores), and herbal medicine (Jin et al. 2002). In Thailand, herbal medicine is used not only for curing disease but also for taking control of one's health. Herbal products, in the form of food sup-plements and health drinks, have gained increasing popularity among people of all ages (Muangman 2000). Thai traditional medicine is regarded as traditional philosophies, bodies of knowledge, and practices to cure diseases and illnesses, and is seen as congruous with the Thai way of life and culture. Practices in Thai traditional medicine involve herbal medicine, steam baths, massaging, and tradi-tional healing of bone injuries with oil and bamboo (Subcharoen 1995). In ad-dition, a number of Thai menopausal women practice alternative medicine for their well-being (Chirawatkul et al. 2002). Subcharoen (1995) also described that the application of Buddhist rites and rituals such as prayer and meditation to cure mental health problems and the use of heat and herbs (or "lying by the fire") are parts of the Thai traditional medicine methodology. The elements of Thai traditional medicine are derived from the systematic accumulation and transfer of knowledge and experiences to others by means of word of mouth, observation, recording, training, and instruction at institutes of Thai traditional medicine.

Ethnohistorical Remarks

As ethnohistory is an important dimension of cultural care, some brief remarks on the history of the Thai people will be presented. Thailand, located in Southeast Asia, has a population of about 62 million people and an area of 514,000 square kilometres. Thai people have defended their country's independence, sometimes fiercely, for more than eight hundred years, and Thailand is the only country in Southeast Asia that has never been a European colony.

The great majority of the Thai people, about 94 percent, are devout Buddhists, but religious tolerance is, and always has been, extended to most other religions. As a philosophy, Buddhism has played a profound role in shaping the character of Thais and particularly their reactions to events. The Buddhist concept of earthly impermanence and the idea of the absurdity of trying to establish certainties in an ever fluid existence have done much to create the relaxed, carefree charm that is a characteristic of Thais. To the average Thai, there is nothing inconsistent about the intermingling of other practices and beliefs with Buddhism. Their broad, easygoing tolerance makes it possible for Thais to accept, or at least to absorb, all such beliefs (Cooper and Cooper 1990).

The Official Statistics of Sweden (2004) show that the number of Thai immigrants has been increasing over the years, up to about 14,600 in 2003. These Thai immigrants constitute one of the relatively small ethnic minority groups in Sweden. Their patterns of behavioral and cultural values differ greatly from those in the West. Most Thais are unfamiliar with Northern Europe and have some difficulty adjusting to European lifestyles, beliefs, religions, languages, education, and so on. Also, most of them are women, often with weak educational backgrounds. Commonly, they have Swedish husbands or are living together with Swedish men without being married.

Methods

This study made use of ethnography (Leininger 1985; 1991), a methodology that provides a systematic process of observing, detailing, describing, documenting, and analyzing the ways of life or particular patterns of a culture in order to understand the meaning people give to particular behaviors. The method supports the goal of becoming knowledgeable about the informants' emic (or insider's) perspective while remaining attentive to the etic (or outsider's) knowledge related to worldviews, professional attitudes, biases, and racial and gender views, as well as other factors that could influence collection and interpretation of data. Both emic and etic data were studied with this method, which was chosen to discover, describe, and analyze the use of herbal remedies by Thai immigrant women in Sweden (Leininger 1985: 35). From the Thai names of the plants mentioned by the informants, English and scientific names were obtained by use of Inglis (1997). These names were also checked by use of the W^3Tropicos database (Missouri Botanical Garden 2006).

Selection of Informants

The informants were a purposive sample of Thai immigrant women living in and around Uppsala, a university town with around 190,000 inhabitants. Thai women were asked by telephone if they would be willing to participate. Inclusion criteria were as follows: (a) born in Thailand and identifying themselves as Thai; (b) living in Sweden for at least two years since immigration from Thailand; (c) familiarity with and history of use of herbal remedies; (d) willingness to participate in the study.

A total of thirty-nine informants (age range 25–65) participated voluntarily. They received information about the study and agreed to participate by sharing their experiences with the researcher. Most of the women were married to or living together with Swedish men. Their duration of residence in Sweden varied from two to twenty-eight years. They were interviewed in-depth twice, each time for one to two hours. In addition, they were observed during their daily activities.

Interviews

The interviews comprised of four open-ended questions. Before the survey, the questions were tested for validity and reliability with five Thai women. They were semi-structured in order to allow either in-depth exploration of a particular topic or change of the topic sequence, depending on the informant's narrative (Leininger 1991; Dyck 1992). The questions were posed in the Thai language by the investigator who, being herself a Thai immigrant in Sweden, had lived in Thailand for more than thirty-five years and in Sweden for more than fifteen years.

Interview questions were as follows: (a) What do herbal remedies mean to you?; (b) What kind of herbal remedies have you used in your life?; (c) What are the expected consequences or benefits of the herbal remedies you have used?; (d) What do you know about these herbal remedies? Further questions were added during subsequent interviews to validate previous data and to explore emerging themes. As the interviews progressed, meanings and observations were checked regularly with the informants. The interviews continued until the information obtained became redundant so that no new meanings or descriptions were expressed. The informants were also asked questions about demographic data. The women were always interviewed at home for their comfort. Interviews were recorded on tape and transcribed verbatim. Field notes were written out shortly after each interview in order to describe the settings, the informants, and the interviewer's thoughts and reactions. Data collection was carried out during the period from February to June 2004.

Observation-Participation

Participation in the daily life of informants, in their natural environment, enhanced the understanding of their culture and use of herbal remedies. Therefore, observation-participation was a vital part of the ethnographic research method used (Leininger 1991). This process included activities at home, at a Thai food shop, and at a place for traditional Thai massage. Preparation and use of herbal remedies at informants' homes was observed as a part of the interview process.

Data Analysis

The criteria used for data analysis included credibility, confirmation, meaning in context, recurrent patterning, saturation, and transferability (Lincoln and Guba 1985; Leininger 1991). Credibility is the truth-value mutually established between the researcher and informants to ensure accuracy of findings. Confirmation reaffirms what the researcher has heard, seen, or experienced with respect to the phenomena under study, and with confirmed informant checks and audit trails. Meaning in context refers to and focuses on the significance of interpretations and understanding of the actions, events, communications, symbols, and other activities within specific or total contexts. Recurrent patterning refers to identifiable sequenced patterns of repeated experiences, expressions, events, or activities over time. Saturation is reached when data are redundant, and there is no more new information, as all has been said or shared. Transferability refers to any general similarity of findings that can be transferred to another similar context or situation in a new research context.

The data analysis involved Leininger's (1991) four-phase method. The four-step process of this method begins with documentation and analysis of all observations and interview experiences. The second phase focuses on identification of code descriptions and categories from the raw data, and the third phase identifies patterns related to the domain of inquiry. The fourth phase, finally, includes formulation of themes. The analysis moved from collection of qualitative raw data, identification of descriptors and structural analysis, to eventual theme formulation. The principal meanings and expressions of herbal remedies and health practices are presented below with supporting verbatim descriptors and patterns.

Findings

Three major themes that were related to the use of traditional Thai herbal remedies by the Thai immigrant women emerged in this study, namely, (1) a supplement (to food) for strength, health, and absence of disease; (2) medicine for treatment of common sicknesses and illnesses; and (3) a means to regain good health in accordance with traditional folk beliefs. Table 5.1 provides an overview of the herbal remedies used by the Thai women as well as the percentage of women citing each remedy.

A Supplement for Strength, Health, and Absence of Disease

The Thai immigrant women expressed that the use of traditional herbal remedies maintains their health by making them strong, healthy, and free from disease. All women mentioned that they used herbal remedies for cooking Thai food. They believed that the herbal remedies and aromatic herbs are forms of natural medicines that support health. They prepared Thai food for their families, guests, and friends. They described that traditional Thai food, which they often made, contains fresh herbs, spices, and a blend of unusual aromatics. It was observed that Thai women used several herbal remedies and aromatic herbs while cooking Thai food. The herbs most commonly used by Thai immigrant women in everyday cooking were found to be garlic, sweet basil, holy basil, wild

Table 5.1. Herbal and Aromatic Remedies Used by Thai Immigrant Women in Sweden

Thai name	English name	Scientific name (Family plant name)	Folk treatment	Percent citing
Kra thiem	Garlic	Allium sativum L. (LILIACEAE)	Reduce blood pressure and cholesterol levels	97.4
Khing	Ginger	Zingiber officinale Roscoe (ZINGIBERACEAE)	Relieve indigestion, anti-flatulence, carminative	87.2
Hom daeng	Shallot	Allium cepa L. var. aggregatum G. Don (LILIACEAE)	Use against colds, flu, and cough, anti-inflammatory, anti-rheumatic	82.1
Ma nao	Lime	Citrus limon (L.) Burm.f. (RUTACEAE)	Use against colds, bleeding, and scurvy	79.5
Wan hang chorakhe	Aloe	Aloe vera (L.) Burm.f. (LILIACEAE))	Leaves used as a laxative, also used on fresh wounds and on skin burns	76.9
Ta khrai	Lemongrass	Cymbopogon citratus (DC.) Stapf (POACEAE)	Relieve indigestion and flatulence, cold, cough, and sore throat	74.4

Fha talai jon	*Chiretta*	Andrographis paniculata (Burm.f.) Nees (ACANTHACEAE)	*Reduce blood pressure, help for cold, sore throat, bronchitis, fever, and useful for all kinds of allergies*	53.8
Mara khinok	*Bitter cucumber*	Momordica charantia L. (CUCURBITACEAE)	*Relieves indigestion and flatulence, purifies the blood, and reduces blood sugar*	48.7
Garn phlu	*Cloves*	Syzygium aromaticum (L.) Merr. & L.M. Perry (MYRTACEAE)	*Relieve indigestion and stomach pain*	46.2
Bai toey hom	*Pandanus palm*	Pandanus amaryllifolius Roxb. (PANDANACEAE)	*Crush leaves and put on the skin to cure measles and skin diseases. Relieve indigestion and flatulence, refresh heart*	43.6
Gra jeeap dang	*Roselle*	Hibiscus sabdariffa L. (MALVACEAE)	*Protect from fat coagulation in blood, good for indigestion and flatulence, and help for ankle, back, and flank pain*	43.6
Dok kham foi	*Safflower*	Carthamus tinctorius L. (ASTERACEAE)	*Reduce blood cholesterol, improve blood circulation, lower blood pressure*	38.5
Hanuman prasan kai	—	Schefflera leucantha R. Viguier (ARALIACEAE)	*Leaves boiled in water and administered to remedy coughing*	38.5
Gah raboon	*Camphor*	Cinnamomum camphora (L.) J. Presl (LAURACEAE)	*Use to treat breathing and sinus problems, poor circulation, and painful joints*	35.9
Kwao kreu khao	—	Pueraria mirifica Airy Shaw & Suvatabandhu (FABACEAE)	*Reduce menopausal symptoms*	
Ho ra pa	*Sweet basil*	Ocimum basilicum L. (LAMIACEAE)	*Reduce gas distension and improve digestion*	30.8
Ka phrao	*Holy basil*	Ocimum sanctum L. (LAMIACEAE)	*Reduce gas distension and improve digestion*	30.8

(continued)

Table 5.1. Continued

Makham pom	*Tamarind*	Tamarindus indica L. (FABACEAE)	*Juice used to remedy diarrhoea and coughing and to remove secretions*	*30.8*
Kramin	*Turmeric*	Curcuma longa L. (ZINGIBERACEAE)	*Digestive complaints, circulatory disorder, anti-rheumatic, and treatment of eczema*	*30.8*
Ma krut	*Kaffir lime*	Citrus hystrix DC. (RUTACEAE)	*Reduce gas distension and improve digestion*	*28.2*
Bai chaplu	—	Piper sarmentosum Roxb. (PIPERACEAE)	*Use against headaches and pain in the bones, coughs and asthma*	*25.6*
Saled pung porn	—	Clinacanthus nutans (Burm.) Lindau (ACANTHACEAE)	*Help for nettle rash, dysentery, and fever*	*25.6*
Yah nuat maoh	*Cat's whisker*	Orthosiphon grandiflorus A. Terracc. (LAMIACEAE)	*Aid for kidney disorders, relieve back and waist pain, reduce blood cholesterol*	*25.6*
Bai bua bok	*Asiatic pennywort*	Centella asiatica (L.) Urb. (APIACEAE)	*Nourish the internal organs, spleen and gall bladder, and good for contusions and bruises*	*23.1*
Kra chai	*Chinese keys, Finger root*	Boesenbergia pandurata (Roxb.) Schltr. (ZINGIBERACEAE)	*Relieve indigestion, anti-flatulence, carminative*	*23.1*
Op churie tet	*Cinnamon*	Cinnamomum verum J. Presl (LAURACEAE)	*Relieve fever, mild paralysis, fatigue, and dyspepsia*	*20.5*
Dok geg huay	*Chrysanthemum*	Chrysanthemum morifolium Ramat. (ASTERACEAE)	*Improve function of the brain, liver, heart, and eyes, thirst quencher*	*20.5*
Ma toom	*Bael*	Aegle marmelos (L.) Corrêa (RUTACEAE)	*Relieve flatulence, help blood digestion, and promote the appetite*	*17.9*
Kha	*Galangal*	Alpinia officinarum Hance (ZINGIBERACEAE)	*Relieve indigestion, anti-flatulence, carminative*	*12.8*

pepper, lemongrass, kaffir lime leaf, ginger, galangal, mint, red onion, and lemon (table 5.1).

> "I use *kra thiem* [garlic] every day when I make Thai food because I know that it can reduce high blood pressure."

> "When we cook Thai food it is common for us to use different kinds of herbs and aromatics, such as *ho ra pa* [sweet basil], *ka phrao* [holy basil], *kra chai* [Chinese keys/finger root], *kha* [galangal], *ta khrai* [lemongrass], *ma krut* [kaffir lime leaf], *khing* [ginger] and *ma nao* [lemon]. We believe that these herbs and aromatics help us to be healthy."

Some women born in the northeast of Thailand mentioned that *lab kai* or *lab muo* is native Northeastern food which contains different kinds of herbal remedies and is usually eaten with vegetables. In addition, some women who had experienced childbirth stated that *kaeng lieng* is useful and a common food for women after childbirth as it is "hot" due to a large amount of pepper, and it also contains a large amount of nutrients from dried shrimps and different fresh vegetables. It is believed that *kaeng lieng* helps to get rid of childbirth blood and that it stimulates the production of breast milk.

> "My friend here made *kaeng lieng* for me after I had childbirth. She told me that it would help me get rid of childbirth blood and that it would stimulate production of breast milk for my baby."

Most of the Thai immigrant women described that they used such herbs as *khing* (ginger), *ta khrai* (lemongrass), or mint in tea. They also used Thai herbal teas such as *dok kham foi* (safflower), *op churie tet* (cinnamon), *dok geg huay* (chrysanthemum), and *bai bua bok* (Asiatic pennywort). They stated that these teas were used for their health problems. The ingredients in the tea varied depending on the type of sickness. For example, they used certain teas against colds, throat problems, or stomach conditions in their families.

> "My mother drinks *chah op churie tet* [cinnamon tea] when she has problems with her stomach. She tells me that it helps against indigestion."

> "I put dry *khing* [ginger] into the tea before drinking. I feel that my cold gets better. I also give it to my family to drink."

> "I know that my friends like to drink *chah dok kham foi* [safflower tea]. They believe that it helps reduce blood cholesterol and improve blood circulation."

In addition, some women indicated that they used Thai herbs to make preventive health drinks. They described that they boiled herbs such as *ma toom* (bael), *toey hom* (screw pine), *bai bua bok* (Asiatic pennywort), *dok geg huay* (chrysanthemum), or *ta khrai* (lemongrass) with water, then took away the herbs and let the water become cold before drinking. They mentioned that they were using herb drinks in order to maintain their health. They believed that such drinks would strengthen their bodies and immune systems and therefore prevent diseases. One of the women made a health drink from *ta khrai* (lemongrass) and offered it to the investigator as a cold drink when she visited her apartment.

> "I boil *ta khrai* [lemongrass] with water to make health drinks. I also boil *bai bua bok* [Asiatic pennywort], *geg huay* [chrysanthemum], or *gra jeeap* [roselle] with water. I make different health drinks and drink them instead of soft drinks."

The Thai women in this study had kept their cultural health beliefs and practices intact concerning the use of herbal remedies as a means to maintain or promote health and prevent or cure sickness. This finding is supported by Lundberg (2000) who found that the practice of traditional health care is one manifestation of cultural care by Thai immigrants in Sweden. Health has been conceptualized as a state, a process of development, an actualization, an outcome, and a lifestyle (Meleis 1990: 108). Health perceptions are the ways people comprehend and reflect on their health (Sholkamy 1996). They also determine people's health behavior and their decisions about when to ask for help (Lundberg and Manderbacka 1996). In this study, the Thai women preferred traditional food because they were accustomed to it. Food beliefs and food taboos are types of eating behavior that people learn from their parents and relatives (Aree et al. 2004). This means that belief and culture still strongly influence these women's ways of life. Like people in every part of Thailand, these women had shared their experiences regarding eating behavior with their forefathers and descendants through storytelling (Suparp 1993). Their use of herbal remedies to maintain health was supported by Factor-Litvak et al. (2001), who found that herbs for medical teas were often used by women from different ethnic groups.

Medicine for Treatment of Common Sicknesses and Illnesses

Most of the Thai immigrant women had experiences from Thailand in using traditional herbal remedies at home when they felt sick or ill. They used these remedies in the same way in Sweden. The women described that they used *ya samon phai* (Thai herbal medicine) for treatment of common sicknesses and illnesses such as cold, cough, sore throat, stomachache, bladder problems, menopause, high blood pressure, and high blood sugar. Nowadays, Thai herbal medicines are mainly obtained from drug stores in Thailand, where they are available as powders, tablets, or capsules. Some mentioned examples of herbal remedies they had used when they were sick, such as *wan hanuman prasan kai* (*Schefflera leucantha*

R. Viguier), *wan hang chorarakhe* (*Aloe vera* (L.) Burm.f.), *saled pung porn* (*Clinacanthus nutans* (Burm.) Lindau), *fha talai jon* (*Andrographis paniculata* (Burm.f.) Nees), *yah nuat maoh* (*Orthosiphon grandiflorus* A. Terracc.), *kwao kreu khao* (*Pueraria mirifica* Airy Shaw & Suvatabandhu), *op churie tet* (*Cinnamomum verum* J. Presl) and *mara khinok* (*Momordica charantia* L.) (table 5.1). Most Thai women described that *wan hang chorarakhe* (*Aloe vera*) was the most common herb used to reduce pain and obtain a cooling sensation for skin burns. They used *fha talai jon* (*Andrographis paniculata*) in the form of tablets or capsules when they had fever, colds, headache, and sore throat. They also mentioned that they had learnt about herbal medicines by listening to their mothers, relatives, and friends, and by reading herbal medicine books, newspapers, and magazines. They used herbs as medicines since they believed herbs give fewer side effects than medicines that make use of chemistry. However, they visited the doctor in case they did not get better.

"When I was coughing heavily, my mother put *wan hanuman prasan kai* [*Schefflera leucantha*] with water into a pot and boiled it. She let me drink this remedy three times a day. After three days my coughing had disappeared, and I felt better."

"My sister had problems with her bladder. It was difficult for her to remove urine. She felt much pain. She told me that she boiled *yah nuat maoh* [*Orthosiphon grandiflorus*] with water and drank it several times. She got better after using the drink."

"My friend told me about *mara khinok* [*Momordica charantia*]. It is believed to reduce blood sugar for people who have diabetes. Her mother takes this capsule three times a day."

Some women mentioned that nowadays *ya samon phai* (Thai herbal medicine) is more and more popular in Thailand. Many Thai people prefer to use herbal medicine for treatment of common sicknesses and illnesses. They brought herbal medicine when they went back to Thailand or asked friends or relatives to buy it for them. They also bought herbal medicine in Thai shops.

"I brought back several *ya samon phai* [Thai herbal medicine] and Thai herbal teas when I visited my relatives in Thailand. Sometimes I bought them in a Thai shop in Sweden."

These findings show that the Thai women paid attention to the use of Thai herbal medicines for the treatment of common illnesses or diseases. They were alert to learning about herbal medicines for the promotion of good health. This is in agreement with several studies (Astin 1998; Zapata and Shippee-Rice 1999; Kraatz 2001; Jin et al. 2002; Ailinger et al. 2004). Also, Hilton et al. (2001) found that many South Asian immigrant women used traditional health practices for small problems or when conventional medicines did not work. In Thailand, Chi-

rawatkul et al. (2002) found that a number of Thai menopausal women practiced alternative health care for well-being. Muangman (2000) also described that
herbal drugs and food have been investigated and used for the treatment of diseases such as diabetes. Several researchers (Kaczmar 1998; Jouad et al. 2001)
showed that herbs such as gymnema, fenugreek, bitter gourd, garlic, and onions
exhibit antidiabetic activity to some degree.

A Means for Regaining Good Health in Accordance with Traditional Folk Beliefs

Thai immigrant women had cultural beliefs, passed on from generation to generation, about using herbal remedies for recovering health or curing diseases.
Some women indicated that there are several customs in Thailand that Thai
women practice after childbirth, such as *yu fai, kao krachome,* and *ya dong.* However, in Sweden *kao krachome* was the only method used after childbirth. *Yu fai*
is a ritual in which a woman who has just given birth lies on a wooden bed over
a warm fire for the thirty days of the confinement period. *Kao krachome* is a traditional confinement practice for new mothers in the Thai culture that often
makes use of herbal remedies. It is believed to help sweat out poisonous water
from the body so that the body may absorb good water, and this makes the
woman healthy. It also helps to get rid of the bad smell of *lochia* (discharge from
vagina after delivery) and to protect from blurred vision, dizziness, headaches,
and fatigue later in life. *Ya dong* is a traditional medicine that is brewed from several traditional herbal medicines in alcohol (rice wine). Some herbal medicines
must be prepared in advance and are difficult to find in Sweden. It is believed
that these medicines are hot and help the woman gain full strength.

> "I used *kao krachome* after my first childbirth in Thailand. I felt that I quickly became
> strong. I still used *kao krachome* for my health after my second childbirth in Sweden.
> My friend here helped me to contact an old Thai woman who knew how to prepare
> herbal remedies."

Some women described that they often used *nuad phaen boran* (traditional
Thai massage) when they felt fatigue, back pain, or headache. In Sweden they
got traditional Thai massage at home or in a traditional Thai massage place. The
women reported they felt better after a massage, and they also met friends and
had some kind of social life at the traditional Thai massage place. One of the
Thai women who worked with traditional Thai massage explained that such
massage is based on the concept of invisible energy lines running through the
body. These lines begin at the toes and go up through the head, and they conduct and carry the energy of the body. Traditional Thai massage is sometimes
combined with herbal remedies or aromatics such as *kramin* (turmeric) to treat
skin diseases and rashes. *Gah raboon* (camphor) can be applied to treat bruises,

sprains, and swelling. The herbs are put in a muslin or linen bag, which has been steamed for at least ten minutes until it is hot and moist. The heat creates a sort of compress, which is then patted on the joints.

> "When I feel fatigue or back pain I visit the traditional Thai massage place. I feel better after such a massage."

Some women mentioned that they had seen some people in their hometown using *ya mor* or *ya tom* (folk medicine) to treat cancers. They explained that people found out about the efficacy of this medicine through word of mouth. It is common that monks or laypersons, living in the local community and functioning as folk healers, give a recipe or prepare *ya mor* or *ya tom* for sick persons. To prepare this medicine, sick persons or relatives boil a mixture of tree roots and herbs. Some informants said that the value of traditional treatment with *ya mor* or *ya tom* is in subjective feeling and visible symptomatic improvement. They also said that Thai people often use *ya mor* or *ya tom* in combination with Western medicine for treatment of diseases such as cancer. The informants had no experience using *ya mor* or *ya tom* in Sweden. They mentioned that they visited the doctor at the health care centre or hospital when they were sick. However, they would make use of traditional treatment as well as Western medicine if they believed that these practices would help them recover from diseases such as cancer.

> "My aunt told me that her friend who is living in Thailand had cervix cancer. She had her cancer treated by a doctor at a hospital. She also went to a *wat* [temple] for treating cancer. A monk gave her a recipe to prepare *ya mor*. She took about four or five pots and felt better. I don't know what kinds of tree roots or herb leaves she used."

These findings show that the Thai immigrant women in Sweden believed in herbal medicines and used them in harmony with traditional folk beliefs. For example, they used *kao krachome* and practiced traditional confinement after childbirth to avoid ill health later in life. This was supported by Liamputtong Rice (2000) and Liamputtong and Nakssok (2003) who found that Thai traditional confinement helps to restore heat lost in childbirth, to flush out retained blood and the placenta from the uterus, and to make the uterus shrink to its normal size. Liamputtong and Nakssok (2003) found that Thai women practiced traditional confinement (the "mother heating" ritual), which is a common practice in a number of Southeast Asian cultures, as well as dietary precautions. The findings also show that *Nuad phaen boran* (traditional Thai massage) is still in use by Thai people. The use of combined therapies such as *ya mor* or *ya tom* and Western medicine indicates that both modern medicine and traditional folk medicine are important for patients in Thailand with diseases such as cancer. They try to find a secondary form of treatment that supports the medical therapies. This is supported by Nilmanat and Street (2004) who found that caregivers moved between modern medicine, traditional/folk medicine, supernatural healing rites, religious performances, and home remedies in their search for a cure for AIDS.

From the findings, the Thai immigrant women's beliefs are culturally determined, and this may affect their health practices and the type of health care sought. Traditional Thai beliefs are a mixture of Indian and Chinese medicine and spiritual/animistic beliefs (Lundberg 2000). Traditional folk medicine practices are still important in contemporary Thai culture, and the Thai immigrant women often used them as they did before migration. Yet, Western or modern medicine was the dominant approach adopted by these women. They tried to make their choices between different types of treatment that they believed would be helpful to them even though they lived in Sweden, and they considered several practices for treatment of their problems. Therefore, their experiences of using traditional herb remedies in their country of origin had an important impact on their behavior in the new country.

Discussion

The present findings show that all the informants had experience with the use of herbal remedies both in their native country and in Sweden. They were strong believers in traditional herbal medicines for maintaining health, treating common sicknesses, and preventing diseases. However, they also used physicians and the professional health care system. Kleinman (1988) suggested that, in looking at any complex society, one can identify three overlapping and interconnected sectors of health care: the popular sector, the folk sector, and the professional sector.

Findings from this study indicate that traditional herbal remedies are an integral part of health care for Thai immigrant women who also use professional health care. Cultural beliefs, values, and meanings associated with expressions of health and illness were found among the Thai immigrant women. Leininger (1991; 1997) emphasizes the importance of meeting the clients' cultural needs to provide care that heals and maintains health. Generic care needs to be discovered and integrated into professional health care in order to offer culturally congruent care practices. In addition, Papadopoulos et al. (1998) identify four main concepts (cultural awareness, knowledge, sensitivity, and competence) of the transcultural skills development model in order to achieve cultural competence, as well as anti-discriminatory and anti-oppressive practice. Therefore, health care providers should be culturally sensitive and open to ways in which different cultures attempt to achieve and maintain health among their members.

Regarding the theory of cultural health care diversity and universality, health care providers should (1) help preserve culture to meet care needs and promote health and well-being of immigrant clients; (2) respect and accommodate cultural beliefs and lifestyles of clients; and (3) consider factors that can be influenced to restructure areas in which caring and generic folk practices are incongruent.

Providing an opportunity for clients to express whether they use or practice herbal medicine, and creating an atmosphere of caring and attentiveness to the whole person, opens a door to establishing a trusted relationship. It is in the context of this trust that health care providers and clients can help each other to learn different or new ways of healing and promoting health. Using the cultural knowledge gained from clients, health care providers can take actions that are culturally sensitive and make decisions with their clients that support cultural beliefs and practices to promote and enhance the healing process.

Since data were obtained from a purposive sample of voluntary Thai immigrant women interested in participating, the results cannot be extrapolated to Thai immigrant women in general. These results reflect the perceptions and experiences of the informants of the current study. In future studies, we will work with larger samples of subjects. A focus on gender differences and the use of a quantitative methodology will provide more details about the perspectives of Thai immigrants. In addition, the attitudes of Thai youth toward using herbal remedies will be addressed.

Conclusion

The findings of the present study show that the Thai immigrant women used traditional herbal remedies to maintain and regain good health and to treat common diseases in accordance with traditional folk beliefs. An implication is that health care providers working with immigrants from different cultures should be aware of immigrants' traditional health beliefs and practices. In order to provide holistic, culturally congruent care, they should also understand the role herbal medicine plays in their clients' health care values and beliefs. Therefore, health care providers are instrumental in helping clients from different ethnic groups integrate the professional systems of care in such a way that they promote an optimal level of health and well-being.

References

Ailinger, R.L., S. Molloy, L. Zamora, and C. Benavides. 2004. "Herbal remedies in a Nicaraguan barrio." *Journal of Transcultural Nursing* 15: 278–282.

Aree, P., V. Tanphaichitr, W. Suttharangsri, and K. Kavanagh. 2004. "Eating behaviors of elderly persons with hyperlipidemia in urban Chiang Mai." *Nursing and Health Sciences* 6: 51–57.

Astin, J.A. 1998. "Why patients use alternative medicine: results of a national study." *Journal of the American Medical Association* 279: 1548–1553.

Chirawatkul, S., K. Patanasri, and C. Koochaiyasit. 2002. "Perceptions about menopause and health practices among women in Northeast Thailand." *Nursing and Health Sciences* 4: 113–121.

Cooper, R., and N. Cooper. 1990. *Culture shock! Thailand.* Singapore: Times Books.

Dyck, I. 1992. "Managing chronic illness: an immigrant woman's acquisition and use of health care knowledge." *American Journal of Occupational Therapy* 46: 696–704.

Factor-Litvak, P., L.F. Cushman, F. Kronenberg, C. Wade, and D. Kalmuss. 2001. "Use of complementary and alternative medicine among women in New York City: A pilot study." *The Journal of Alternative and Complementary Medicine* 7: 659–666.

Helman, C.G. 2000. *Culture, health, and illness.* 3rd ed. Oxford: Butterworth-Heinemann.

Hilton, B.A., S. Grewal, N. Popatia, J.L. Bottorff, J.L. Johnson, H. Clarke, L.J. Venables, S. Bilkhu, and P. Sumel. 2001. "The desi ways: traditional health practices of south Asian women in Canada." *Health Care for Women International* 22: 553–567.

Hufford, D.J. 1997. "Folk medicine and health culture in contemporary society." *Primary Care* 24: 723–741.

Inglis, K., ed. 1997. *Tropical herbs & spices of Thailand.* Singapore: Periplus Editions (HK) Ltd.

Jin, X.W., J. Slomka, and C.E. Blixen. 2002. "Cultural and clinical issues in the care of Asian patients." *Cleveland Clinical Journal of Medicine* 69: 50–61.

Jouad, H., M. Haloui, H. Rhiouani, J. El Hilaly, M. Eddouks. 2001. "Ethnobotanical survey of medical plants used for the treatment of diabetes, cardiac and renal diseases in North centre region of Morocco (Fez-Boulemane)." *Journal of Ethnopharmacology* 77: 175–182.

Kaczmar, T. 1998. "Herbal support for diabetes management." *Clinical Nutrition Insights* 6: 1–4.

Kleinman, A. 1988. *The Illness narratives: Suffering, healing, and the human condition.* New York: Basic Books.

Kraatz, E.S. 2001. "The structure of health and illness in a Brazilian favela." *Journal of Transcultural Nursing* 12: 173–179.

Leininger, M. 1985. *Qualitative research method in nursing.* Orlando: Grune & Stratton.

———.1991. *Cultural care diversity and universality: A theory of nursing.* New York: National League for Nursing Press.

———.1997. "Overview of the theory of cultural care with the ethnonursing research method." *Journal of Transcultural Nursing* 8: 71–78.

Liamputtong Rice, P. 2000. "Nyo dua hli—30 days confinement: traditions and changed childbearing beliefs and practices among women in Australia." *Midwifery* 16: 22–34.

Liamputtong, P., and C. Nakssok. 2003. "Perceptions and experiences of motherhood, health, and the husband's role among Thai Women in Australia." *Midwifery* 19: 27–36.

Lincoln, Y.S., and E.G. Guba. 1985. *Naturalistic inquiry.* Beverly Hills: Sage.

Lundberg, O., and K. Manderbacka. 1996. "Assessing reliability of a measure of self-rated health." *Scandinavian Journal of Social Medicine* 24: 218–224.

Lundberg, P.C. 1999. "Meanings and practices of health among married Thai immigrant women in Sweden." *Journal of Transcultural Nursing* 10: 31–36.

———. 2000. "Cultural care of Thai immigrants in Uppsala: A study of transcultural nursing in Sweden." *Journal of Transcultural Nursing* 11: 274–280.

Meleis, A.F. 1990. "Being and becoming healthy: the core of nursing knowledge." *Nursing Science Quarterly* 3: 107–114.

Missouri Botanical Garden. 2006. *W3 Tropicos database.* Retrieved 9 August 2006 from http://mobot.mobot.org/W3T/Search/vast.html

Muangman, V. 2000. *Use of herbal remedies for treatment of adult diseases.* Bangkok: Author.

Muller, G.J. 2003. "Traditional medicines and acute poisoning," *CME* 21: 481–484.

Nilmanat, K., and A. Street. 2004. "Search for a cure: Narratives of Thai family caregivers living with a person with AIDS." *Social Science & Medicine* 59: 1003–1010.

Official Statistics of Sweden. 2004. *Statistical yearbook of Sweden 2005.* Stockholm: Elanders Novum AB.

Papadopoulos, I., M. Tilki, and G. Taylor. 1998. *Transcultural care: A guide for health care professionals.* Dinton, UK: Quay Publication, 175–211.

Sholkamy, H. 1996. *Women's health perceptions: A necessary approach to an understanding of health and well-being.* Egypt: Giza.

Subcharon, P. 1995. *The history and development of Thai traditional medicine.* Bangkok: Thai Traditional Institute.

Suparp, S. 1993. *Thai society and culture: Values, families, religions, and customs.* 8th ed. Bangkok: Thammasat University.

Zapata, J., and R. Shippee-Rice. 1999. "The use of folk healing and healers by six Latinos living in New England: A preliminary study." *Journal of Transcultural Nursing* 10: 136–142.

Chapter 6

Medicinal Plant Use by Surinamese Immigrants in Amsterdam, the Netherlands
Results of a Pilot Market Survey

Tinde van Andel and

Charlotte van 't Klooster

Introduction

Almost 90 percent of the population of developing countries relies largely on traditional herbal medicine to meet their primary health care needs (WHO 2002). The frequently inadequate supply of Western medicine, the cost of consultations and pharmaceuticals, the distance to the nearest primary health care center, as well as cultural and religious practices, contribute to the widespread and continued use of herbs as medicine in developing countries (Slikkerveer 1990; Ososki et al. 2002; Hamilton 2004; Vandebroek et al. 2004). It is often assumed that the demand for wild plants will decrease with increasing welfare, because they will be replaced in time by cultivated or synthetic products (Homma 1996; Ros-Tonen 1999). Moreover, the reliance on medicinal plants may decline in the long term as modern health care facilities become available (Cunningham 1993).

Recent studies of urban ethnobotany, however, contradict these assumptions. In fact, research on the use of medicinal plants by ethnic minorities in Europe and the United States has shown that immigrants generally adhere to their culture and continue their traditional medical practices after emigration (Ososki et al. 2002; Reiff et al. 2003; Pieroni et al. 2005). Instead of being replaced by conventional medicine as part of the process of cultural adaptation, the demand for medicinal plants remains, even when modern health care facilities are available (Balick et al. 2000). There are a number of reasons for this. For example, traditional healers are often less expensive than practitioners of conventional medicine, and they share the immigrant's language and cultural background. In addition, immigrants are often faced with quite different health problems in their new surroundings than in their homeland. The question is: What aspects of traditional health care are maintained in the new environment, and what aspects are lost through acculturation?

With more than 90 percent of its surface covered with dense tropical forest, Suriname numbers only 450,000 inhabitants. The population is concentrated around the capital city of Paramaribo and consists mainly of Creoles, East Indians, Javanese, and Chinese. The country's interior is inhabited by various tribes of Amerindians and Maroons, the latter being descendants of escaped African slaves who created their own traditional societies. The civil war of the 1980s, road improvement, and the lack of employment and education opportunities in the interior have caused an ongoing migration of Amerindians and Maroons to Paramaribo (Kambel and MacKay 1999). Today, more than 60 percent of the population lives below the poverty line (Pan American Health Organization 2004). Health care facilities in Suriname are limited, with only 2.5 physicians per 10,000 people.

Suriname's high biological and cultural diversity, however, corresponds with a wealth of ethnobotanical knowledge and the widespread use of medicinal plants (Raghoenandan 1994; DeFilipps et al. 2004). The traditional health care system in Suriname consists of folk medicine (orally transmitted within households) and shamanistic (spiritual) medicine. The latter can only be applied by specialized practitioners like Afro-Surinamese *winti* priests and Amerindian shamans. Several studies have been conducted on the traditional *winti* religion, a mixture of Afro-American beliefs and Christianity that includes possession by supernatural beings, rituals, and herbal baths (Wooding 1972; Stephen 1998; van Wetering 2003). Although plants are essential ingredients in *winti* rituals, the species involved have been poorly documented.

Almost half of the Surinamese population migrated to the Netherlands between 1972 and 1996, after which Dutch immigration rules became more restricted (Pan American Health Organization 2004). Surinamers (particularly Creoles and East Indians) now form the largest immigrant group in the country. Compared to ethnocultural groups from Turkey and Morocco, Surinamers have integrated fairly well into Dutch society (Musschenga 1998). Nevertheless, even after decades of living in the Netherlands, many Surinamese immigrants still have a "transnational identity." They maintain strong social and economic ties with their home country, which is generally seen as the major cause of their weak socioeconomic position (Gowricharn and Schuster 2001).

Unlike those in the United States (Reiff et al. 2003), most legal immigrants in the Netherlands have a health insurance that covers visits to general practitioners (biomedical health care providers), hospitalization, mental health care, and nearly all prescription medicine. This means they are able to take full advantage of Dutch health care facilities. Surinamers, however, seem to have retained their cultural concepts of health and illness. In fact, several social scientists have reported that the *winti* religion still plays a key role in the mental health care of Surinamese immigrants (Stephen 1998; Musschenga 1998; Leenders et al. 2001; van Wetering 2003). Hence, the following questions have arisen: Do these immigrants also continue to use herbal medicine from their homeland? And if so, for which ailments do they use these plants? In this chapter we discuss the results of a pilot study on urban ethnobotany carried out in the Netherlands.

Methodology

Fieldwork took place in July 1999 as part of a Master of Science research project at Utrecht University and the Free University of Amsterdam (van 't Klooster 2000). It was decided that the venue of this pilot market survey would be in Amsterdam, where large numbers of Surinamese immigrants reside. To start with, the researchers visited a number of Surinamese shops in Amsterdam and asked the shopkeepers whether they sold herbal medicine. Most sold only vegetables, fruits, and frozen and salted fish from Suriname; just a few offered a small assortment of dried herbs. After a lengthy search, the authors discovered "Piking Nanga Grang Vanoedoe Oso" (PNGVO, "warehouse for big and small (spiritual) necessities"), a store specializing in medicinal plants and *winti* ingredients from Suriname. The owner and his family were Saramaccan Maroons, whose roots were in the village of Goejaba along the Suriname River and transmigration settlements near the Brokopondo reservoir, a day's travel from Paramaribo by boat and road. Traditional Maroon textile and woodcrafts were displayed in the shop window, while three large freezers packed with medicinal plants occupied the back of the shop. Also in the back were a couple of bulky rice bags stuffed with dried plants.

The authors made an inventory of the fresh, dried, and frozen medicinal plants offered for sale and tried to identify as many as possible on the spot using the illustrated pharmacopoeia of French Guiana (Grénand et al. 1987). Twenty-six of the plants that could not be identified to the species level, but did contain fertile or woody material, were bought and taken to the Utrecht branch of the National Herbarium of the Netherlands (NHN-U) for identification. Only twenty-six items could be purchased because of time and budget constraints. The researchers then copied the list of vernacular plant names in the Saramaccan and Surinamese ("Sranan Tongo") languages that hung above the freezers. The scientific names were later sought among those mentioned in the (ethno)botanical literature available for Suriname (e.g., Görts-van Rijn 1985–1996; Görts-van Rijn and Jansen-Jacobs 1997–2001; Jansen-Jacobs 2002–2003; van 't Klooster et al. 2003; DeFilipps et al. 2004). The shopkeeper and his family were interviewed with regard to the store's customers and the provenance, quantities, uses, and cultivation status of the plants on sale. The authors also questioned a Dutch customs officer at Schiphol International Airport to get an idea of the quantity and legal status of the import of medicinal plants from Suriname. Plant uses recorded in the shop were compared with an unpublished report by DeFilipps et al. (2004) that was compiled from over 240 literature references and that listed medicinal plant species from the Guianas and their uses.

Results

Although not always in stock, PNGVO sold more than 149 species of medicinal plants from Suriname at the time of the inventory (see table 6.1). Forty-eight well-known cultivated medicinal species could be identified directly (e.g., *Cymbopogon citratus, Aloe vera,* and *Ixora coccinea*). Of the sample of twenty-six plants taken to the Utrecht herbarium, twenty-two could be identified to species level (CK numbers, see table 6.1). The majority of the plant material in the store, however, was in such poor condition that it was virtually impossible to identify the genus or even the family. The plants had been rolled tightly in newspapers before being frozen or dried so that leaf shapes had become unrecognizable. Moreover, flowers and/or fruits were often lacking. Fortunately, the vernacular name list provided by PNGVO allowed the authors to trace the scientific names of another seventy-six species. More than sixty of the Saramaccan names of plants sold in the shop had never before been mentioned in the literature. Only because some of the plants were also listed by their Sranan Tongo name could they be linked to a scientific name. In the end, thirty-eight local (mainly Saramaccan) plant names could not be matched to a scientific name.

The species offered for sale in PNGVO included herbs, ferns, woody lianas, seeds, bark, resin, roots, fruits, and vegetable oil. Plants or pieces of bark were sold in bundles for about €5 a piece or in more expensive ready-to-use mixtures. The shopkeeper told that, two to three times a week, a cardboard box (0.05 m^3) full of fresh plants arrives at Schiphol from Suriname on regular KLM flights. Although most of the species were said to be harvested in the vicinity of Paramaribo, some come from the hinterland forests near Brokopondo Lake. If customers ask for specific plants that are not generally in stock, a request is sent to the shopkeeper's family to find and send the species to Amsterdam. The demand from clientele in the Netherlands strongly determines the supply of herbal medicine.

According to the store owner, the species on sale were not deliberately cultivated for their shop. Of the 149 species offered for sale at PNGVO, 91 (61 percent) are found only in the wild, 38 (26 percent) are cultivated, and 21 (14 percent) occur in the wild, but are also domesticated in the Guianas (Boggan et al. 1997; van Andel 2000). Although the majority of the Surinamese herbal medicine encountered during this pilot study appears to be harvested in the wild, about two-

Table 6.1. Plants Offered for Sale by "Piking Nanga Grang Vanodoe Oso" (PNGVO), Amsterdam, the Netherlands

CK numbers correspond with collected voucher specimens. Species marked with # are indentified on the spot. Others are traced back by local their names.

w = wild, c = cultivated, w/c = in process of domestication, c/w = introduced, naturalised, ? = uncertain

tea = decoction and infusion; bath = ingredients soaked in hot or cold water; on skin = entire leaf, poultice or oil rubbed on skin; in alcohol = extract used internally or externally

* Use confirmed by literature from the Guianas

1 = Stephen (1998); 2 = van Andel (2000); 3 = DeFilipps (2004); 4 = Titjari (1985); 5 = ter Steege et al. (2003); 6 = Lachman-White et al. (1987)

Vernacular names are spelled following van 't Klooster et al. (2003). Sr = Sranang tongo, Sa = Saramaccan, - = no information available

Family	Species	w/c	nr	Names used in PNGVO	Part used	Preparation	Uses mentioned by PNGVO (bold, italic) Uses in Guianas from literature (not bold)
Acanthaceae	*Justicia calycina*	w		Brudu wiwiri (Sa)	whole plant	bath?, in alcohol?	winti (1), charm for enemies (2)
Acanthaceae	*Justicia pectoralis*	w		Papa wiwiri (Sa)	whole plant	tea	***stomach problems, diarrhoea**, *too frequent menstruation***
Amaranthaceae	*Amaranthus sp.*	c	CK2	Redi mowa (Sa), Redi klaroeng (Sr)	whole plant	bath	winti (1), herbal baths (3)
Anacardiaceae	*Anacardium occidentale* #	c		Kasyu (Sr)	leaf, bark	tea, bath	***stomach ache, vaginal steam bath****
Anacardiaceae	*Spondias mombin* #	w/c		Mope (Sr)	leaf	bath	***"to clean spirit", winti****
Annonaceae	*Annona muricata*	c		Alukutu (Sa), Sunsakka, Zuurzak (Sr)	leaf, flowers	tea	fever, headache, hypertension, nervousness, heart problems (3)
Annonaceae	*Xylopia discreta*	w	CK25	Kunge (Sa), Pegrekoe, Pedreku wiwiri (Sr)	fruits, leaf	tea, bath	***stomach problems**, *diarrhoea, asthma, cold**, *vaginal steam bath****
Apiaceae	*Eryngium foetidum* #	w/c		Sneki wiwiri (Sr)	whole plant	bath	herbal bath for fright and convulsions (3)
Apocynaceae	*Allamanda cathartica*	w/c	CK17	Ingi broemtjie (Sr), Wilkens bita (Sa)	leaf, root	tea, bath	***bowel problems**, *cold, headache, sleeplessness, ritual body cleaning****
Apocynaceae	*Asclepias curassavica*	c		Konu broemkie (Sa), Frederiks katoen, Koningsbloem (Sr)	whole plant?	bath	***luck charm***
Apocynaceae	*Geissospermum sericeum*	w		Bergi bita (Sr)	bark	tea, in alcohol	venereal diseases, diabetes, diarrhoea, fever, worms, itches (3)

Family	Species			Local name	Part used	Preparation	Uses
Araceae	*Montrichardia arborescens* #	w		Mokomoko (Sr)	leaf?	on skin	eye infection, hypertension, rheumatism, aphrodisiac, sores (3)
Arecaceae	*Astrocaryum vulgare*	w/c		Awara koko (Sr)	oil (seeds)	on skin	rheumatism, pain, laxative, earache, fever, aching feet (3)
Arecaceae	*Euterpe oleracea*	w/c		Pina wiri tongo (Sr)	inflorescence	charm	ward off evil spirits (2)
Arecaceae	*Oenocarpus bacaba*	c		Kumbu lutu (Sa)	root	tea	infertility
Aristolochiaceae	*Aristolochia consimilis* #	w		Loangotitei (Sr)	leaf	tea	*blood purifier, ritual body cleaning*
Asteraceae	*Ayapana triplinervis*	w		Sekrepatu wiwiri, Schildpadblad (Sr)	leaf	tea	*hypertension**
Asteraceae	*Clibadium surinamense* #	c	–	Kunami (Sr)	leaf	tea	winti (1), colds (3)
Asteraceae	*Eclipta prostrata*	w		Rosa wiri, Luisa wiwiri (Sr)	leaf	tea, bath	*stomach ache, diarrhoea, too frequent menstruation, winti*
Asteraceae	*Rolandra fruticosa*	w	CK13	Bokokosu, brokopanjie (Sa)	whole plant	bath	*painful joints, stomach ache, scabies, back ache, ritual body cleaning, winti*
Asteraceae	*Struchium sparganophorum*	w	CK26	Baka mujeh (Sa), Blaka oema, Scegroto (Sr)	whole plant	bath	*rheumatism, sprained limbs, winti**
Asteraceae	*Tagetes sp.*	c		Koelibroentjie, Afrikaanjes (Sr)	flower	bath	*winti, ritual use during marriages**
Asteraceae	*Unxia camphorata* #	w		Campfer bita (Sr)	leaves	tea	colds, bitter tonic (2)
Asteraceae	*Wulffia baccata*	w		Sukrutanta (Sr)	leaves	bath	skin spots, blotches (3)
Begoniaceae	*Begonia glabra*	w/c	CK4	Korto hati (Sa)	leaf (juice)	bath	*anxiety, rheumatism, ritual body cleaning, luck charm, winti**
Bignoniaceae	*Crescentia cujete* #	c		Kuya (Sa), Krabasi, Kalebas (Sr)	leaf, fruit	bath, as cup	*asthma, bronchitis, ward off evil influences, winti**
Bignoniaceae	*Jacaranda obtusifolia*	w		Kandra udu, Gobo-gobo wiri, Goebai (Sr), Dia makka (Sa)	leaf?	bath	skin infections, ward off evil spirits (3)
Bignoniaceae	*Mansoa alliacea*	w	CK5	Abonenge tatai (Sa), Knofroekoe-tete (Sr)	leaf	tea, bath, in alcohol	*rheumatism*, hypertension, anxiety, fever*, laxative, ward off evil spirits*, stomach problems, to lose weight*

Table 6.1. Continued

Family	Species	w/c	nr	Names used in PNGVO	Part used	Preparation	Uses mentioned by PNGVO (bold, italic) / Uses in Guianas from literature (not bold)
Bixaceae	Bixa orellana #	c		Kuswe (Sa), Kusewe (Sr)	fruit	on skin?	**ear ache, winti***
Bombacaceae	Ceiba pentandra #	w		Kankantri, Dia makka (Sr)	leaves	bath	*winti**
Boraginaceae	Cordia curassavica #	w/c		Blaka oema (mama) wiwiri (Sr)	whole plant	bath	*vaginal steam bath*
Boraginaceae	Heliotropium indicum	w		Kaka fo kangkang (Sa)	whole plant	tea	venereal diseases, diabetes, excessive vaginal discharge, clean out uterus, abortive (2)
Bromeliaceae	Ananas sp.	w		Busi nanasi (Sr)	?	?	abortive (3)
Burseraceae	Protium sp.	w		Busie kadea (Sa), Buskandra (Sr)	resin	in water	winti (1), chest colds, bronchitis, asthma (3)
Caricaceae	Carica papaya #	c		Papaya (Sr)	leaf?	tea	hypertension, aphrodisiac, venereal diseases (3)
Cecropiaceae	Cecropia peltata	w		Papantii (Sa), Busi papaya (Sr)	leaf	tea	*laxative, to loose weight*
Cecropiaceae	Cecropia sciadophylla #	w		Womi pantii (Sa), Man/ bofro busi papaya (Sr)	leaf	tea	kidney and bladder problems, hypertension, nervousness (3)
Chenopodiaceae	Chenopodium ambrosioides	w/c		Woronmenti, Tiki menti, Fukufuku menti (Sr)	whole plant		—
Combretaceae	Terminalia amazonia	w	CK16	Anangoswiti (Sa)	leaf, root, bark	bath	*vaginal steam bath, womb infections, wounds*
Combretaceae	Terminalia catappa #	c/w		Amandra, Amandelboom (Sr)	leaf	tea, bath	*diabetes, vaginal steam bath, improve vitality, winti*
Commelinaceae	Tripogandra serrulata	w		Gado dede (Sr)	whole plant	bath, on skin?	winti (1), venereal diseases, warts (2)
Costaceae	Costus arabicus	w		Fiko-fiko (Sa), Sangrafu (Sr)	leaf	bath	*ritual body cleaning, winti**
Crassulaceae	Bryophyllum pinnatum	c		Kalabana (Sa), Wonderblad (Sr)	leaf	on skin?	colds, skin problems (3)
Cuscutaceae	Cuscuta americana	w		Limki wisi, Duivelsnaaigaren (Sr)	whole plant	on skin?, tea?	sores, dysentery, urinary tract disorders (3)

Family	Species	Code	w/c	Local name	Part used	Preparation	Uses
Cyperaceae	*Cyperus sp.*		w	Ningre kondre (adru) (Sr)	whole plant, root	tea, in alcohol	stomach problems, muscular pain, painful joints (3)
Cyperaceae	*Scleria secans*		w	Babunnefi (Sr)	whole plant		—
Euphorbiaceae	*Chamaesyce hirta*		w	Bobi bobi (Sa)	whole plant	tea	cholagogue, fever, colds, flatulence, filaria, rheumatism (3)
Euphorbiaceae	*Euphorbia cotinifolia*	CK1	c	Redi alasa (Sa)	leaf?		laxative (3)
Euphorbiaceae	*Euphorbia thymifolia*		w	Tjembe wiwiri (Sa), Merki wiwiri (Sr)	whole plant	tea	***excessive vaginal discharge*, vaginal yeast infection*, winti, herpes***
Euphorbiaceae	*Hura crepitans #*		w	Possentri (Sr)	leaf	tea	*diabetes*
Euphorbiaceae	*Jatropha curcas*		c/w	Po-oka (Sa)	leaf	bath	pain, fever, wounds, headache, swelling, abscess, cold, heart problems, sores, tooth ache (3)
Euphorbiaceae	*Manihot esculenta #*		c	Casaba, Cassave (Sr)	root	in alcohol?	abscess, skin eruptions, skin fungi (3)
Euphorbiaceae	*Phyllanthus amarus*		w	Bita, Fini bita, Djari-bita (Sr)	leaf	bath, tea	***vaginal steam bath, blood purifier****
Euphorbiaceae	*Ricinus communis #*		c/w	Amioamio (Sa), Krapata, Castorolie (Sr)	oil, leaf	on skin	laxative, pain, swelling, ulcers (6)
Fabaceae	*Abrus precatorius #*		w	Kokriki (Sr)	seed	tea	stomach ache, thrush, colds, asthma, cancer, sores, eye infection (3)
Fabaceae	*Caesalpinia pulcherrima*		c	Krerekrere (Sr)	flower	bath	***luck charm***
Fabaceae	*Copaifera guianensis #*		w	Opro-oli, Hoepel olie (Sr)	oil	?	***too frequent menstruation***
Fabaceae	*Desmodium adscendens*	CK19	w	Mapinda pinda (Sa), Torban (Sr)	whole plant	tea, bath	***stomach ache, luck charm****
Fabaceae	*Hymenaea courbaril #*		w	Loksi (Sr)	leaf	tea	***anaemia, blood purifier***
Fabaceae	*Hymenolobium flavum*		w	Kabbes, Wormbast (Sr)	bark?		sores (2)
Fabaceae	*Lonchocarpus sp.*	CK9	w	Neku-udu (Sa, Sr)	leaf, stem	tea, bath	***asthma, bronchitis, cancer*, 'to give positive energy'***
Fabaceae	*Lonchocarpus sp.*	CK12	w	Hogi paw (Sa)	leaves	bath	***black magic, winti, to ward off dark forces***
Fabaceae	*Machaerium sp.*		w	Branti makka (Sr)	leaf	tea?	***hypertension***
Fabaceae	*Mimosa pudica*		w	Sehmai (Sa), Sinsin (Sr)	whole plant	bath	alcoholism (4)

Table 6.1. Continued

Family	Species	w/c	nr	Names used in PNGVO	Part used	Preparation	Uses mentioned by PNGVO (bold, italic) Uses in Guianas from literature (not bold)
Fabaceae	*Parkia pendula*	w		Kwatakama (Sr)	bark	bath	**ritual body cleaning, stomach ache, winti**
Fabaceae	*Senna alata* #	w		Slabriki (Sr)	leaves	tea, bath	winti (1), uterus problems, filaria, ringworm (3)
Fabaceae	*Senna occidentalis*	w/c	CK7	Komanti sangu (Sa), Jorkapesi (Sr)	root	tea, bath	**womb problems, absence of menstruation*, stomach ache*, heartburn, spleen problems, to ward off evil spirits, winti**
Fabaceae	*Tamarindus indica* #	c		Tamarinde (Sr)	leaf, fruit?	bath	swelling, pain, skin problems (3)
Fabaceae	*Tephrosia sinapou*	c	CK21	Wanapu (Sa)	leaves	bath	**"against negative influences"**
Fabaceae	*Voandzeia subterranea* #	c		Awogogo pinda (Sa)	whole plant		—
Gentianaceae	*Irlbachia alata* #	w		Dagoe jesie (Sr, Sa)	leaf	tea?	sores, foot fungus, colds, fever, biliousness, malaria, ward off evil spirits (2)
Lamiaceae	*Hyptis atrorubens*	w		Fuku fuku menti (Sr)	leaf	tea	*diabetes*
Lamiaceae	*Hyptis lanceolata*	w	CK24	Dyana faya (Sa), Faya djang, Knoppo wiri (Sr)	leaf	tea, bath	**cold*, shortness of breath, stomach ache, winti**
Lamiaceae	*Leonotis nepetifolia*	w		Bradi bita (Sr)	leaf, stem	?	*eczema*
Lamiaceae	*Ocimum campechianum* #	c		Bonu wiwiri (Sa), Smeri wiri (Sr)	leaf	bath, tea	winti (1), fever, headache, cold, sedative, swollen groin (3)
Lauraceae	*Persea americana* #	c		Afkatie, Advocaat (Sr)	leaf	tea?	hypertension (3)
Lecythidaceae	*Lecythis chartacea*	w		Sneki bita (Sr)	?		—
Liliaceae	*Aloe vera*	c		Aroeweh (Sa), Aloe, Semprefisi (Sr)	leaf	in alcohol?	**shortness of breath*, womb cramps, too frequent menstruation**
Loganiaceae	*Spigelia anthelmia*	w		Drunguman, Kromantikankan (Sr)	whole plant	bath	**winti***
Loganiaceae	*Strychnos melinoniana*	w		Dobrudua (Sr)	stem, bark, root	tea, in alcohol	**male aphrodisiac***
Loganiaceae	*Strychnos sp.*	w		Gran babadu (Sa)	stem, bark, root	tea, in alcohol	**male aphrodisiac***

Family	Species		CK	Local names	Plant part	Preparation	Uses
Loganiaceae	*Strychnos sp.*	w		Pikin babadu (Sr)	stem, bark, root	tea, in alcohol	*male aphrodisiac**
Lycopodiaceae	*Lycopodiella cernua*	w	CK8	A moh ma (Sa), Adrangaman, Pratilobi (Sr)	whole plant, root	tea, bath	*fits, to break up love affairs*, winti*, 'herbal baths for good and evil'*
Lythraceae	*Lawsonia inermis*	c		Reseda (Sr)	leaf		feet fungi, wounds, mouth wash (3)
Malvaceae	*Abelmoschus esculentus*	c		Okro, oker (Sr)	leaf	bath	*to ward off evil spirits*
Malvaceae	*Abelmoschus moschatus*	w		Okro, oker (Sr)	leaf	bath	*to ward off evil spirits*
Malvaceae	*Gossypium barbadense* #	c		Maau (Sa), Katoeng, Katun (Sr)	leaf, fibre	tea	*stomach ache, winti**
Malvaceae	*Hibiscus sabdariffa*	c		O sele (Sa)	root?	tea	bitter tonic, hypertension (3)
Melastomataceae	*Aciotis purpurascens*	w	CK18	Sombo wiwiri, Beminja wiwiri (Sa)	leaf	tea	*back ache, excessive vaginal discharge, fresh breath, hair improvement*
Melastomataceae	*Miconia lateriflora*	w	CK20	Matu bonu wi (Sa), Busie smerie wiwiri (Sr)	leaf	bath	*winti*, ritual body cleaning*
Melastomataceae	*Nepsera aquatica*	w	CK23	Ingi wiwiri (Sr)	root	tea, bath	*cancer, stomach infections, winti*
Meliaceae	*Azadirachta indica*	c		Niem (Sr)	leaf	tea, in alcohol	abortion, fever, intestinal worms, diabetes, jaundice, measles, itching (3)
Meliaceae	*Carapa guianensis*	w		Krapa (Sr)	oil?	bath	*to ward off evil spirits*
Moraceae	*Ficus schumacheri*	w	CK14	Finoe wiwi katu (Sa)	leaf	tea, bath	*eczema, to cool down body, winti*
Musaceae	*Musa sp.* #	c		Bananen bladeren (Sr)	leaf	bath	*winti**
Musaceae	*Musa sp.* #	c		Groeng bana, bakbananenblad (Sr)	leaf	bath	*winti**
Myrtaceae	*Campomanesia aromatica*	w		Andoja (Sa), Adoja kers (Sr)	leaf	bath	vaginal steam bath (5)
Myrtaceae	*Syzygium cumini* #	c/w		Djamu (Sr)	leaf, bark	bath, tea	*vaginal steam bath, diabetes*
Nymphaeaceae	*Nymphaea amazonum*	w		Tookooghagba (Sa), Pankuku wiri, Waterlelie (Sr)	leaf	bath?	*winti*

Table 6.1. Continued

Family	Species	w/c	nr	Names used in PNGVO	Part used	Preparation	Uses mentioned by PNGVO (bold, italic) Uses in Guianas from literature (not bold)
Phytolaccaceae	*Microtea debilis*	w		Eiwit wiri, Eiwitblad (Sr)	leaf	tea	**excessive vaginal discharge during pregnancy**
Phytolaccaceae	*Petiveria alliacea*	w		Alatatere (Sr)	whole plant	tea	womb problems, abortive, fever, toothache (3)
Piperaceae	*Peperomia pellucida*	w	CK3	Konsaka wiwiri (Sa, Sr)	leaf	bath, tea, in alcohol	**eye infection*, hypertension, depression, anxiety, head lice, gonorrhoea, fits, luck charm, winti**
Piperaceae	*Peperomia rotundifolia*	w		Pikin fowru sopo, Tien sensi wiwiri, dubbeltjesblad (Sr)	whole plant		—
Piperaceae	*Piper arboreum*	w	CK10	Malembe toko, Gaa maa udu (Sa), Man anesi wiwiri (Sr)	leaf	bath	**vaginal steam bath, ritual body cleaning, winti***
Piperaceae	*Piper marginatum*	w	CK22	Malembelembe (Sa), (oema) Aneisi wiwiri (Sr)	leaf, stem, root	bath, tea	**to calm down, recover from caesarean, luck charm, winti, vaginal steam bath*, male aphrodisiac**
Piperaceae	*Pothomorphe peltata*	w		Switi aneisi (Sr)	leaf	tea	womb problems, diuretic, headache (3)
Poaceae	*Bambusa vulgaris*	c/w		Bambusie (Sr)	leaf	bath, tea	winti (1), fever, clean out uterus, expulsion placenta (2)
Poaceae	*Cymbopogon citratus* #	c		Citroengras, Stroeng grasi (Sr)	leaf	tea, bath	**fever**
Poaceae	*Eleusine indica* #	w		Gaan womi hati (Sa), Mangrasi (Sr)	leaf	tea, bath	**winti*, 'many medical purposes'**
Poaceae	*Oryza* sp.	c/w?		Busi alesi, Blakka alesi (Sr)	seeds	tea	winti (1), bedwetting (3)
Poaceae	*Saccharum officinarum* #	c		Suikerriet, Tjing (Sr)	stem (juice)	juice	winti (1), rheumatism, worms, digestive, prevent caries (3)
Poaceae	*Zea mays**	c		Karu (Sa)	leaf, seeds	tea	**male aphrodisiac, winti***

Family	Species		w/c	Local name	Part used	Preparation	Medicinal use
Polygonaceae	*Triplaris weigeltiana**		w	Mira udu (Sr)	leaf	bath	**general body strengthening**
Portulacaceae	*Portulaca oleracea*		w/c	Gronposren (Sr)	whole plant	on skin	**back pain, neck pain**
Pteridophytae	*Campyloneurum repens?*		w	Makoko tabaku (Sa)	whole plant	?	—
Pteridophytae	*Pityrogramma calomelanos*		w	Kapuaweri, Kapilairi wiri (Sa)	whole plant	tea, on skin	bronchitis, wounds, stomach ache, colds (3)
Pteridophytae	*Asplenium serratum*		w	Makoko tabaku (Sa)	wholeplant		
Rubiaceae	*Ixora coccinea**		c	Faya lobi (Sr)	flower	bath	**love charm**
Rubiaceae	*Psychotria poeppigiana*		w	Apuku roos, Dangdang (Sa)	whole plant	juice?, tea?	earache, whooping cough (3)
Rubiaceae	*Psychotria ulviformis*	CK11	w	Saw sapatu (Sa), Kibri wiwiri (Sr)	whole plant?	bath?	**winti**
Rubiaceae	*Uncaria guianensis*		w	Popokainangra (Sr)	whole plant	bath	**to ward off evil influences**
Rutaceae	*Citrus aurantium**		c	Limki, Lemmetje (Sr)	fruit?	bath	**to ward off evil influences**
Rutaceae	*Ertela trifolia*		w	Koffimisa (Sr)	whole plant	?	—
Salicaceae	*Casearia javitensis*		w	Kanusubi (Sa)	leaves	?	
Sapindaceae	*Matayba sp.*		w	Gawetri (Sr)	bark?	in alcohol?	bitter tonic (3)
Sapindaceae	*Serjania paucidentata*		w	Fefifinga, Fefi fina wiwiri (Sr)	whole plant	tea	**back ache, stomach ache** *
Sapotaceae	*Pouteria cf. caimito*		w	Kwata bobi, Atakamara, Lauriekers (Sr)	bark?	in alcohol?	bitter tonic (3)
Scrophulariaceae	*Scoparia dulcis**		w	Sisibi wiwiri (Sr)	whole plant	tea, bath	**hepatitis, vaginal steam bath, winti** *
Selaginellaceae	*Selaginella sp.**		w	Oko kowa (Sa)	whole plant	tea?	headache (3)
Simaroubaceae	*Quassia amara**		c	Kwasibita (Sr)	leaf, wood	tea, in alcohol	**blood purifier** *
Simaroubaceae	*Simarouba amara**		w	Sumaruba (Sr)	bark?	in alcohol	bitter tonic, malaria, dysentery, fever, stomach ache, diarrhoea (3)
Siparunaceae	*Siparuna guianensis**		w	Yarakopi (Sr), Fajapau (Sa)	leaf	tea, bath	**fever, vaginal steam bath, winti** *
Smilacaceae	*Smilax cumanensis*		w	Abago maka (Sa)	root	in alcohol	male aphrodisiac, venereal diseases, blood purifier (3)
Solanaceae	*Capsicum frutescens**		c	Alakaka pepre (Sa)	whole plant	bath	winti (1)

Table 6.1. Continued

Family	Species	w/c	nr	Names used in PNGVO	Part used	Preparation	Uses mentioned by PNGVO (bold, italic) / Uses in Guianas from literature (not bold)
Solanaceae	Nicotiana tabacum*	c		Tabaka (Sr)	leaf	tea	winti (1), liver problems, headache, eyewash, hallucinogenic (3)
Solanaceae	Physalis angulata	w		Batoro bita (Sr)	leaf	on skin?	*eczema*
Solanaceae	Solanum americanum	w		Aguma (wiwiri) (Sr)	leaf	tea, on skin	*rheumatism*
Solanaceae	Solanum subinerme			Blaka maka (Sr)	leaf?	?	fever (3)
Sterculiaceae	Theobroma cacao	c		Cacao pitten (Sr)	seeds	tea	*anaemia**
Urticaceae	Laportea aestuans	w		Azo wiwiri (Sa), Krasi wiwiri, Surinaamse brandnetel (Sr)	leaf	tea?	to ease childbirth, cough (3)
Verbenaceae	Lantana camara*	w/c		Maka maka (Sa), Koorsoe wiwiri (Sr)	leaf	bath	*vaginal steam bath, fever, to ward off evil influences*, spiritual cleaning**
Verbenaceae	Lippia alba	w/c	CK6	Piji Piji pau (Sa), Malva (Sr)	leaf	tea, bath	*blood purifier, fever*, anaemia*, ritual body cleaning*
Verbenaceae	Stachytarpheta spp.	w		Isri wiwiri (Sr)	whole plant	bath	winti (1), flu, headache, liver problems, diarrhoea, dysentery, diabetes, low blood pressure, fractures, venereal diseases (3)
Viscaceae	Phoradendron piperoides	w		Fowru doti (Sr)	whole plant	tea, bath	herbal bath to gain strength (5)
Vitaceae	Cissus verticillata	w		Bun hati mama (Sr)	whole plant	on skin?	snakebite, thrush, ulcers (3)
Zingiberaceae	Aframomum melegueta*	c		Nengre kondre pepre (Sr)	seeds	tea, on skin, in alcohol	winti (1), head colds, rheumatism, abdominal pains, constipation, arthritis, menstrual pain (3)
Zingiberaceae	Renealmia alpinia	w		Grang masusa (Sr)	leaves, rhizome	tea?	Fever, pulmonary problems, typhoid, dandruff, dysentery, 'strengthening the nerves' (3)
Zingiberaceae	Renealmia sp.	w	CK15	Piki masusa (Sa, Sr)	leaf, rhizome	tea	*blood purifier, fever, fits, bowel problems, paralysis*
Zingiberaceae	Zingiber officinale	c		Gember (Sr)	rhizome	raw, in alcohol	asthma, cramps, stomach ache, abscess, rheumatism (3)

thirds of the plants are common weeds or shrubs and tree growing in disturbed habitats. Just over twenty medicinal products, such as the wood of *Strychnos* and *Lonchocarpus* lianas and the bark of *Geissospermum sericeum*, are collected in primary forest. All of the medicinal plants sold at PNGVO are imported from Suriname; none are cultivated or wild-harvested in the Netherlands.

According to the Dutch customs officer, approximately 2,000 kg of plant material enter the Netherlands from Suriname each week, either by passenger flight to Amsterdam or container boat to the Rotterdam harbor (van der Pluijm, pers. comm.). The bulk of this material consists of vegetables and fruits. An import permit is needed only if the plant material is destined for commercial use. No permit is needed for the import of plants for personal use as long as they do not involve protected species. It is forbidden to import pieces of wood, even for personal use. If the customs officers come across protected species or wood, the material is confiscated and national botanical gardens are consulted for identification (van 't Klooster 2000).

Although no CITES-listed species were found during this market survey, wood of several lianas and tree bark were being sold (e.g., *Strychnos* sp., *Lonchocarpus* sp., *Parkia pendula*). Neither the shopkeeper nor his family was very keen on informing the authors about the uses of the plants they were selling. They said they were afraid of losing their income if outsiders learn their "family recipes." Moreover, they would feel responsible if the plants were used the wrong way. Because it took quite some patience and time to gain the trust of the shopkeeper, the authors were only able to document the uses of 79 species. These plants were used for a total of 204 different purposes (see table 6.1). The literature confirmed one or more uses for 42 of the species. The other 37 species were either used differently or not (yet) recorded as useful in the ethnobotanical publications for the Guianas. Although some herbs were used in single-species remedies, most treatments were prepared as mixtures of the species mentioned in table 6.1.

A great number of the recorded plant uses were applied in magical practices (see figure 6.1). In fact, 46 of the plant species were used in *winti* rituals to ward off evil spirits or "negative influences," for ritual cleansing of the body, and/or as lucky charms for "happiness" or love ("to attract a man". According to Surinamese lore, one of the plants (*Psychotria ulviformis*) can even make you invisible. Several of the vernacular names (e.g., *kromanti obia,* meaning "black magic of a certain spirit") indicated their magical purposes. The majority of these plants were used in herbal baths, mostly mixed with other plants and various peculiar ingredients like rusty nails, chalk, afoetida gum, and cork. Other *winti* attributes that were sold over the counter included calabashes, traditional maroon fabrics, and kauri shells, also known as *papa moni,* the money of the African ancestors.

At least twenty of the species were used to treat gynecological disorders such as womb infections, menstrual pains, or delayed or extremely heavy periods. Plants used to treat excessive vaginal discharge (leucorrhoea, often caused by yeast

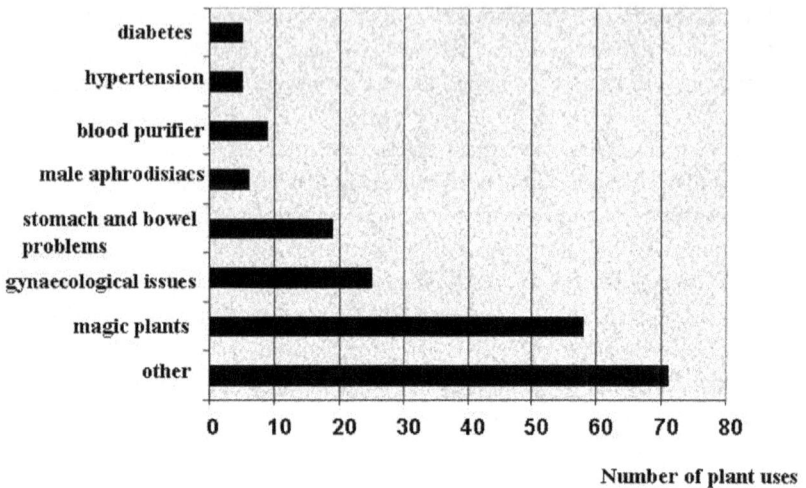

Figure 6.1. *Number of plant species sold for different purposes in Amsterdam*

infections during pregnancy) were usually prepared as a tea and taken orally. The other "gynecological" plants were used as mixtures in antiseptic genital baths. Leaves and/or bark of different species (e.g., *Terminalia amazonia*) were first soaked in a bucket with hot water. The women were then required to sit over the bucket, wash their genitals with the water, and let the steam enter their vagina. The authors were told that this treatment contracts and "cleanses out" the uterus after childbirth. It may have the same effect as dilatation and curettage. The shopkeepers also noted that because these steam baths narrow the vagina, they are widely used by Saramaccan women to improve their sex life. Several fresh smelling herbs (e.g., *Piper* spp.) are added to these baths to avoid bad odors.

Fifteen of the species were used to treat stomach problems and bowel disorders. Most of these plants were prepared as a decoction (boiling the leaves or bark) or infusion (soaking the ingredients in hot water). Seven species were boiled and drunk to "purify the blood." Finally, four species were each used to treat hypertension and diabetes. The other seventy-one plant uses varied from preventing bedwatering to lowering fever (see table 6.1).

The ten most popular plants sold at PNGVO are listed in table 6.2. Although their medicinal uses varied from eye infection to gonorrhea, every plant in the list was also used in *winti* rituals or in herbal baths for ritual cleansing of the body. The authors were not able to document the use of the last plant (*Psychotria poeppigiana*); nevertheless, the vernacular name ("rose of the Apuku forest ghost") indicated a spiritual use. All but one of the species were used in herbal baths: *Euphorbia thymifolia* was only administered as a tea. With the exception of *Ocimum campechianum*, all of the species are wild or domesticated.

Table 6.2. The Ten Most Popular Medicinal Plants Sold at PNGVO

Nr.	Species
1	*Begonia glabra*
2	*Lippia alba*
3	*Ocimum campechianum*
4	*Peperomia pellucida*
5	*Costus arabicus*
6	*Euphorbia thymifolia*
7	*Lycopodiella cernua*
8	*Lantana camara*
9	*Eleusine indica*
10	*Psychotria poeppigiana*

While the authors were conducting their survey, a traditional healer ("luku-man") was preparing mixtures for curative baths by pounding herbs and barks with a traditional Afro-Surinamese wooden mortar and pestle. He said he worked one day a week in the shop, gave advice on the use of medicinal plants, and prepared medicines on request. Most of the store's clientele, however, purchased bundles of plants to make their own baths. In contrast to the genital baths mentioned earlier, these baths are prepared by soaking the plant material in a bucket of hot water and pouring a calabash full of water over one's head and body. The PNGVO customers were mostly Saramaccan, and their native language was often spoken in the shop. However, since Creoles and members of other Maroon tribes also visited the store, the plants were also listed by their Sranan Tongo names. Occasionally, Moroccan, Dutch, and Surinamese customers of non-African descent frequented the shop.

Discussion

This pilot study reveals only a tip of the iceberg. More shops in immigrant neighborhoods of the larger Dutch cities (Amsterdam, Rotterdam, The Hague, and Utrecht) sell Surinamese herbal medicine, yet no other inventories have been conducted. Moreover, extra loads of medicinal plant material are flown in during Surinamese festivities like the Kwakoe festival in Amsterdam. There are also strong indications that Surinamers of East Indian, Javanese, and Chinese descent have similar shops serving their own ethnic communities. However, this still needs to be verified. These shops represent only a fraction of the real trade in medicinal plants. Large numbers of Surinamers travel weekly between the Netherlands and Paramaribo, and many bring back herbal medicine for personal use or for their family or friends.

Because many languages are spoken in Suriname, local plant names vary among and within ethnic communities (van 't Klooster et al. 2003). Several attempts were made in the past to document Suriname's ethnobotanical knowledge (Stahel 1944; Ostendorf 1962; May 1982; Griffith 1985; Plotkin 1986; Heyde 1987; Tyong-Ayong 1989; Sedoc 1992; Raghoenandan 1994). Unfortunately, most of these studies were never published or are no longer available. Moreover, many did not verify local plant names by collecting botanical specimens, which has resulted in listings of common names or highly questionable scientific names. DeFilipps et al. (2004) drew up a list of medicinal plants of the Guianas, though incomplete, based on several of the aforementioned publications. Although the first author deceased before the manuscript could be published, the information was recently put on the Internet. Bioprospecting (the search for new chemicals in plants that will have some pharmaceutical or commercial use) was initiated in Suriname in the 1990s by the US-based International Cooperative Biodiversity Group (ICBG). Although they isolated several new compounds (e.g., Gunatilaka et al. 2001), the results of their ethnobotanical research were never published. This, together with the fact that we discovered many previously undocumented plant names and uses, indicates that there is still a need for thorough ethnobotanical research in Suriname, not only among the Maroon and indigenous tribes in the interior, but also among the citizens of Paramaribo. Collecting fertile botanical vouchers of medicinal plants is essential to obtain reliable scientific plant identifications.

A shift from using traditional medicines to consulting medical doctors (if the latter are available) only occurs with socioeconomic and cultural change, access to formal education, religious influence, and economic growth (Cunningham 1993). Since Suriname's economy has been in a depression for decades and the country's health care standards are not likely to improve in the near future, the authors do not expect the demand for medicinal plants to decrease. Moreover, the present study confirms the findings of Balick et al. (2000) that the demand for traditional medicine remains after emigration and despite improved access to modern health care.

In addition to contributing to their health, the sale of wild-harvested medicinal plants provides a significant income for rural people of Suriname (van Andel et al. 2003). This pilot study confirms the assumptions of Duplaix (2001) that considerable amounts of medicinal plants are being exported from Suriname. Yet, hardly any information on the scale and ecological effects of this trade exists. Since the majority of the herbal medicine encountered during our inventory is harvested from the wild, the importance of Suriname's biodiversity for the well-being of the country's citizens and overseas immigrants should not be underestimated. Traditional cultures and biological resources are becoming increasingly vulnerable to the pressure of market economies (Sheldon et al. 1997). Sustainable harvesting, therefore, is essential, not only for the conservation of medicinal plant species, but also for the livelihoods of many forest-dwelling communities.

Ethnobotanical studies of the indigenous peoples of the Guianas have reported a wide variety of plant species used to treat common tropical diseases like malaria, skin sores, diarrhea, snake bites, and intestinal parasites (Lachman-White et al. 1987; Grénand et al 1987; van Andel 2000). Finding herbal cures for these illnesses is of lifesaving importance, especially for people in the remote interior where health care facilities are limited. In contrast, herbal pharmacies in Guyana's capital Georgetown specialize in cures for venereal diseases, impotence, and gynecological problems (van Andel 2000). Although living closer to the city means improved access to better hospitals, people seem to prefer to buy herbs for sexual problems instead of consulting a doctor.

When Surinamese citizens migrate to the Netherlands, they become surrounded by a new culture and a different health care system. How does this change in environment affect their use of medicinal plants? Obviously, malaria and other tropical infections no longer threaten their lives. The change of climate, food, and society, however, may bring about new health problems. We expect that when people migrate from rural areas to the urban environment of Paramaribo, and then to the Netherlands, their focus of herbal medicine shifts from treating the basic health problems of a forest dweller to the sexual and gynecological disorders and psychosocial ailments typical of immigrants.

When a large number of plant species are used to treat a certain ailment, one can assume its frequent occurrence among community members or its cultural importance as perceived by the community (Slikkerveer, pers. comm.). Because most of the plants found during this pilot market survey were used for gynecological conditions and in *winti* rituals, it must be concluded that these applications are considered important by the Dutch Surinamese community. It has been mentioned before that, especially in the case of drug addiction and sexual and psychological problems, Surinamese patients prefer to visit a traditional healer rather than consult a conventional physician (Stephen 1998; Leenders et al. 2001). Language problems may contribute to this phenomenon, since not all Surinamese immigrants speak fluent Dutch. Nearly all of the *winti* ingredients mentioned by Stephen (1998) were sold at PNGVO. A similar trend has been noted among Dominicans in New York, where traditional healers mostly treat patients suffering either from gynecological problems or stress, anxiety, and feelings of alienation caused by economic insecurity, minority status, and cultural and personal loss (Reiff et al. 2003). Witchcraft or "Santería" forms an essential part of the traditional healing practices among Latino immigrants in New York (Balick et al. 2000).

Most ethnobotanical studies in the Guianas mention the use of astringent bark and leaf decoctions to stop hemorrhages, treat womb infections, "clean out" the uterus and tubes after birth or in cases of infertility, or to diminish menstruation flow and leucorrhoea. However, there are no literature reports of plant remedies referring to the use of vaginal steam baths as a means to improve a woman's sex life. Perhaps this practice was not considered appropriate for publication by the authors of articles on Surinamese herbal medicine, given the hesitation by some of them to list plants used for abortion (Griffith 1985). Nonetheless, these herbal baths appear to be used frequently by Maroon women as was shown by a controversial documentary made by the Dutch television program "Waskracht" (VPRO). It is also common practice among Saramaccan women in Pikien Slee (Brokopondo district, Suriname) to wash their genitals with a bucket of fresh smelling herbs in the morning (van 't Klooster, pers. comm.). Herbal steam baths for gynecological disorders are also common among Mayan Indians (Burke and Arvigo 2005), the Patamona of Guyana (Tiwari 1999), and several other Latin American communities (Ososki et al. 2002). Obviously, indigenous and immigrant women suffering from gynecological problems prefer to treat themselves with herbal medicine rather than consult a (often male and white) medical doctor.

The widespread practice of using plants as "blood purifiers" has been documented throughout the Guianas (Griffith 1985; Lachman-White et al. 1987; Grénand et al. 1987; van Andel 2000). Bottles are filled with small pieces of bark, roots, wood, and herbs and then soaked in alcohol. A small glass of this tonic is drunk on a daily basis to "bitter the blood", i.e., to protect oneself from parasites and skin sores and for one's general well-being. A major ingredient of these tonics is the bitter wood of *Quassia amara* (Simaroubaceae).

According to Griffith (1985), one of the main reasons that people in Suriname prefer to use herbal medicine lies in the fact that conventional medical health care is much more expensive than medicinal plants. This is also the case in the United States, where many immigrants cannot afford hospital treatment and where an extensive parallel health care system of traditional healers and herbal pharmacies exists (Balick et al. 2000; Ososki et al. 2002; Reiff et al. 2003). In the Netherlands, where almost all (legal) immigrants have health insurance and thus access to modern health facilities, these individuals still do not make optimal use of the national health care system. The Netherlands Organization for Health Research and Development (ZonMw) is aware of this problem and has funded several projects to improve the newcomers' access to health care, disease prevention, and socioeconomic causes of typical immigrant diseases (van der Veen et al. 2003). Little attention, however, has been paid to the immigrant's use of herbal medicine and traditional health care. Conventional health care providers need to become familiar with the diversity of health beliefs and practices among immigrant communities, consider the implications of combining modern medical treatment with other therapies and acknowledge the potential role of traditional practitioners in their patient's mental and physical health care (Reiff et al. 2003).

Surinamese herbs and traditional healing sessions in the Netherlands are quite costly and usually not covered by insurance companies. Factors other than economic ones must, therefore, play a role in the choice for herbal medicine in the

Netherlands. More in-depth research is needed to clarify the reasons why Surinamese immigrants continue to use plants from their homeland to cure ailments. Furthermore, little is known about the traditional health care practices of other ethnic minorities that are much less integrated into Dutch society, such as Turks, Moroccans, Somalis, Antilleans, and West Africans. The recent discovery of smoke-dried bats from Ghana, sold at an Amsterdam vegetable market as medicine to treat asthma and bronchitis (AT5 News Archive 2005), suggests that future studies of urban ethnopharmacy in the Netherlands may reveal quite interesting results.

As a follow up of the present pilot study, a postdoctoral research project started in 2005 at the National Herbarium of the Netherlands that aims to discover how cultural and geographic transplantation affects the use of herbal treatments by Surinamese immigrants in the Netherlands.

Conclusions

The wide variety of medicinal plants offered for sale in Amsterdam and the frequent import of fresh material indicate that traditional health care still plays an important role in the well-being of Surinamese immigrants in the city. It is not known, however, if this is true for all ethnic Surinamese groups, or if this use prevails among newcomers and/or illegal immigrants as compared to long-time (and legal) ethnic minorities.

The majority (61 percent) of the herbal medicine encountered during this market survey represents a wild-harvested resource. Nothing, however, is known about the scale of this trade or its impact on natural plant populations. The conservation and sustainable use of medicinal plants in a poor country like Suriname is of great importance, not only because these plants are the only available (and affordable) source of health care for many Surinamese citizens, but also because of the demand from the country's emigrants.

The medicinal plants that were inventoried in the present study were predominantly used for gynecological problems and *winti* rituals. The authors presume that no adequate or custom-made medical treatment is yet available from conventional physicians for these issues.

The number of previously unrecorded plant names and uses found in this inventory indicates the need for thorough ethnobotanical studies in Suriname. In addition, more research is needed to assess why people continue to use traditional medicine, alone or in combination with conventional medicine, after emigration.

Acknowledgements

The authors wish to thank Richard Cullimore, Ruud Parami, and their families working in "Piking Nanga Grang Vanoedoe Oso," 1e Oosterparkstraat 101, Amsterdam, for their time, trust, and valuable contribution to this research. Gaan tangi fii! We are also very grateful to Marion Jansen-Jacobs, Jan Lindeman, Paul Maas, Bep Mennega, and the other staff members of the Utrecht branch of the National Herbarium of the Netherlands for their help with identifying the plant specimens. Mr. Giel van der Pluijm of the Schiphol custom office is thanked for his assistance as well. A grant from the Alberta Mennega Foundation facilitated the purchase of the voucher specimens.

References

AT5 News Archive Amsterdam. 2005. *Inspectie vindt gerookte vleermuis.* Retrieved 23 March 2005 from www.at5.nl/archief.

Balick, M.J., F. Kronenberg, A.L. Ososki, M. Reiff, A. Fugh-Berman, B. O'Connor, M. Roble, P. Lohr, and D. Atha. 2000. "Medicinal plants used by Latino healers for women's health conditions in New York City." *Economic Botany* 54: 344–357.

Boggan, J., V. Funk, C. Kelloff, M. Hoff, G. Cremers, and C. Feuillet. 1997. *Checklist of the plants of the Guianas (Guyana, Surinam, French Guiana).* Washington D.C.: Smithsonian Institution.

Burke, P., R. and Arvigo. 2005. "Healing wombs: Maya abdominal massage." Retrieved 21 June 2007 from http://www.ofspirit.com/patburke8.htm

Cunningham, A.B. 1993. "African medicinal plants: setting priorities at the interface between conservation and primary health care." *People and Plants working paper* vol. 1.

DeFilipps, R.A., S.L. Maina, and J. Crepin. 2004. "Medicinal plants of the Guianas (Guyana, Surinam, French Guiana)." Retrieved 21 June 2007 from http://www.mnh.si.edu/biodiversity /bdg/medicinal/index.html

Duplaix, N. 2001. *Evaluation of the animal and plant trade in the Guiana Shield eco-region: preliminary findings.* Paramaribo: World Wildlife Fund.

Görts-van Rijn, A.R.A., ed. 1985–1996. *Flora of the Guianas.* Koenigstein: Koeltz Publishers.

Görts-van Rijn, A.R.A., and M.J. Jansen-Jacobs, eds. 1997–2001. *Flora of the Guianas.* Kew: Royal Botanic Gardens.

Gowricharn, R.S. and J. Schuster. 2001. "Diaspora and transnationalism: the case of Surinamese in the Netherlands." In *20th century Suriname: Continuities and discontinuities in a New World society,* ed. R. Hoefte and P. Meel. Kingston: Ian Randle Publishers/Leiden: KITLV Press, 155–173.

Grénand, P., C. Moretti, H. Jacquemin, and M.F. Prévost. 1987. *Pharmacopées traditionelles en Guyane: Créoles, Palikur, Wayãpi,* Paris: ORSTOM.

Griffith, W. 1985. *Famiri-encyclopedia Fa da Natoera Dres-Fasi. Gezinskruidenboek van de natuurgeneeswijzen.* Amsterdam: Titjari.

Gunatilaka, A.A.L., J.M. Berger, R. Evans, J.S. Miller, J.H. Wisse, K.M. Neddermann, I. Bursuker, and D.G.I. Kingston. 2001. "Isolation, synthesis and structure-activity relationships of bioactive benzoquinones from *Miconia lepidota* from the Suriname rain forest." *Journal of Natural Products* 64: 2–5.

Hamilton, A.C. 2004. "Medicinal plants, conservation and livelihoods." *Biodiversity and Conservation* 13: 1477–1517.

Heyde, H. 1987. *Surinaamse medicijnplanten.* Paramaribo: Granma.

Homma, A.K.O. 1996. "Modernisation and technological dualism in the extractive economy in Amazonia." In *Current issues in non-timber forest product research*, ed. M. Ruiz-Pérez and J.E.M. Arnold. Bogor: CIFOR, 59–81.

Jansen-Jacobs, M.J., ed. 2002-2003. *Flora of the Guianas.* Kew: Royal Botanic Gardens.

Kambel, E., and F. MacKay. 1999. *The rights of indigenous peoples and maroons in Suriname.* Leiden: KITLV Press.

Lachman-White, D.A., C.D. Adams, and U.O. Trotz. 1987. "A guide to the medicinal plants of coastal Guyana." *Commonwealth Science Council Technical Publications Series* 225.

Leenders, F.R.J., R.V. Braam, H.T. Verbraeck, and G.F. van de Wijngaart. 2001. *Winti en de verslavingszorg.* Utrecht University: Centrum voor Verslavingsonderzoek.

May, A.F. 1982. *Surinaams kruidenboek, Sranan oso dresie.* Paramaribo: de West.

Musschenga, A.W. 1998. "Intrinsic value as a reason for the preservation of minority cultures." *Ethical Theory and Moral Practice* 1: 201–225.

Ososki, A.L., P. Lohr, M. Reiff, M.J. Balick, F. Kronenberg, A. Fugh-Berman, and B. O'Connor. 2002. "Ethnobotanical literature survey of medicinal plants in the Dominican Republic used for women's health conditions." *Journal of Ethnopharmacology* 79: 285–298.

Ostendorf, F.W. 1962. *Nuttige planten en sierplanten in Suriname.* Amsterdam: Gebroeders van Leeuwen.

Pan American Health Organization (PAHO). 2004. *Country health profile for Suriname.* Retrieved 21 June 2007 from http://www.un.org/esa/agenda21/natlinfo/wssd/suriname.pdf.

Pieroni, A., H. Muenz, M. Akbulut, K. Hüsnü Can Ba_er, and C. Durmu?kahlya. 2005. "Traditional phytotherapy and transcultural pharmacy among Turkish migrants living in Cologne, Germany." *Journal of Ethnopharmacology* 102: 69–88.

Plotkin, M.J. 1986. "Ethnobotany and Conservation of the Tropical Forest with special reference to the Indians of Southern Suriname." Ph.D. Dissertation. Boston: Tufts University.

Raghoenandan, U.P.D. 1994. *Etnobotanisch onderzoek bij de Hindoestaanse bevolkingsgroep in Suriname,* Paramaribo: Anton de Kom University.

Reiff, M., B. O'Connor, F. Kronenberg, M.J. Balick, P. Lohr, M. Roble, A. Fugh-Berman, and K.D. Johnson. 2003. "Ethnomedicine in the urban environment: Dominican healers in New York City." *Human Organization* 62: 12–26.

Ros-Tonen, M.A.F., ed. 1999. *NTFP research in the Tropenbos Programme: Results and perspectives.* Wageningen: Tropenbos.

Sedoc, N.O. 1992. *Afro-Surinaamse natuurgeneeswijzen.* Paramaribo: VACO Press.

Sheldon, J.W., Balick, M.J., and Laird, S.A. 1997. "Medicinal plants: Can utilization and conservation coexist?" *Advances in Economic Botany* 12.

Slikkerveer, L.J. 1990. *Plural medical systems in the Horn of Africa: The legacy of 'Sheikh' Hippocrates.* London: Kegan Paul International.

Stahel, G. 1944. "De nuttige planten van Suriname." *Departement Landbouwproefstation in Suriname Bulletin* 59.

Stephen, H.J.M. 1998. *Winti culture: Mysteries, Voodoo and realities of an Afro-Caribbean religion in Suriname and the Netherlands.* Amsterdam: Karnak.

Tiwari, S. 1999. "Ethnomedicine of the Patamona Indians of Guyana." Ph.D. Dissertation. New York: Lehman College, City University of New York.

Tyong Ayong, G. 1989. "Het gebruik van medicinale planten door de Javaanse bevolkingsgroep in Suriname." Master's Thesis, Department of Biology, Free University of Amsterdam.

van Andel, T.R.. 2000. "Non-Timber Forest Products of the northwest district of Guyana." Ph.D. Dissertation. Utrecht: Utrecht University.

van Andel, T.R., A. Mackinven, and O.S. Bánki. 2003. *Commercial Non-Timber Forest Products of the Guiana Shield: An inventory of commercial NTFP extraction and possibilities for sustainable harvesting*. Amsterdam: NC-IUCN.

Vandebroek, I., P. Van Damme, L. Van Puyvelde, S. Arrazola, and N. De Kimpe. 2004. "A comparison of traditional healers' medicinal plant knowledge in the Bolivian Andes and Amazon." *Social Science & Medicine* 59: 837–49.

van der Veen, E., C. Schrijvers, and E. Redout. 2003. Bewijs voor verschil? The Hague: ZonMW.

van't Klooster, C.I.E.A. 2000. "Medicinale planten gebruikt door Surinamers in Amsterdam." Master's Thesis, National Herbarium of the Netherlands, Utrecht University / Free University of Amsterdam.

van't Klooster, C.I.E.A., J.C. Lindeman, and M.J. Jansen-Jacobs. 2003. "Index of vernacular plant names of Suriname." *Blumea Supplement* 15.

van Wetering, I. 2003. "Winti en voodoo." In *Multiculturaliteit in de strafrechtspleging*, eds. F. Bovenkerk, M. Komen, and Y. Yesilgöz. Den Haag: Boom Publishers, 89–99

Wooding, C.J. 1972. *Winti: een Afro-Amerikaanse godsdienst in Suriname*. Meppel.

WHO (World Health Organization). 2002. "Traditional medicine-growing needs and potential." *WHO Policy Perspectives on Medicines* 2: 1–6.

Chapter 7

The Use of Home Remedies for Health Care and Well-Being by Spanish-Speaking Latino Immigrants in London

A Reflection on Acculturation

Melissa Ceuterick, Ina Vandebroek,

Bren Torry, Andrea Pieroni

Introduction

Today Britain is one of the most multicultural of societies, encompassing traditional Commonwealth immigrant groups from Indian, Caribbean, African, and Irish descent, as well as increasing numbers of people originating from South America and Central Europe (Kyambi 2005). Despite the government's aim of rendering the National Health Service (NHS) more culturally appropriate for the ethnically diverse population, little attention has been paid to study traditional health care practices of immigrant communities in the United Kingdom (UK) (Green et al. 2006). When people migrate to urbanized centers, they often bring along their medical traditions. Balick et al. (2000) describe how immigrant communities in New York City continue to import, buy, and utilize traditional remedies. Urban ethnobotany studies these plants used as medicine by ethnic communities in an urban environment, and also focuses on the changes that traditional medicine undergoes when it is transplanted from one culture to another. The present study aspires to contribute to this new discipline, by exploring the plants that are used for health care by Latino immigrants in London, and the influence of migration on these medical practices. Doing so, the study also aims to provide a better insight into particular health care patterns of one of Britain's least documented communities, the Latino population. The results that are presented are based on nine months of fieldwork among the Spanish-speaking Latin American community in London (September 2005–May 2006).

This chapter starts with a general outline of the community in London and will then focus on the subject of the illness narratives of Latinos and their choice of plants for health care. Medical anthropological theories on the concepts of "illness and disease" will clarify the issue of the incorporation of both English terms into one word in Spanish (*enfermedad*), and its particular difference with the commonly used concept of discomfort (*malestar*). Theories on medicalization will allow exploring this linguistic discrepancy more profoundly. In addition, Kleinman's (1980) model on health care systems, and the updated version of Stevenson et al. (2003) will be put forward as a framework for Latinos' choice of health care. This theoretical explanation will provide the fundaments for discussing the different types of natural home remedies used by Latinos in London. It will be shown that, in general, people use plants for self-treatment of minor ailments and for maintaining well-being in a home context. Yet, the medicinal plants reported by the participants represent only a part of what they used formerly. Furthermore, there appears to be a substantial difference between passive and active knowledge, i.e., between past uses and maintained practices. Hence, it will also be explored what the main reasons for this difference are, and if these can be ascribed to acculturation processes. It will be hypothesized that the use of herbal remedies by Latinos is mainly influenced, and even dominated, by practical, material factors, such as import regulations, and is not always a matter of "free choice."

Methods

The methodology draws on a range of qualitative anthropological interview techniques and participant observation methods. The results presented here, are based on thirty-five semi-structured interviews with immigrants mostly from Andean countries (i.e., Ecuador, Colombia, Bolivia, Peru, and Chile). All but one interviews were conducted in Spanish by the first author (M.C.). Participants differed in age (between 18 and 65), social background, and length of stay in the United Kingdom (between 9 months and 26 years). Open-ended interviews and casual conversations were conducted at Latin American shops, restaurants, in community centers, and with two London-based, Latin American practitioners of Complementary and Alternative Medicine (a homeopath and a holistic doctor) who offered further data to sketch a solid ethnographic background. Information on importation laws was gained through interviews with representatives of the Food Standards Agency and the Department for Environment, Food, and Rural Affairs (DEFRA), both part of Her Majesty's Customs and Excise. In addition, several months of participant observation and two group-interviews with the associates of the Latin American Elderly Project (LAEP) were conducted. Finally, several visits were made to a Latino herbal shop in the Elephant and Castle shopping center in south London. Apart from casual conversations and an open-ended interview with the Ecuadorian shopkeeper, an inventory was made of all herbs sold there and samples of all available remedies were bought. A reference collection of bagged plant samples, consisting of pharmacognostic specimens, is deposited at the herbal drug collection in the Herbarium of Pharmacognosy of the University of Bradford. The species mentioned in this paper are identified using Heywood (1979), Maas and Westra (1993), Ody (1993), Mansfeld's Encyclopaedia of Agricultural and Horticultural Crops (2001) and the International Plant Names Index (IPNI) (2006).

Ethnographic Background

Ever since the tumultuous 1970s, people from all over Latin America have come to the United Kingdom either for political or economic reasons. Generally speaking, the first Latin American immigrants were refugees fleeing the military dictatorships in their respective countries, many of them coming from Chile and Argentina. This first wave is sometimes perceived as a brain drain of the educated middle class (Berg 2004). Nowadays, the gloomy image of the continent has changed and so has the "profile" of the Latin American migrants. During the early 1980s, British immigration rules were relaxed in order to fill in labor shortages in the service industry. At the same time, several Latin American countries were facing deep economic crises. This further encouraged immigration to the United Kingdom, especially from Colombia (Berg 2004). Unlike the first political refugees, these economic immigrants often come from a more deprived socio-economic background. However, the line between political and economic immigrants is not always easy to draw since, in several countries, years of ongoing political conflicts aggravated already poor economic situations.

Who belongs to the Latino community? Broadly speaking, Latin America includes all Spanish speaking South and Central American countries, as well as Mexico (geographically part of North America) and Brazil (where Portuguese is the official language). The Latino community in London embraces people from all these countries. According to the 2001 National Census, 0.62 percent of London's population was born in a South American country. In terms of composition by nationality, Colombians represent the largest group in London. Bermúdez Torres (2003) estimates that approximately 50,000 Colombians reside in London, where they constitute about half of the Latino community.

This research focuses on the Spanish-speaking Latino community. One might wonder whether this choice is artificial or disproportional. From an emic point of view, the answer is: no.[1] The Latino community does not embody the "classical" elements of an ethnic minority and is not recognized as such by the British government. Yet, ethnicity is not a fixed given, nor does it have anything to do with "colonial" concepts of either racial or even national boundaries, or as Baumann (1999: 21) puts it: ethnicity is "not given by nature, but an identification created through social action[,] . . . not the character or quality of an ethnic group, but an aspect of a relationship, constituted through social contact." In this way, Latin American migrants do not tend to act as "classic immigrants" (Block 2005); i.e., instead of emphasizing their national background, they take on a transnational or pan-Latino identity in their new environment.[2] The different Latin American organizations, community centers, and newspapers in London reinforce this pan-Latino identity. Subsequently, a division based on country of origin might seem slightly artificial or "etically" imposed. Language, on the other hand, while being a unifier, also creates borders. Brazilians are often considered a separate group somehow. While officially the Latino organizations welcome people from all Latin American countries, in reality almost all come from Spanish-speaking countries. Furthermore, Brazilians also have their own specialized shops, bars, and restaurants and tend to socialize more with people from other Portuguese-speaking countries.

Latinos' self-representation as a single group might be interpreted as a form of acculturation, not in the sense of assimilation toward the host society, but rather as identification with a shared pan-Latino distinctiveness. Or in Baumann's (1997: 219) words: "internal divisions" are replaced "by a shared external distinctiveness." According to Perna (1996), this centrifugal effect is visible in the community's use of *salsa*, a shared Catholic faith and the Spanish language.

No exact statistics exist on the real number of (Spanish-speaking) Latin American immigrants in the United Kingdom. This can partly be attributed to a conceptual gap in the (2001 and previous) National Census. As mentioned previously, the Latino community is not recognized as an official ethnic minority in the United Kingdom. None of the Latin American countries ever belonged to the British Commonwealth, a political reality that is echoed in the absence of a separate "Latino" option in former Census documents. So, some Latinos probably got lost in the broad "(white) others" category, while some refugees and illegal residents are not counted in these statistics. Furthermore, the London-based Latin American organizations do not agree on exact numbers either. Estimations of numbers for the Spanish-speaking Latino community in London vary from upward of 80,000 to as many as 300,000 (Block 2005). The absence of a clear number reflects both the mobility and so-called invisibility of the community. The Latino community is partly a community in flux. Some, usually younger people, do not intend to stay permanently and move on after their visa expires, when they have done a few jobs, or when they finish their studies. Others have been living and working in the United Kingdom for many years (Román-Velázquez 1999). Some migrants however, remain illegally, i.e., they enter with a tourist visa and, after the visa's term has passed, fall through the holes of the bureaucratic net. Moreover, these people do not have the opportunity to rely on the National Health Service. Subscribing with a biomedical health care provider can jeopardize one's invisibility considerably and applying for treatment through the National Health Service sometimes even leads to arrest, since an individual's details can be given to the Home Office (Román-Velázquez 1999).

The invisibility of the community is further reflected in the fact that there are no specific Latino boroughs in London. Unlike Southall (for the Bangladeshi communities), Hackney (for the Turkish and Kurdish communities), or Brixton (for the Caribbean communities), there is no Latino area as such. Rather, the Latino community in London is dispersed throughout the whole city and thus less visible than some other ethnic minorities (Román-Velázques 1999). Recently, some Latino shops and restaurants opened in Brixton where many new immigrants—mainly from Ecuador and Colombia—reside. Yet, the exceptions that prove the rule are the shopping centers of Elephant and Castle (south) and Seven Sisters (north) (Block 2005). Both enclose several Latino shops, ranging from hairdressers to food stalls and grocery stores (Román-Velázquez 1999).

Results and Discussion

The Use of Plants for Health Care Equals Home Remedies for Self-Treatment: Latinos' Explanatory Model of Health

Before exploring which plants are actually used for health care by Latinos in London, it remains to be elucidated what specific form health care takes in this context. Although the primary focus of the project was on the use of *plants* for *health care,* research made clear that neither of these terms apply literally. In what follows, this methodological riddle will be explained, using various theories based on the concepts of illness and disease and the sectors of health care.

One of the key questions in the semi-structured interviews originally aimed at finding out what plants people use in London *to treat a disease* (translated into Spanish: *para tratar una enfermedad*). During the pilot testing, interviewees surprisingly answered they could not think of any disease they had been treating with herbal remedies. Instead many came up with several remedies for what they described as *malestar.* This term can best be translated in English as "discomfort." And in and of itself, this linguistic twist offered an important insight into Latinos' explanatory model of health and illness.

The Spanish word *enfermedad* can be translated both as "illness" or "disease." In medical anthropology, a clear difference is stipulated between both terms (Kleinman 1988; Helman 1990). "Illness" is generally described as a subjective "expression of unhealth" from the patient's own experience, sometimes when no disease can be found, whereas disease is "the pathological process, the deviation from a biological norm," as seen from a medical perspective (Boyd 2000: 10). According to Pelto and Pelto (1990: 275), culture is the main determinant responsible for this distinction. Illness then refers to "the culturally defined feelings and perceptions of physical and/or mental ailment in the mind of people in specific communities." "Disease" on the other hand "is the formally taught definition of physical and mental pathology from the point of view of the medical profession" (Pelto and Pelto 1990: 275). Since there is only one word that stands for both "illness" and "disease," a linguistic difference between both concepts does not exist in Spanish (Zapata and Shippee-Rice 1999; Collins Spanish Dictionary 2004). Instead, there is the difference between *malestar* and *enfermedad,* which was pointed out by participants during this study. This difference is summarized in table 7.1.

In the explanatory model offered by Latinos, the concept of "discomfort," or *malestar,* has a strong connotation of being less severe than *enfermedad.* This implies that somebody suffering from *malestar* usually does not consult a general practitioner. *Enfermedad,* on the other hand, is mostly perceived as something that does require professional medical help (table 7.1.). In general, there seems to be a consensus on what is considered "just" a discomfort. Stomachaches, headaches, colds, chills, and flu symptoms, for example, are all seen as forms of *malestar,* as illustrated by the following quote:

> "A disease. . . . I cannot think of a name of a disease, but I can think of discomforts. I drink for example an infusion of celery when my stomach hurts[;] . . . for menstrual pain, a decoction of cinnamon[;] . . . for sore throat, ginger, orange, honey, and a spoonful of butter is a relief." (Colombian woman, 24 years old).[3]

This can be interpreted as a demedicalization of certain disorders. Some conditions that would be described by a practitioner as "diseases," such as flu, are demedicalized by Latinos in that they are not considered "medical" problems. This is rather the opposite of what Myllykangas and Tuomainen (2006) describe as "a tendency to detect medical problems everywhere, turning ordinary ailments into medical problems, seeing mild symptoms as serious, and treating personal problems as medical ones." Modern allopathic medicine has often been criticized for turning nonmedical phenomena into medical problems (Illich 1976; Helman 1990; Illich 2003; Myllykangas and Tuomainen 2006). The umbrella of biomedicine now encompasses all normal phases of the female life–cycle such as menstruation, childbirth, and menopause, as well as, for example, unhappiness (Helman 1990). This expansion of the medical sphere is not part of the explanatory model of health and unhealth for Latinos.

When people say, for example, that *toronjil* (*Melissa officinalis* L.) is "good for the heart" (*bueno para el corazón*), the word "heart" has a symbolic meaning. "Good for the heart" stands for "uplifting" (or a more revealing English synonym: "heartening"). Lemon balm, as *toronjil* is called in English, is indeed described as a sedative and antidepressant (Ody 1993). Participants did not consider this a proper disease (illness/disorder), but rather a form of distress or discomfort. Similarly, all kinds of female problems are not considered medical disorders either.

> "Types of discomfort, for example [are], what a woman does when she has her periods, ... cinnamon, *la canela,* cinnamon tea is good when you have too much discomfort, discomfort during menstruation, when it hurts. But diseases, I do not know, no." (Colombian woman in her fifties, living in the United Kingdom for 26 years).[4]

Quite a few plants are used to alleviate pain related to menstruation (infusions of *Apium graveolens* L., *Cinnamomum verum* J. Presl., *Ocimum basilicum* L.) or as emmenagogues (infusions of *Petroselinum crispum* Nyman., *Origanum vulgare* L., *Ruta* spp.). They are all described as remedies to alleviate female discomforts.

Equally, certain folk illnesses are not considered medical problems either. As one woman remarked, with *mal de ojo* (the evil eye) "you do not go to a "normal" doctor. He would tell you, you came to the wrong shop."[5]

Table 7.1. Summary of the Main Differences between *Enfermedad* and *Malestar*

Enfermedad	Malestar
Can be translated as both "disease" and as "illness"	Discomfort
Tratar (to treat)	*Aliviar* (to alleviate)
Connotation of being severe, needs professional help	Connotation of being less severe, "minor" disorders
Medicalized, treated within the professional sector	Self-treatment, treated within the home context (popular sector)
As opposed to *salud* (which is also a very medicalized term)	As opposed to *bienestar*

Moreover, remedies to treat a form of *malestar* did not only include medicinal herbs, but also a range of health foods and food medicines. The term *plantas* which is often interpreted as a synonym for *hierbas,* herbs, was thus in a way insufficient to cover the whole range of natural remedies. Moreover, these remedies were not only used to alleviate discomforts, but also for cosmetic home treatments, for magico-religious purposes, and in some cases also to prevent illnesses. These uses are included because they are all applied in a home context. They can be summarized as treatments for attaining *bienestar,* or well-being, as opposed to *malestar.*

These results can be explained further by employing Kleinman's (1980) model on the three sectors of health care systems, which has been recently updated and applied to the British context by Stevenson et al. (2003). According to this model, treatment for a disorder can be sought in one of the three sectors that form every health care system. To begin with, there is the professional sector, dominated by biomedicine. The second is the folk, or nonprofessional, specialist sector, which ranges from osteopathy to faith healing. Nowadays, the distinction between both sectors has become vaguer. Since Kleinman first published his model, many Complementary and Alternative Medicine practices have received official recognition (Stevenson et al. 2003). The final sector is the popular sector, "the lay, non-professional, and non-specialist arena" (Stevenson 2003: 513). This lay sector includes health maintenance activities such as diet and home remedies (Stevenson 2003). It is the domain where "ill health is first recognized and defined, and where health care activities are initiated" (Helman 1990: 55).

The plants used by Latinos in London that are described in this article, are all utilized within the popular sphere: the household, or the more extensive social network including family, relatives, and friends. Sometimes, knowledgeable people in the community are also consulted, such as the owner of the Latino herbal shop. All interviewees (n=35) agreed that within Latino families, women (mothers and grandmothers) are the key "sources" when it comes to home remedies.

Stevenson et al. (2003) point out that in the United Kingdom, most of the natural home remedies that were formerly used within the popular sphere are now replaced by manufactured medicines. Nowadays, biomedicine dominates all sectors. These authors declare that decades of overprescribing allopathic medicine might have undermined people's confidence in self-treatment, which is not in concurrence with health care seeking habits of Latinos.[6] In general, self-treatment with natural remedies was preferred by all interviewees (n= 35) for a whole range of illnesses and disorders that are considered forms of *malestar.*

The propensity to stick to self-treatment can also be interpreted as a form of critique on biomedicine as a whole. Latinos' views on the British medical system are not always unanimously positive. The most common complaints include disapproval of doctors' tendencies to an "impersonal and short treatment" and to "only prescribe aspirins or paracetamol." Language can be a barrier as well. For others, bad experiences with the National Health Service instigated their distrust. This critique on the biomedical sector is a summary of what came out of the thirty-five interviews conducted so far and does not claim to be applicable to the whole community.

During recent years the UK government has started to emphasize and encourage the use of self-treatment in its health care policy (Stevenson et al. 2003). Hence, paradoxically, by trying to maintain existing customs, Latinos might be more "acculturated" towards this objective than anyone else.

Home Remedies

Home remedies used by Latino immigrants in London can be roughly divided into the following categories: (1) medicinal herbs (used preventively or curatively);

(2) functional food, food medicine, food (plants) used in a multifunctional way; (3) "ritual" plants and remedies; and finally (4) cosmetic home remedies or *tratamientos*. Each type of use will be discussed shortly by means of representative examples.

MEDICINAL HERBS AS CURE AND PREVENTION:
"THE DOCTOR HERE TENDS TO CURE, AND NOT TO PREVENT."[7]

Medicinal herbs are used either preventively or curatively. Among all mentioned remedies that are used to cure or alleviate a certain ailment, most (13 out of 60) are used to alleviate an upset stomach, to treat the symptoms of flu and common cold (11), or as laxatives (8). None of these ailments typically is considered a disease and thus none is "worth going to a doctor" for consultation. When participants did mention seeking professional help, they often criticized the British practitioners for focusing only on curing a disease and not drawing enough attention to preventive care.

Several preventive usages came up during interviews. Preventive remedies are mostly ingested through food, as juices and smoothies (i.e. a blended nonalcoholic beverage made from natural ingredients), or as herbs mixed with a meal. Smoothies of tropical fruits (e.g., mango, *Mangifera indica* L., Anacardiaceae) and plants such as aloe (*Aloe vera* L., Liliaceae) belong to often-cited preventive measures, used to "regulate the digestive system." They are mostly drunk in the morning. Common examples of species used preventively and ingested as herbs are garlic (*Allium sativum* L., Alliaceae) and ginger (*Zingiber officinale* Rosc., Zingiberaceae). The following quote is about the use of both.

> "Above all, these are remedies that are not used when you are already ill; it's because it is April and it is going to rain a lot, so you change your diet, and you use more ginger or you put more garlic in it, so they would give you more defenses . . . In reality it has to be before. This is something that you adapt to your lifestyle, I suppose. It gives you a better defense system . . . before the virus comes. . . . As I said, I use more things preventively." (Colombian woman, living in the United Kingdom for more than 20 years).[8]

This also shows that the preventive use of food differs from season to season. In Colombia it is related to the rainy season; in Britain this is transformed into a more frequent use during autumn and winter. Preventive uses often overlap to a certain extent with functional foods, which will be discussed next.

REMEDIES ON THE FOOD-MEDICINE CONTINUUM: "YOU CANNOT AVOID ILLNESSES,
BUT YOU CAN PREVENT THEM THROUGH A GOOD DIET."[9]

Some natural remedies are used, sometimes concurrently, as food and medicine. we draw on Pieroni and Quave's (2006) model—here of course transposed onto cultivated species—to schematize the different degrees of correlation between food and medicines. They differentiate among functional foods, food medicines, and remedies with multifunctional uses.

Functional foods are species used in food preparations with a nonspecific action assumed to benefit health. They are said to be "good for the blood," to promote a good circulation of the blood, or to purify the blood (e.g., parsley, *Petroselinum crispum* Nyman, Apiaceae). Others are said to "enhance the defense system." So, in this respect, functional foods are sometimes used preventively, as is the case with garlic. Celery (*Apium graveolens* L., Apiaceae) is one of the most commonly used examples. Eaten raw, or prepared in soups and stews, *apio* as it is called in Spanish, is said to be "good for the blood," to provide vitamins and to "clean the liver." Another, typically Colombian, example is *chontaduro* (hearts of palm, *Bactris gasipaes* Kunth, Arecaceae). In London, the fruits are only available in cans, preserved in salted water. They are sold in every Latino shop and highly appreciated. The fruits are said to be "rich in proteins" and are considered a natural energizer and aphrodisiac. Finally, almost every interviewee mentioned that eating carrots (*Daucus carota* L., Apiaceae) is supposed to be "good for the eyes." Pieroni and Quave (2006) found that in the Italian Basilicata region the consumption of functional foods is often seasonal. This is in accordance with what was pointed out earlier about the seasonal use of garlic and ginger in the Latino community.

Food medicines, on the other hand, are "ingested in a food context" as well, but are assigned specific medicinal properties; or they are "consumed in order to obtain a specific medicinal action" (Pieroni and Quave 2006: 110). In the Latino community, food medicines are used curatively more often than preventively. Freshly pressed orange juice (*Citrus sinensis* (L.) Osbeck, Rutaceae) is used to alleviate symptoms of common cold, flu, and a sore throat. It is also drunk as a laxative. A soup of chili peppers (*Capsicum frutescens* L., Solanaceae) is prepared as a relief for various diseases typically contracted in wintertime, mainly to treat flu because it promotes sweating. Sweating is considered a means to cure more quickly, because it purifies the body, "expels" the disease, or makes one literally "sweat it out."

Some examples like flax or linseed (*Linum usitatissimum* L., Linaceae) and oats (*Avena sativa* L., Poaceae) are somewhere in between food medicines and functional foods. On the one hand, both are used curatively (as laxatives), but they are also eaten because they "regulate the digestive system" (preventive). Another "borderline example" is beetroot (beets), *remolacha* or *betabel* (*Beta vulgaris* L., Chenopodiaceae). Beetroot is eaten because it "enhances the level of red blood cells," producing a general effect (and therefore a functional food). However, it is also consumed by some to prevent and even to treat anemia, which makes it a food medicine.

Finally, some species are used in a multifunctional way. This means they are simultaneously used as food and medicine, "without any relationship between both uses" (Pieroni and Quave 2006: 108). A good example here is potato (*Solanum tuberosum* L., Solanaceae). Apart from being part of the everyday diet of most Latinos, potatoes are also used in several medicinal applications. The sap is ingested as a sedative. Slices are applied topically on the skin as an anti-inflammatory or on the forehead to alleviate fever and related headaches. Finally, the water in which potatoes have been cooked is drunk to expel kidney stones.

It should be mentioned that a few non-vegetal foodstuffs could be situated within the food-medicine continuum as well. Some products of animal origin, such as milk (drunk as a laxative), honey (mixed with infusions or juices to alleviate sore throat), chicken stock (a flu remedy), and eggs (gargled against sore throat) are widely used. Water and salt are also claimed to have several medicinal properties. Water is considered a general cleanser and is drunk against headaches, whereas salt is used as an antiseptic to rinse minor wounds. A mixture of lukewarm water and salt is gargled to alleviate a sore throat. Finally, even beer and Coca-cola® are both common home remedies (respectively as a diuretic and against diarrhea).

RITUAL PLANTS: BUENA SUERTE AND ZAHUMERIOS

A few plants are used for what might be called magico-religious purposes. The ones that are still in use in London can be subdivided into two categories. On the one hand, there are plants that are used *para buena suerte,* to "bring good luck," "to protect against bad spirits" or "bad energy." They are put next to the front door of the house, usually in a flowerpot (since the majority of the interviewed Latinos live in apartments). Common examples are *sábila* (*Aloe vera*) and *ruda* (*Ruta* spp., Rutaceae). This is not an isolated use; almost all participants mentioned it or had an aloe at home. Some Latino shops or bars also have an aloe plant at their entrance. These uses are not tied to one specific country, but interpretations may change a bit from country to country. The use of aloe to avert bad luck is widespread among Colombians, whereas the use of rue is commonly known by people from all Andean countries.

Outside the popular sphere, rue is often used in spiritual healing rituals, performed by *curanderos* (specialized healers). The plant is said to absorb the negative energy from a person struck by certain cultural diseases, such as the evil eye. While this practice is common in different home countries, nobody reported this use in the United Kingdom. The use of rue might have been introduced to Latin America by the Spaniards and rue is described by San Miguel (2003: 239) as a typical plant "to counter supernatural evil in Western traditions."

Apart from just keeping a plant in the house to protect against misfortune, people also burn certain herbs. This is done initially to clean the house "from bad energy" and "to bring good luck." The most widespread example is eucalyptus (*Eucalyptus* spp., Myrtaceae). Some people who follow this habit explained it with a modern twist. Some say burning eucalyptus cleanses the air of germs and that doing so prevents diseases. This can be interpreted as a medicalized explanation of a traditional ritual. Others claimed to use eucalyptus "to make the house smell good" or as somebody said: "as a sort of hippie-like incense" without a specific symbolic meaning. It is interesting to see how this ritual is explained using Westernized narratives that turn it into a health conscious or a fashionable practice. In the Ecuadorian *botánica* one can also obtain *zahumerios* (incense). This is a mixture of herbs that have been blessed by a *curandero* or specialized healer. To attract good luck, they should be burned inside the house.

Latinos do not see these practices as promoting health in a medicalized way, but they do belong to the broader field of well-being, since they are indeed used to avoid *malestar*. A final category that fits under the umbrella of general well-being is *tratamientos de belleza*.

Cosmetic Home Remedies: Tratamientos de Belleza

Every woman interviewed, young or old, came up with certain home remedies for cosmetic purposes, also called *tratamientos (de belleza)* (literally: "[beauty] treatments"). They are used because they are cheaper, and said to work better, than products bought from a shop, because they "are more natural." Treatments range from emollients for skin and hair, to natural peelings made from a mixture of sugar and lemon. In total, eleven different *tratamientos* were reported. Aloe, egg (yolk) mixed with honey, olive oil, and even mayonnaise are all used as conditioners or as treatments for dry hair. They are put on the hair, as any other conditioner would be, and washed out afterwards. Another *tratamiento* often heard about is a mixture of lemon (*Citrus* spp.) juice and sugar, applied as a peel to scrub the skin.

Most Frequently Mentioned Remedies

Table 7.2 gives an overview of the ten most frequently mentioned remedies, how they are used and for what purpose, where they are obtained, and the percentage of interviewees (n = 35) that use them in a certain treatment. Herbal remedies are mostly purchased in supermarkets and markets. Often, Latino shops also have an assortment of herbal infusions. Some remedies are brought or imported from home countries. Finally, some people mentioned the Latino herbal shop at the Elephant and Castle shopping center. The significance of this shop cannot be underestimated, not only because it seems to be unique of its kind, but also because its supply is completely based on the demands of customers (90 percent of the clientele are Latinos, the other 10 percent are either Spanish or Portuguese). Furthermore, the shop is also representative of the pan-Latino identity, i.e., of people from different countries who buy herbs at the shop imported from different countries. There is one thing that stands out from table 7.2 and the examples discussed in the previous paragraphs. Most of these remedies are "common" herbs or natural products that can be found in any supermarket. Every typical brand of English tea has developed an assortment of herbal infusions that commonly include chamomile and mint, which are also sold fresh in some markets and supermarkets. Aloe has become a panacea and is nowadays used virtually in everything, from beauty products and washing powder to "healthy" juices sold in retail shops. Fresh leaves are sold per piece in almost every Caribbean market, and even international chains like Ikea sell whole plants. In short: except for aloe, the home remedies used most often by Latinos in London are not at all indigenous Latin American plants. The question that arises is how this use is related to the debate on acculturation. In the following and final section of this chapter we will analyze this question more in detail.

Acculturation

The plants mentioned in the text and table 7.2 provide an overview of the ten most frequently mentioned plants that are still in use after migration. However, this is only a fraction (5 percent) of all the remedies that people mentioned. Depending on the geographical region of origin (urban or rural), there exists a huge difference between the number of remedies people used previously in their home countries (the total of different remedies used in home countries is 196) and the ones that are still in use after migration to the United Kingdom (the total of different remedies used in London is 66). This difference could be measured because interviewees were asked to distinguish between the two. Often only a small segment of their knowledge is still actively applied. In general, interviewees used at least three times as many natural remedies back in their home country than they use currently in the United Kingdom. Whether or not this can be seen as a sign of acculturation depends on the interpretation of the concept.

First of all, acculturation is often interpreted as assimilation toward the so-called host society that results from a simultaneous process, encompassing deculturation from the original culture and enculturation towards the host culture (Kim 2001). This model works when the host society is culturally rather homogenous (Pieroni et al. 2005). However, in multicultural London it is hard to speak of a fixed, homogenous substrate. One would have to become multicultural, in order to "assimilate to the host society," and this, of course, is a contradiction in itself.

In addition, there is Baumann's (1997) theory, already mentioned, that acculturation is a conformation toward a pan-Latino identity. While this model works on a political and sociological plan, the question remains whether the formation of a cross-diasporic alliance has any influence on the use of plants. The answer is hard to determine and could be somewhere in the middle. There may be both a continuation of former "national" uses and a pan-Latino exchange. Some trivial examples include the following. A Colombian interviewee mentioned a flu remedy he started using in London after he got it from a Bolivian shaman (also living in the United Kingdom, but not practicing anymore). A Guatemalan woman

Table 7.2. Ten Most Frequently Mentioned Remedies among the Latinos Living in London

Scientific name	Spanish name	English name	Part(s) used	Administration	Claimed medicinal use(s)	Provenance	Frequency of quotation
Allium sativum L. (Alliaceae)	Ajo	Garlic	Bulbs, fresh	Functional food	"Good for the heart," "cleanses the blood," "enhances the immune system" Anthelmintic, depurative	bs	More than 50%
				Chewed or tincture (with *agua ardiente*)			
				Infusion	Relieves symptoms of common cold, flu, and sore throat		
				Suspension of all garlic in diet	Antidiarrheal		
				Rubbed on nails	To make nails grow faster		
				Hung on door	Protection against evil spirits (Colombia)		
Allium cepa L. (Alliaceae)	Cebolla, Cebolla cabezona	Onion	Bulbs, fresh	Food medicine: prepared as *chango caballuna* (soup made of water, salt, and onion)	To alleviate an upset stomach, gastritis	bs	Between 30 and 50%
				Functional food	Energizer, aphrodisiac		
				Boiled, steam inhalant	Treatment against nasal congestion		
				External application	To heal wounds, antiseptic		
Aloe vera L. (Liliaceae)	Sábila, Aloe (Arg.)	Aloe	Leaves, fresh	Mixture (of the gel or "crystal") with (raw) eggs, honey, ingested in spoons	Expectorant (emergency home remedy, used to alleviate acute asthmatic problems)	bs	More than 50%
				Mixture with ginger (fresh), ground turmeric (*curcuma* powder), black pepper and honey, ingested in spoons (has to be taken during 45 days)	Detox for the liver		
				Smoothie with mango	Stimulates the bowels, purgative		
				External application	Vulnerary, cicatrizant for minor wounds/cuts/burns, skin emollient (cosmetic use), applied on hair as conditioner or against lice		
		Whole plant		Put at entrance of house	To prevent bad spirits from entering the house, brings "good luck"		

Scientific name	Spanish name	English name	Part used	Preparation	Use	Availability	Percentage
Citrus sinensis (L.) Osbeck (Rutaceae)	Naranja	Orange	Fruits, fresh	Juice, smoothie Food medicine: juice	"Fortifies the defense system," contains vitamin C To alleviate symptoms of common cold/flu/sore throat, laxative	bs	Between 30 and 50%
Citrus limon (L.) Burm.f. Citrus aurantifolia Swingle (Rutaceae) Both species are used interchangeably, depending on availability	Limón, Lima limón	Lemon, lime	Fruits, fresh	Juice, hot, with honey or panela Juice with salt, gargles Juice (drunk pure) Juice, mixed in teas or infusions Juice, external use: eye drops Juice, topically applied	Relieves symptoms of common cold and flu Against sore throat To heal gastric ulcers, stomach aches (caused by salmonella) Digestive, to alleviate nausea "To whiten the eyes" Antiseptic for small wounds, astringent	bs	More than 50%
Eucalyptus spp. (Myrtaceae)	Eucalipto	Eucalyptus	Leaves, dried	Infusion (with hot milk or with cane sugar (panela) and cinnamon) Boiled Burned	To alleviate symptoms of common cold or flu Inhalation of steam to alleviate symptoms of common cold, sore throat, problems with sinus, nasal congestion To (ritually) clean the air, house from evil spirits or germs	bs, bl	Between 30 and 50%
Matricaria recutita L. (Asteraceae)	Manzanilla	Chamomile	Flowers, dried or fresh	Infusion or tea (in bags) Infusion Infusion, used as eye-bath Infusion	Digestive, calming effect (reduces nausea) Sedative Antiseptic Mouth antiseptic, to alleviate toothaches	bs, bl	More than 50%
Mentha piperita L. (Lamiaceae)	Hierba buena, menta poleo	(Pepper) mint	Leaves, dried or fresh	Infusion or tea (bags)	Digestive, calming effect, to heal gastritis, to heal sore throats	bs, bl	
Honey of Apis mellifera	Miel	Honey		Sweetener in teas or hot milk with whisky Externally applied	To alleviate sore throat, flu symptoms, "honey cures" ("la miel cura") Hair conditioner	bs	Between 30 and 50%
Salt	Sal	Salt		Dissolved in hot water, gargled or externally applied	Antiseptic, to heal sore throat, to clean minor wounds, salt is a remedy ("sal es medicina")	bs	Between 30 and 50%

bs: bought in British supermarkets or markets in London   bl: bought in Latino shops in London   ih: imported or sent from home countries said she regularly drinks mate (an Argentinean social beverage) with her other Latino, Spanish, and even English friends. Recipes get spread among the whole community through Latino newspapers that dedicate a separate section to herbal remedies. Many Colombians shop at the Latino botánica in Elephant and Castle and buy either Peruvian or Ecuadorian herbs. Yet again, this only reveals something relating to the social networks in which herbal remedies are shared, and nothing about acculturation.

Acculturation is often interpreted as a cultural transformation and adaptation toward the host society, resulting from a mental negotiation process within the migrant community (Kim 2001). Newcomers seem to adopt values, language, beliefs, and traditions of the dominant group, or alternatively, they reject these and stick to their cultural customs, as a deliberately chosen ethnic marker. The use of traditional food is often seen as a symbol in the maintenance of ethnic identity and a cultural trait most resistant to change (Nguyen 2003). However, the model that explains the use of food as an element in the negotiation of cultural or ethnic identity does not pay sufficient attention to the fact that adaptation is often a practical matter, based either on financial considerations or availability of ethnic products. During this survey, participants often uttered complaints about unavailability or cost of traditional products (some herbs are available only in specialized shops and are often expensive). Thus, in the first place, acculturation—seen from the point of view of using less herbal remedies in the host society than in the home country—seems to be an inevitable practical adaptation to a new environment. Because, for practical reasons, it is not possible for people to use all the remedies previously known and used by them in their home country, knowledge gets lost, irrespective of whether they are willing to use these remedies or not. And this loss of knowledge is reflected in the difference between what people know and what they are still using. This loss also results in the fact that, despite a willingness to use herbal remedies, immigrants use only what is available, and those are mostly global species, as shown in table 7.2. Or, as Balick et al. (2000) put it, "studies in an urban setting require a specific awareness and understanding of the limits of a city environment." Acculturation is very much related to an acceptance of those limits, which is often not a negotiation of ethnic identity, but rather a practical adaptation.

Another practical problem that influences the diminished use of traditional remedies is summarized in the following quote from Octavio Paz' famous novel, *El laberinto de la soledad*. It describes how a Mexican immigrant explains to a friend how she cannot really enjoy nature anymore, thereby revealing both her homesickness and the linguistic difficulty that numerous migrants experience. It grasps the fact that many people often do not know a plant's English name. For many species known by Latinos an English translation may not exist, which makes looking for these plants in London even more difficult.

"How do you want me to like the flowers, if I do not know their real name, their English name, a name based on the colors and petals, a name that is the flower itself . . . because those who say plum and eucalyptus, they do not say it to me, nor does it mean anything to me." Paz (1994: 21).[10]

Furthermore, "time," or the lack of it, is often cited as a main reason why people do not use herbal remedies in general (meaning not limited specifically to Latin American herbs). Some interviewees who only had been living in the United Kingdom for a short period, and who said to have uncertain prospects about whether they are going to stay in the United Kingdom permanently, claimed they tend to use fewer herbal remedies than they did before when they were in their home countries. Most of these immigrants stay in London primarily for economic reasons and accordingly work very long hours. Some of these interviewees claimed that while preferring natural remedies in their home country, in London "they cannot wait" for a certain remedy to work, because they simply cannot stay away from work. Most of these migrants are younger people (less than thirty years old). So age is a major determinant as well. Permanent residents, on the other hand, often make more efforts to buy remedies in specific shops or markets, and even to cultivate their own remedies, as well as to find new alternatives for plants used formerly but that are now no longer available. In addition, these people frequently have a more solid social network on which they can rely for purchasing herbal remedies.

However, the main reason, and the most often cited reason why people hardly use any native Latin American remedies lies in the harsh importation laws:

"It is prohibited for us, Latinos, because of the drugs and everything, we, as Latinos, as Colombians, we cannot import fruits or plants. It is stupid, for one person, we all pay the bill. . . . I could not bring a plant with me from my country, officially I could, but it is something very risky." (Colombian woman, living nine years in the United Kingdom).[11]

During different group interviews at the Latin American Elderly Project, there was a general consensus that Latinos (especially Colombians) cannot import herbal material: "We can import some, but it is not worth taking the risk." In addition, stories about thorough—sometimes physical—inspections, applied particularly to people entering the United Kingdom directly from a Latin American country (especially from Colombia and Ecuador), further discourage the importation of potentially prohibited remedies. While some of these stories might be seen as exaggerated urban legends, there are enough common narratives to have a basis in reality.

What then, are the official rules and restrictions on the importation of herbal material for personal use? A representative of International Trade of Her Majesty's

Revenue and Customs who was contacted several times summarized the answer to this question as follows: "before herbal remedies can be posted or brought to the United Kingdom, the importer or traveler needs to check with all relevant (British) authorities, to make sure the herbal remedy will be allowed into the United Kingdom." This means that people first have to check with the Department for Environment, Food and Rural affairs (DEFRA) and with Her Majesty's Revenue and Customs.

A leaflet on the importation of plants entitled *"If in doubt, leave it out,"* published by Her Majesty's Customs and Excise (May 2004), says: "There are strict penalties for smuggling prohibited and restricted items and this can include limited fines, the possibility of imprisonment or both." It further affirms that most of the plant material imported from non-European Union countries consists of controlled items, meaning that they "require a phytosanitary certificate issued by the plant protection service of the exporting country or a license issued by DEFRA or the forestry commission." A telephone call to DEFRA made clear that they "have no concerns" about the importing of *dried* herbs for personal use, only about fresh plant material, in order to prevent diseases from entering the country. They advised, however, to double-check with the Food Standards Agency as well. Here, a short interview with a representative revealed a different point of view. The import of herbal remedies for home and personal use, according to the Food Standards Agency is a "dodgy area." People from Latin America are "not allowed to bring in medicinal herbs," or actually "some are allowed, some are not." No clear official guidelines exist about this matter, people "just have to bring in their herbs and have to let them be checked by an officer, who then decides whether or not the material will be allowed into the country." Yet, not many migrants want to face the risk of "limited fines" and "imprisonment, or both," for just a couple of innocent teabags. However exaggerated this may sound, it is a very common assumption about the importing of herbal material among the Latino community, and especially among Colombians. Those who do want to make the effort to get permission often get lost in the bureaucratic web. Therefore, people may turn instead to what is available, and that is what we would like to call "globalized species," plants that nowadays are available worldwide.

Conclusion

This paper has looked into the different types of herbal remedies used by Spanish-speaking Latin American immigrants in London. In doing so it has tried to address the issue of immigrants' health care patterns. By focusing on interviewees' views of health and unhealth, it has been shown that Latinos tend to use herbal remedies in a home context, and for minor ailments, described as forms of *malestar.* These home remedies can be divided roughly into the following categories: (1) medicinal herbs (used preventively or curatively); (2) functional food,

food medicine, food (plants) used in a multifunctional way; (3) "ritual" plants/ remedies; and finally (4) cosmetic home remedies or *tratamientos*. However, these plant remedies only represent a fraction of the number of remedies that people used before in their home countries. Hence, an answer has been sought to the question of how this difference between former and actual uses might be related to processes of acculturation. It was found that acculturation first and foremost is probably the result of practical adaptation and plant availability.

Acknowledgements

This project received financial support from the Leverhulme Trust (Grant number F00235D). I (M. Ceuterick) would also like to thank all interviewees and other participants for letting me be part of their lives for a while and David Block for kindly sharing his data.

Notes

1. "*Emic*" stands for the insider's point of view, or the community's own standpoint, whereas the terms "*etic*" and "*etically*" refer to the researcher's or outsider's viewpoint.
2. This does not mean that people simply put their (former) nationality aside. On the contrary, when one goes further into the subject, people often show a huge sense of pride about their home country, and first and foremost feel that they are, for example, Colombian, Peruvian, or Mexican. Narratives about this Latin identity are not always very clear or logical. I overheard more than once: "Of course I'm Latino, I'm Colombian."
3. "Alguna enfermedad. . . . No viene el nombre de enfermedad, pero si malestar. Por ejemplo, yo tomo agua de apio si tengo dolor de estomago . . . para cólicos mensuales, agua de canela. . . . para el dolor de garganta, es jengibre, naranja, miel, y una cucharadita de mantequilla es un alivio."
4. "Tipos de malestar, por ejemplo, lo haga la mujer con los periodos, . . . el "cinnamon," la canela, el té de canela es bueno cuando tiene demasiado malestar, malestar durante el periodo . . . te aprieta, te duele, no? Eso, sí. Pero enfermedades, no sé, no."
5. Free translation of this quote: "Con mal de ojo se dice: [reproachful]: te equivocaste de lugar; aquí es para gripa, dolores de cabeza, algo así. Pero es mal de ojo lo te cura un curandero." Mexican woman, twenty-nine years old.
6. Self-treatment is defined as any treatment or therapy without a professional (folk or biomedical) practitioner's intervention (Stevenson et al. 2003).
7. ". . . el médico aquí tiende a curar, y no a prevenir." Quote from a Colombian woman, living in the United Kingdom for over twenty-six years.
8. "El ajo y ginger. Sobre todo son remedios que . . . no . . . se usa cuando ya estás enfermo . . . porque dice: ah, viene abril, va a llover mucho, entonces cambian las comidas y a toda la comida le ponen más "ginger," o le ponen más ajo, para que te den más defensas. . . . En realidad tiene que ser antes. Pero es una cosa que adaptas a tu vida . . . supongo. Te da defensas, te da defensas . . . antes de que venga el virus. Como te digo, yo uso cosas más de prevención."
9. "No se puede evitar enfermedades, pero si prevenir, con una buena alimentación." Quote from a fifty-three year old Colombian man, living eight years in the United Kingdom.
10. "¿Comó quieres que me gustan las flores si no conozco su nombre verdadero, su nombre ingles, un nombre que ha fundido ya en los colores y a los pétalos, un nombre que ya es la cosa misma? . . . porque lo que dicen el ciruelo y los eucaliptos no lo dicen para mí, ni a mí me lo dicen."

11. "Es prohibido [para] nosotros, los Latinos, con las drogas y todo, nosotros como Latinos, como Colombianos, no podemos importar frutas ni plantas. Es tonto, por uno, pagamos todos. . . . Yo no podría traer una planta de mi país, oficialmente puedo traer, pero es algo muy riesgoso."

References

Balick, M.J., F. Kronenberg, A.L. Ososki, M. Reiff, A. Fugh-Berman, B. O'Connor, M. Roble, P. Lohr, and D. Atha. 2000. "Medicinal plants used by Latino healers for women's health conditions in New York City." *Economic Botany* 54: 344–357.

Baumann, G. 1997. "Dominant and demotic discourses of culture." In *Debating cultural hybridity. Multi-cultural identities and the politics of anti-racism,* eds. P. Werbner and T. Modood. London: Zed Books, 209–225.

———. 1999. *The Multicultural riddle: Re-thinking national, ethnic and religious identities.* London: Routledge.

Berg, M. 2004. "Migration and development in Latin-America and the Caribbean." The Centre on Migration, Policy and Society (COMPAS). Working Paper No. 7. Oxford: Oxford University.

Bermúdez Torres, A. 2003. *Navigation guide to refugee populations in the United Kingdom: Colombians.* London: ICAR.

Block, D. 2005. *Multilingual identities in a global city: London stories.* London: Palgrave.

Boyd, K. 2000. "Disease, illness, sickness, health, healing and wholeness: Exploring some elusive concepts." *Journal of Medical Ethics: Medical Humanities* 26: 9–17.

Collins Spanish Dictionary. 2004. New York: Harper.

Green, G., H. Bradby, A. Chan, and M. Lee. 2006. "We are not completely Westernized: Dual medical systems and pathways to health care among Chinese migrant women in England." *Social Science & Medicine* 62: 1498–1509.

Helman, C. 1990. *Culture, health and illness.* 2d ed. London: Wright.

Heywood, V. 1979. *Flowering plants of the world.* Amsterdam: Elsevier.

Illich, I. 1976. *Medical nemesis. The expropriation of health.* London: Calder and Boyars.

———. 2003. "Medical Nemesis." *Journal of Epidemiology and Community Health* 57: 919–922.

International Plant Names Index (IPNI). 2006. Retrieved July 2006 from http://www.ipni.org

Kim, Y. 2001. *Becoming intercultural. An integrative theory of communication and cross-cultural adaptation.* Thousand Oaks: Sage.

Kleinman, A. 1980. *Patients and healers in the context of culture.* Berkeley: University of California Press.

———. 1988. *The illness narratives—suffering, healing and the human condition.* New York: Basic Books.

Kyambi, S. 2005. *Asylum in the United Kingdom, an IPPR fact file.* London: Institute for Public Policy Research.

Maas, P., and L. Westra. 1993. *Neotropical plant families.* Koenigstein: Koeltz Scientific Books.

Mansfeld's Encyclopaedia of Agricultural and Horticultural Crops. 2001. Retrieved July 2006 from http://mansfeld.ipk-gaterleben.de/mansfeld/start.htm

Myllykangas, M., and R. Tuomainen. 2006. "Medicalization." *Health values, media publicity and citizens society.* Retrieved 4 June 2006 from http://www.uta.fi/laitokset/tsph/health/society /medicalisation.html

Nguyen, M. 2003. "Comparison of food plant knowledge between urban Vietnamese living in Vietnam and in Hawai'i." *Economic Botany* 57: 47–80.

Ody, P. 1993. *The complete medicinal herbal.* New York: DK Publishing.

Paz, O. 1994. *El laberinto de la soledad. Postdata. Vuelta a el laberinto de la soledad.* 2d ed. Mexico: Fondo de Cultura Económica.

Pelto, P. and G. Pelto. 1990. "Field methods in medical anthropology." In *Medical anthropology, Contemporary theory and method,* eds. T. Johnson and C. Sargent. New York: Praeger, 269–297.

Perna, V. 1996. "Latin lovers: Salsa musicians and their audience in London." In *Music, culture and society in Europe,* ed. P. Rutten. Brussels: Soundscapes, pp.104–115.

Pieroni, A., H. Münz, M. Akbulut, K.H.C. Baser, and C. Durmuskahya. 2005. "Traditional phytotherapy and trans-cultural pharmacy among Turkish migrants living in Cologne, Germany." *Journal of Ethnopharmacology* 102: 69–88.

Pieroni, A., and C. Quave 2006. "Functional foods or food medicines? On the consumption of wild plants among Albanians and Southern Italians in Lucania." In *Eating and healing. Traditional food as medicine,* eds. A. Pieroni and L. L. Price. New York: Food Products Press, 101–130.

Román-Velázquez, P. 1999. *The Making of Latin London.* Aldershot: Ashgate.

San Miguel, E. 2003. "Rue (*Ruta* L., Rutaceae) in traditional Spain: Frequency and distribution of its medicinal and symbolic applications." *Economic Botany* 57: 231–244.

Stevenson, F.A., N. Britten, C.A. Barry, C.P. Bradley, and N. Barber. 2003. "Self-treatment and its discussion in medical consultations: how is medical pluralism managed in practice?" *Social Science & Medicine* 57: 513–527.

Zapata, J., and Shippee-Rice R. 1999. "The use of folk healing and healers by six Latinos living in New England." *Journal of Transcultural Nursing* 10: 136–142.

Chapter 8

Hackney's "Ethnic Economy" Revisited

Local Food Culture, Ethnic "Purity," and the Politico-Historical Articulation of Kurdish Identity

Sarah Keeler

Introduction

It has long been noted within anthropological discourse that absorption of food-stuffs can serve integrative and appropriating functions socially and symbolically, with the old adage "we are what we eat" serving as a recurrent everyday reminder of this (Van den Berghe 1984; Douglas 1997). Increasingly since the emergence of theories surrounding globalization, researchers have also turned their attentions to the cross-cultural consumption of food, and the ways in which this can act as a site for the construction of identity and reiterate social relations at the global, national, and local levels (Caglar 1995; James 1996; Warde 1997; Bell and Valentine 1997). The ways in which foodstuffs are produced, presented, thought about, distributed, and consumed, particularly within a migratory context of growing mobility of goods and people, now also require us to pose the question: if "we are what we eat," in what ways does the *how, where,* and *with whom* of our eating habits contribute to our sense of self and other? This chapter, based on fourteen months of fieldwork with Kurdish migrants and refugees in a multiethnic London community, looks at food culture from this perspective.[1] In particular, it explores the ways in which food becomes a marker of ethnic difference and a means of imagining and maintaining identity boundaries within the Kurdish diaspora;[2] and by consequence, of narrating their history and migratory experience, including the changing contours of social relations in the diaspora. Of particular significance for the Kurdish community is the way these processes are played out vis-à-vis their relations with the sizable Turkish community with whom they coexist in the same inner-city area of London. It indicates how foodstuffs and cuisine culture are revealed within these narratives to be emblematic of a politico-historical experience in the homeland in which Kurdish ethnic identity was perceived as being obscured and aspects of Kurdish culture appropriated by Turkish antagonists, and how this is reproduced in local political economic relations in the position of exile. Also, it looks at how the reclamation of this "stolen" and degraded identity is enacted through foodways that memorialize imaginings of the homeland and assert a continued presence and resistance through embodied practices of food production and consumption.

It is not concerned therefore with the substantive features of the ethnic cuisine or foodstuffs used by Kurds in London so much as it is with how they are perceived and utilized through social interactions within these processes of identity construction and negotiation. How is food as a cultural marker called upon to create solidarity, signpost shared heritage, demarcate difference, and even facilitate the political aims of nationalist discourse? In exploring these questions, it becomes possible to understand also how food as a cultural marker in these contexts can serve to assert, in very fundamental and tangible ways, the presence of particular groups in multiethnic public spaces which are open to contestations over access and use. I consider the traditional concepts of use value and exploitation of resources in symbolic rather than material terms, and more specifically, how this exploitation of resources within public space is used as a means to manage expressions of identity.

Food, Migration, and Ethnic Identity Constructions

While in the last decade work on migration and ethnicity has turned its attentions increasingly to the cross-cultural consumption of food, this work—based typically in the study of colonial relations, social inequality, or ethnic enclave economics—tends to assume a hierarchical framework whereby the dominant culture of the host society engages in a process of commodifying and appropriating the "authentic" food culture of the subjugated minority (Parker 1994; Caglar 1995). However these approaches tend to overlook the nuances which form interactions between the concepts of dominant/minority, place and consumption, positionality and identity, and which are enacted in everyday encounters in public space (Howes 1996). What is more, there is little account of the fluidity and multiplicity of the identifications with which migrants affiliate themselves as social actors in contingent social interactions, all the more so as second and third generations contribute their own experiences of enacting hybridity to the settlement process. Rather, "migrant groups" are often viewed as bounded cultural entities struggling to manage the contradictory and often incompatible experience of several fixed "cultures" and, as they move through social spheres, usually from the vulnerable position of a subject, not as active agents engaged in the construction and continual renegotiation of place, space, and identity. The anthropology of food and consumption thus needs to be imbued with the theoretical reevaluations which have informed more recent work on other areas of migration studies; to explore transnational phenomena and their influence on situational identities which are constantly in flux, just as efforts to reconstruct "pure" identities form a response to these conditions (Hall 1997).

Further, in a cosmopolitan context such as London, where the research for this paper was carried out, we cannot conceive of a simple dichotomy of host/ migrant cultures, but must take into consideration interactions across and within minority groups as well as the internal diversity of migrant populations and the often subversive manifestations of overarching power relations that exist within supposedly homogeneous "groups." In the case of this research, the unity of the "Turkish speaking community" in Hackney as perceived by researchers, local council officials, and consumers alike not only overwrites a vast plurality of ethnic, political, religious, and gender identifications, but has helped in reshaping those distinctions and divisions for community members, as well as their interaction in the diaspora setting. More than 40 percent of Hackney's recorded population belongs to ethnic minorities, in particular large numbers of Afro-Caribbeans, Vietnamese, Bangladeshis, and of course Turkish nationals. There are believed to be approximately 10,000 Kurds, (Hackney Council 1999a; 1999b) almost all from southeastern Turkey. The dynamic of coethnic interaction in the "ethnic economy" has a particular relevance in terms of power relations and hegemonic systems among the Turkish speaking communities. Previous waves of migration from Cyprus and mainland Turkey are an extremely salient issue in terms of local policy and the delivery of service designed for the "Turkish speaking community" in Hackney. From the perspective of local government, perceptions about issues facing these particular populations have been informed by exposure over several decades to migrants arriving from Cyprus and western Turkey. Thus, despite fairly significant differences in the makeup, needs, and aims of each population in the London context, the provision of "culturally sensitive" services for Kurds has been grafted onto existing programs designed for other Turkish speaking populations. Taking into account the significant differences in background and culture between Kurds and other Turkish speakers, this has had an impact on economic adaptation, and concomitantly, on the development and maintenance of distinct ethnic identities.

Returning for the moment to theoretical concerns, however, we may look to evolutionary models commonly used for explaining the nexus of food, ethnicity, and collective identity. Van den Berghe uses such an approach, looking at the role of food sharing in strengthening bonds of kinship which he sees to be microcosms of ethnic identity, transformed into increasingly bounded categories in the face of multiethnic interaction in urban space. For him, food works within this process to become an easily accessible and maintained "badge of ethnicity" (1984: 392). This can be observed most visibly as a symbolic expression of group solidarity within what he calls the "culinary complex" consisting of "the inventory of food items, the repertoire of recipes and the rituals of commensalisms" (ibid.: 392). This notion is useful for our ethnographic account here as a way of thinking about how foodways, or the "culinary complex," not only demarcate borders of ethnic difference, but are also tied in strong and continuing ways to practices of exclusion, the maintenance of insider status, or sharing on the basis of degrees of intimacy. What is interesting in this case, though, is the fact that, while the elements that make up the "culinary complex" are predominantly shared by Turks and Kurds, in practice there is much debate about their real or perceived "ownership."

The concept of incorporation has been another anthropological device employed frequently to analyze the role of food in culture. In the research presented here, incorporation is within taken-for-granted and quotidian forms of everyday consumption rather than in the ritualized contexts traditionally addressed. These everyday forms, often tied to explicit and self-conscious manipulation of the central role of food in the development of identity, allow us to understand the significance of food in the lives of Kurdish refugees in London. L. Harbottle finds the relevance for migration research in the concept itself, stating, "incorporation is the basis for collective identity and thus of marking a distinction from 'others'" (2000: 24). She indicates that an important way in which foods can demarcate identity boundaries is through the variable classification of edibility; that is, foods that may be prized in one cultural setting will be considered inconsumable in others. However, my interest here is in areas in which common use of foodstuffs, or the entire food culture, have become loci around which pivot contestations over ownership, identity, and appropriation. This is not only a recurrent theme in personal narratives of migration but also a very publicly expressed point of debate.

In the course of my fieldwork among Kurds in Hackney, where the close physical, social, and cultural proximity of non-Kurdish fellow nationals from Turkey integrates within the local ethnic economy but also provides the impetus for differentiation from a perceived antagonist, the culture of food—actual content of the cuisine, its preparation and consumption, and the social and symbolic relations which guide this—emerged within discourse to serve two important functions. First, within the multiethnic context of Hackney, food and foodways were identified as important markers of ethnic identity, articulating both the distinctiveness of "Kurdish cuisine" from culturally contiguous Turkish or Arab cuisines as well as the oppression and appropriation of any version of distinctive "Kurdishness" by those same groups. The migratory process and the changing context of political, social, and material relations to which it gives rise can become particularly salient, for it has been noted that a coherent and bounded sense of identity (individual, ethnic, or national) comes to be articulated in response to exposure to various forms of otherness (Van den Berghe 1984), and often when a perceived threat is seen to be present (Bowman 2003). Second, foodways represent a cultural site characterized in this case by informality, making them both more subjective and transitory (and thus less susceptible to official sanctioning by dominant systems), but also potentially more "vulnerable"; a space in which culturally "pure" practices and identities may be threatened by hegemonic forces. As such, there is the potential for food culture to become a highly politicized and intensely felt aspect of ethnic identity formation and power negotiation.

The Refugee Context

Traditional food and foodways serve an important memorial function in the lives of displaced peoples, often being the only tangible link to the world left behind (Harbottle 2000; Sutton 2001). For refugees whose experiences of migration may have been sudden and traumatic, for practical reasons the reconstruction of homeland cannot find its basis within a storehouse of material objects, as is the case for some migrant groups. But recipes, techniques for the preparation and serving of certain foods, if not the actual foodstuffs, need not be present in physical repositories but are transmitted by other means, particularly social relations. As such there are important psychological, social, and symbolic dimensions making food an extremely salient feature of daily life and collective experience in exile. Food serves as a basic resource in this respect, momentarily soothing feelings of loss, celebrating aspects of identity and cultural rites which continue to be practiced in the new country, maintaining social relations and the boundaries which define them, and informing images of the reconstructed homeland.

Equally however, food comes to serve new functions in the dynamic process of migration and identity negotiation in a diverse setting. For Kurdish refugees in Hackney, for example, the local ethnic economy of "Little Turkey," comprised largely of businesses in the catering sector (including cafes, restaurants, kebab takeaway shops and small greengrocers), means that the meanings and uses of food are driven into public space and increasingly commodified; what is more, a public space in which encounters with Turkish people—as employers, competitors, or customers—form the basis of commercial activity and social exchange. The presentation of food and foodways within this context is a highly condensed cultural site reflecting ideas about and contestations over ethnicity, homeland, and political events taking place both locally and transnationally.

As informants frequently indicated to me, the denial of a distinctly Kurdish culture, identity or population, followed by the emergence over time of a sense of "Kurdishness" within the population (especially strong in the diaspora), is a reminder of nearly a century of brutal oppression of the Kurds at the hands of the Turkish state. When the modern Turkish republic was founded in 1923, the

Kemalist doctrine sought to establish a secular and unifying national identity which left little room for the expression—in cultural, linguistic, or religious terms—of the myriad ethnic identities present in the former Ottoman lands. The imposition of this nationalist ideology on the lives and consciousness of the people is epitomized by the phrase *Ne Mutlu Turkum Diyene* ("Happy is he who calls himself a Turk"), "decorative" versions of which can be found emblazoned in massive letters painted across mountainsides throughout the country and visible from every village for miles around. In sociocultural terms, this ideology has sought historically to obscure the presence of distinct Armenian and Greek ethnic identities (and their contributions to Turkish culture), and in recent decades, with uprisings in the region, the oppression of ethnic identity has been carried out against the fifteen million strong Kurdish populace indigenous to the southeast of the country. From the inception of the Turkish Republic, and increasingly in later decades as Kurdish resistance movements gained in strength and frequency, all aspects of Kurdish cultural identity were subject to negation, often violently, by state forces; a campaign which included the criminalization of the Kurdish language and traditional cultural celebrations such as *Newroz*, the Kurdish new year.

Thus, whenever Kurds in the diaspora perceive attempts to obscure their identity in any way, cultural practices and traditions call upon this history of oppression and inform the relations with Turks in the migratory setting, with whom there is a considerable degree of interaction (particularly in commercial encounters within the ethnic economy) and an equally high level of social antagonism. However, given the pragmatic demands for solidarity required to access resources made available through agencies such as local government, the Home Office, or the Refugee Council, Kurds do at times ally themselves with their better-established Turkish counterparts.[3] This means that, rather than resulting in open conflicts at the community level, the social antagonisms tend to be somewhat latent and usually expressed informally through verbal asides and digressive social encounters. In the course of conducting fieldwork, the most commonly used vehicle for expressing such resentments was through the language of food.

Although these interethnic social antagonisms are a salient feature of relations between the groups within Hackney at the informal level, they are largely obscured from the public eye and are certainly not phenomena of which "outsiders" would be aware. Indeed, given the nature of their expression, one would have to understand the entire complex of historical relations not only in Turkey but with respect to overlapping migration histories in the United Kingdom, to say nothing of the nuanced social cues, linguistic forms, and behavioral patterns which shape mutual resentments at the local level in Hackney. For these reasons, it is easy to see why local government services, popular perceptions, and visitors' imaginings have led to the creation of a supposedly homogeneous "Turkish speaking community." However, this too has had an impact on the dynamic and ever changing nature of relations, as a reactive form of oppositional identity emerges in resistance to such labeling by politicized members of the Kurdish diaspora in London.

This being the case, informants frequently expressed that their personal sense of "Kurdishness," and their identification with a transnational Kurdish diaspora, was heightened or discovered in the experience of migration and exile.[4] The reconfiguration of social relations between Kurds and Turks in the diaspora setting, and especially the increasing visibility of Kurdish identity as a distinct and separate ethnic category in Hackney and internationally, has helped to shape the self-conscious representation of Kurdish cuisine within the context of the local ethnic economy. The opening in the spring of 2005 of the first Hackney restaurant claiming to serve "traditional Kurdish cuisine" marked a significant and much lauded event within the community. Further, its physical placement amid several successful and well known Turkish restaurants was noted by community members. Thus, while the process of social differentiation is certainly facilitated through the experience of displacement, this may in fact be viewed by refugees themselves as an empowering process, contrary to the frequent assertions that displacement and exile are likely to give rise to a conception of membership in a "victim diaspora" (Cohen 1999), a group whose identity is formed from what they have lost or how they suffer. Kurds in the UK diaspora setting experience an increased connection to their ethnic identity, a growing sense of openness in expressing it, and often an increasing resistance to being classified as "Turkish" (Wahlbeck 1999; Keeler forthcoming). This is reflected in the dynamics of interaction within Hackney's ethnic economy.

THE "ETHNIC ECONOMY," SUBALTERN POSITIONING AND PUBLIC SPACE

Hackney, the east London neighborhood discussed herein, is an inner city area of more than one hundred thousand, with a sizable minority of migrants and refugees from Turkey among its ethnically diverse populace. While Afro-Caribbean, Vietnamese, Bangladeshi, and Eastern European communities are also present in large numbers, it is the "Turkish speaking community" that most visibly occupies public space, particularly in commercial arenas, leading to the designation of Hackney as a "Little Turkey" of sorts. This is strongly in evidence in the many shops, cafes, and restaurants lining the main shopping streets, their signage in Turkish. The Turkish music heard emanating from these places, as well as the often large groups of men to be found socializing outside the cafes and restaurants, represents an observable reproduction of public space typical of towns in southeastern Turkey.

I focus on food in these public spaces as opposed to the more private domestic sphere—though I view the two as often overlapping points on a continuum, rather than as mutually exclusive—since it is within this sphere, and particularly the ethnic economy, that food comes to speak about relations of power, the management of identity, and its many functions within a diverse social context. For within this public space, by becoming visible consumers (by acts as simple as the choosing of one particular café or restaurant over others), customers also identify themselves as members of a distinct community—including all the ideological, political, and social dimensions that implies. The recent proliferation of explicitly "Kurdish" establishments, for example, signifies for Kurdish consumers simultaneously a memorial to a beloved and now lost homeland and an emerging commitment to the rediscovery, through such consumer practices as eating particular foods, of the ideals and ethnic identity which for them forms their political struggle. One informant described the history of oppression of Kurdish identity in Turkey, and its impact on resettlement in the United Kingdom, as follows:

> "Because of the eighty years of assimilation policies the Turkish government imposed on the Kurds . . . [they] were assimilated. And this is reflected in their business and daily life in London. Up until recently, you know, a lot of Kurds gave businesses Turkish names saying "Turkish Restaurant" or "Turkish Supermarket" or even presented themselves as Turks, thinking to improve their business. They put food in there that maybe was Kurdish but they wouldn't say so, people wouldn't eat it then. Another factor is that when we arrived . . . there was a large settled community of Turks and Turkish Cypriots living in north London. But the authorities, and customers, well, assume these people [Kurdish business owners] are Turkish. They don't understand the idea of a Kurd. It causes a self-esteem problem, we feel like hiding who we are, so it's just like being in Turkey again. But what can you do? People needed to make a living, right?"

Thus, despite recent assertions of "Kurdishness" visible within an upwardly mobile segment of the population, Kurdish refugees employed in the local ethnic economy remain aware of the potentially degrading and essentializing impact of expressions of identity through foodways. This is observable increasingly in louder resistance to inclusion in the so-called "kebab culture" seen as representing Turks in Hackney. At a public meeting held to discuss labor market segregation and educational underachievement among members of London's Turkish speaking community, one individual spoke out to declare:

> "I am not a member of the "Turkish speaking community" and I am not Turkish. I was forced to learn Turkish because of the conditions in Turkey before I came as a refugee to this country. Now I work in the same industry, I make food and I am in catering industry for a living, people just say "they are Turkish, it's the kebab culture." I do not wish to be called Turkish, culturally or as a part of the "kebab culture." My identity is my own, and I do not choose to be Turkish, I am Kurdish, and so is the food I make."

While the life experiences of refugees are frequently characterized by themes of dislocation and loss, the above narratives and the wider politico-historical processes to which they point for their tellers reveal underlying continuities—of interethnic power relations and socioeconomic inequalities, and their influence on the opportunity structures available for Kurds in London's "Little Turkey." The pervasiveness of food culture as a device for the narration of migratory experience is thus only partially explained by its sensual and behavioral elements, for it is also true that economic realities, systemic discrimination in the labor market, and the particular trajectories of migration history in Hackney tie Kurdish refugees from Turkey to industries involving the handling of food. With few educational resources and, as a resultant, limited employment opportunities in London, a preponderance of Kurds find themselves shifted into the "kebab culture" of the Turkish speaking community's ethnic economy, with as many as 70 percent of recent arrivals taking up work in the catering industry (Keeler forthcoming). Thus, their everyday encounters with other social actors in public space are often predicated on exchanges involving food, which becomes a sort of language in and of itself. What is more, these are often exchanges the success of which demand a particular form of packaged ethnicity, whether to a fellow Kurdish clientele or to customers of other ethnicities in search of a safe and accessible taste of the exotic. In his description of the demands for ethnic identity in the restaurant trade to be advertised as a part of the commodity being sold, and its relationship to histories of domination and subordination in global perspective, Parker suggests, "discourses about food and servility are mobilized in interactions across the counter . . . customers draw on the archive of images . . . usually exoticized expressions of cultural difference . . . operating together in highly condensed moments of encounter" (1994: 627; cf. Mandel 1996).

Given the convergence of historically embedded inequalities with economic conditions local to the Kurds in Hackney, we begin to understand the process of (re)construction of ethnicity through food and its reactive nature expressed in terms of perceived antagonisms. Equally, however, these dimensions can serve to fragment the possible responses on the part of members of Hackney's Kurdish community. The realities of "people needing to earn a living" can often be seen to conflict with possible desires to express loyalty to, or assert identification with, a particular sense of "Kurdishness." Placement within the ethnic economy, and variable positionality according to differences of age, gender, or social class internal to the so-called "Kurdish community", can also lead to considerable ambiguity in the ways and means considered possible or desirable for expressing ethnic identity. At such times, the lives of informants became interstices for crosscutting sentiments and pragmatic concerns relating to livelihoods and beliefs about identity, with food culture serving to bridge these tensions.

Harbottle's study (2000) of Iranian migrants in the catering industry in the United Kingdom revealed similar equivocal sentiments regarding the display or concealment of ethnic identity through food within the context of public space. However, unlike the politicized Kurds on whom this research was based, the Iranians in Harbottle's study were often observed to obscure their ethnicity intentionally while employed in the catering industry, preferring instead a generic "Middle Eastern" label for their productive activities. In this respect, the catering industry may particularly suit the needs of this and other ethnic minority groups by providing an unobtrusive niche on the fringe of the British economy, allowing interaction with the host culture as far as possible on their own terms (cf. Staub 1981; Harbottle 2000: 78).

However, with respect to interactions internal to an ethnically demarcated economic setting, such as that found in Hackney within the Turkish speaking community, we find an equal degree of tension between various segments of the population, and "for restaurant owners, this raises the dilemma of whether to broadcast their own political or ethnic affiliations" (Harbottle 2000: 94). Certainly, among Kurds in Hackney, only the most well established entrepreneurs can "afford" to display highly visible signifiers of particularly politicized "Kurdishness," or, as one informant explained to me:

"Most people wouldn't risk putting this [flags or maps of Kurdistan] up, because they'd be afraid to lose their Turkish customers, who they need for business." (See figure 8.1).

Figure 8.1. *A Kurdish restaurant in Hackney decorated with political paraphernalia, including prominently displayed flags of Kurdistan*

It is for this reason that mundane acts of consumption like eating out, as well as the informal social exchanges which accompany them, can serve as such an important vehicle for the expression of political values and identity in this multifaceted ethnic economy. As transgressive acts, which resist established patterns of domination and subordination, the choice of a particular restaurant, the consumption of certain foods, and the discourses which surround both of these, can have subversive and empowering functions for people who feel themselves limited in their options.

Food as Narrative Device in Histories of Subjugation and Resistance

One example of this can be found in discussions of *lahmacun,* the distinctive snack food of flatbread topped with minced lamb and spices, which is then filled with salad and eaten rolled up, often "on the go." As one of the cheapest and easiest to prepare foods available in kebab shops, *lahmacun* are eaten across Hackney's Turkish speaking population, popular with both Kurds and Turks, though rarely consumed by English or other ethnic customers. On one occasion while sitting down with three informants to a restaurant meal that included *lahmacun,* a lively discussion broke out (which was joined later by restaurant staff) regarding its origins. It was a point of much aggravation that in "Little Turkey" both Turkish and Kurdish entrepreneurs were advertising *lahmacun* as "Turkish pizza," comparing it to a more familiar fast food product in order to entice customers with a less-than-adventurous appetite. Statements about the "lie" perceived to be inherent in this label led to discussions of cultural diffusion through food and further, to experiences of cultural assimilation at the hands of the Turkish state. As one informant finally summarized referring to the food debate itself:

> "It's not Turkish, it's Kurdish. The Turks don't have anything actually, don't have any culture of their own. They just steal from everyone else who was there first. They use the Greeks, Kurds, Arabs . . . none of it is their own culture. They say *lahmacun* is Turkish but it's Kurdish, it's a Middle Eastern food. They say that to try and get customers, and if people aren't familiar with the culture, well they don't know. But it's not what they say it is. We invented it [*lahmacun*] and they took it."

The statement of this informant powerfully points us toward the subjective interpretations through which food can make claims, not only about who we are but about *how* we got there; in other words, food and the practices surrounding its production and consumption are receptacles and transmitters of historical experience and identity which social actors are constantly reinterpreting and with which they are constantly engaging. For Kurdish refugees in exile, the experiences of displacement and the awakening of nationalist consciousness in the diaspora are paralleled by an increasing salience of cultural sites, such as purveyors of food for the articulation of a distinct and increasingly politicized ethnic identity. Such statements about the dearth of "genuine" Turkish culture and cuisine, therefore, also contain within them assumptions about the political history of the Turkish state, the suppression and obfuscation of Kurdish identity within that state, and the Kurdish struggle for autonomy; a struggle which reproduces itself in the local ethnic economy and food culture of Turkish speakers in Hackney.

Zubaida (1994: 40–41) points to the often close relationship among nationalism, cultural "ownership," and discourses about food, noting that:

> "[N]ationalist pride in food is eminently historical. Aware that similar genres of food are shared by other nations, it is concerned to claim that these are originally "ours" . . . the logic behind these narratives is one of national-historical essentialism . . . the facts of historical transformations and synthesis are excluded from this logic . . . This supposed historical antiquity and continuity are cited as, somehow, a confirmation of the authenticity and the superiority of the present-day national cuisine. History, then, becomes the measure of national virtue, including food."

While Zubaida refers here to discourses pertaining to the national level, including those taken up by the state, what we see in the Kurdish case in Hackney is the transplantation of these tendencies to a migratory setting, where cultural or ethnic identities, perceived as denied, now become "liberated." This is an inherently productive process, as we see not only the rewriting of historical facts and actors ("they," the Turks, are without culture or cuisine, having apparently existed in a static cultural void) but also the denial in turn, of any forms of historical hybridity or diffusion. Events become politicized instead within contemporary apprehensions of interethnic contact (the current antagonisms implied by the suggestion that "we" invented something that "they" took), which are situated directly in the migratory context. Again, the increased awareness of their *otherness from* dominant Turkish cultural narratives, together with an even greater sense of freedom in expressing this, has been channeled into the highly condensed cultural site of foodways, which becomes articulated in informal but nonetheless very public encounters. As Sutton (2001: 86) beautifully states in discussing the amplification of such distinctions for migrant groups:

> "Local divergences . . . become more intensified in the migrant context, where cooking is not simply an everyday practice, but an attempt to reconstruct and remember synesthetically, to return to that whole world of home, which is subjectively experienced both locally and nationally."

In the case of Kurds in the diaspora, many of whom are refugees (as opposed to the Greek migrants of Sutton's work), the traumatic nature of their migratory experience, together with the facts of and reasons behind their flight, their forced dispersal across nation-states, and the denial of the land which they view as "Kurdistan,"[5] renders any attempts at reconstruction and remembrance all the more problematic, intense, and often quite politically charged. Thus we see assertions of difference and purity, if not outright superiority, subjectively expressed and experienced through the medium of food and positioned within a context of interethnic contestation, which recalls but simultaneously diverges from historical relations in the homeland. The negation of these expressions of difference, either through the "mislabeling" of foods or groups of people as "Turkish," is often viewed not only as a reproduction of politico-historical experiences of domination, but also as an outright threat to cultural integrity in the diaspora. While walking me through Hackney to point out major landmarks in the local ethnic economy, one informant powerfully articulated such sentiments:

> "What they do in these places [shops in Hackney], I call it food genocide. They take foods that are originally Kurdish or whatever . . . and they call them by some other name, they say they are Turkish. And then they change everything, put things together in a different way. Look at this [shop signage] for example: In Turkish, it says "Turkish Specialties," the Kurdish one says "Kurdish Specialties," and in Persian that will say "Persian Specialties." They are just trying to please all their customers, but for people who don't speak the languages or know the culture . . . they will just see "Turkish Food Market" and think it is Turkish, and also because it is Hackney. Most of what they sell in there is not Turkish, so what they are doing is wrong. . . . They are misleading people. It's a kind of genocide."

What all informant statements in this article have touched upon—whether inadvertently or not—is the need for small-scale entrepreneurs, if they are ever to break out of a strictly ethnic economy that relies primarily on coethnics as a customer base, to operate in a number of different "cultural registers" and to provide services or products which cater and adapt to a variety of cultural tastes. This phenomenon has been widely observed within the literature on ethnic entrepreneurship (Staub 1981; Pecoud 2000; Kloosterman and Rath 2001) as a positive example of the ways in which hybridity can serve as a valuable cultural resource for enterprising business owners trying to resettle in a new homeland. But what might be interpreted by researchers and analysts as a marketing trend in catering to a diverse clientele and lauded for its resourcefulness, the above informants, and many others I spoke with in the context of shared meals and walks through Hackney's ethnic economy have observed within their own community as a negative factor, and ultimately as a culturally threatening phenomenon, which serves to reinforce the obliteration of all signifiers of "Kurdishness" by those in positions of power; as a phenomenon which is adopted unconsciously by the very people who have been oppressed, serving as further proof of the ubiquity of their subjugation. In fact, while the rhetoric of 'genocide' and threats to purity employed by these individuals is far more dramatic in its expression, researchers into phenomena of ethnic economy and integration have also pointed toward the exploitation and perpetuation of imperialist historical traditions which are often played out in subtle ways in the context of seemingly innocuous commercial exchanges (Mandel 1996). Parker's study of Chinese takeaway restaurants in Britain, for example, illustrates how the essentialization of "ethnic culture" in the form of ethnic food mass-produced for a wide clientele draws on racialized historical interactions and popularized views about the "grotesque" nature of Chinese culture and cuisine (1994; cf. Caglar 1995) Or, as Harbottle reminds us, the success of such ethnic entrepreneurship in the food industry is dependent upon the acceptance by takeaway staff of a continuing relationship of service and subordination, mobilized and reinforced by each face-to-face interaction across the counter (2000: 82).

As the above interview excerpts reveal, Kurds in Hackney are acutely aware of this potential reality, of the tensions which necessarily exist between the self-conscious exploitation of an ethnic identity for the purposes of commercial gain, and the risk of corruption of a strongly-held and culturally "pure" subjective identity, which remains nonetheless an inherent aspect of this livelihood. Perceptions of shared culinary complex are then contingent not only upon sociopolitical relations in the homeland but equally on processes of forced migration, perceptions of refugees and multiculturalism, and local conditions therein. Sutton's recognition (Sutton 2001) that "local divergences" are more salient in the migratory context can then be seen to have special significance for the parallel migrations of ethnic Turks and Kurds into Hackney, which have served to juxtapose members of a dominant elite (ethnic or otherwise) with those subject to a hegemonic system. And while conflicts over such historical relations can be diverted or dissipated through the perhaps wider scope of channels available and officially sanctioned in the receiving context, they can resurface equally in unexpected ways in the diaspora, including, for example, in the considerable importance placed on food culture in the local context of Hackney.

Food, Embodiment, and Assertions of Presence through the Senses

One important way in which we see the various disjunctures from and continuities with politico-historical experiences managed within the lives of Kurdish refugees in London is through embodied practices strongly related to the sensual properties of food. Although such practices are usually observed in the context of production and consumption in public space, their powerful emotional resonance for participants makes them a deeply personal means of conceiving themes which are otherwise loudly articulated in political, and therefore collectivized, discourses.

For example, the scent of innumerable *ocak*[6] that one encounters when walking in Hackney's "Little Turkey" are a powerful reminder of the past and of the position of the self within another system, a way of being and relating to other social actors and to the landscape itself. Visually, the configuration of groups within the cafes—their relaxed demeanor, free movement within the designated space (where typically there is much movement, visiting, and conversation from one table to another)—and the aural stimulation of hearing ones' language spoken that accompanies these experiences—all allow participants opportunities for a reclamation of identity, or what Sutton sees as "a chance to reflect on the past and provide meta-narratives on the felt losses of "modernity" or the exile of migration" (2001: 99). Informants often related to me their desire to reconnect with such collective and memorializing experiences in their daily lives, indicating that regular contact with such experiences was in fact necessary to their well-being in London; echoing that the perception of food and the social behaviors of production and consumption which surround it are integral to "the reimagining of "worlds" displaced in time and/or space" (ibid. 102). In the words of one such individual:

> "When I smell the food and see the people together . . . it reminds me of my mother, and of the town where I grew up, when I was a boy . . . the places I played. Ah, beautiful! You know, when I sometimes don't come to Hackney . . . I need to be here at least once a week to remember those things, eat some good Kurdish food. Also I can talk to people about it. I get sad if I forget, or get too busy."

Clearly, such sensual experiences can conjure emotions and a sense of place tied to an idealized homeland, now lost, and in so doing serve to empower those who identify with an ethnicity perceived as degraded or denied by dominant narratives.

As collective display and experience, however, they serve equally to assert the presence of this ethnic identity within a system of power that historically has not recognized the existence or participation of distinct cultural groups, either in the homeland or in the migratory setting. As such, this presence is a counterhegemonic one, particularly in light of the fact that it is articulated through such vehicles as food culture, which is not subject to official sanctioning. The social, spatial, and political practices which accompany the preparation and consumption of food, in particular the sensual aspects of these processes, become means through which to exercise this counterhegemonic presence in public spaces that represent sites open to contestations over ownership and use by various other groups in the multiethnic context.

Kurdish refugees in London are conscious of changes to their foodways, or what they would characterize as "threats" in the context of migration, exile, and diaspora, and frequently articulate the changing contours of social relations and sensory experience associated with the production and consumption of food in their narratives. As embodied practice, for example, many will refer to the behaviors surrounding the drinking of tea, its differing significance in the receiving culture, and thus as a "Kurdish" practice which they must actively work to preserve. The consumption of tea by the British is seen to be a matter primarily of quenching of ones' thirst, whereas Kurds view their own consumption of tea to be fundamentally about social exchange, a time to share news, to see and be seen in public space. As one informant exclaimed succinctly—"in London, they just order it and go. At home, it would take us two hours to drink a cup of tea!"—meaning that the etiquette and behavioral patterns involved change the focus necessarily from a straightforward act of consumption to a shared social experience.

The corollary of this thinking in the position of exile is often represented in attempts to reconstruct the very material context of spatial and social practices surrounding food. Nowhere is this more striking than in the production of bread, especially when one considers public space and the expression of Kurdish identity through food, for most people do not produce homemade breads for individual household consumption. Rather they shop regularly in the restaurants which are well known in the community for producing handmade versions of these favorite treats. Inside these restaurants, an area within the main dining room (in sight of the customers) is set aside for the preparation of fresh bread, which is baked in the customary *tandur* using traditional instruments and brought warm to the tables. Often, particularly when women are involved, those engaged in the breadmaking will be dressed in a version of traditional Kurdish dress to enhance the scene (see figure 8.2). The most interesting feature with respect to spatial and embodied practices of reclamation of homeland, however, is the art-

Figure 8.2. *Women in Kurdish dress prepare traditional bread in a restaurant*

work that typically serves as a backdrop for these productive activities. In all the restaurants I visited in the course of fieldwork, paintings or photographs depicting the same activity, taking place in a village context in Kurdistan, were placed conspicuously alongside these embodied reproductions of food culture. The laborers and even the public spaces where they work are actively engaged, therefore, not only in the production of food items for everyday consumption, but also in the production of ideas and memories of place and space depicted in the paintings adorning the walls which serve as a backdrop for their reproduction.

For Kurdish customers this tableau represents a synthesis of the sights, flavors, and smells of their idealized and "pure" homeland, or in the words of Appadurai, "a highly condensed social fact . . . a marvelously plastic kind of collective representation (1981: 494)." In putting forward a collective representation in public space that is overwritten as "Turkish" by virtue of its participation in the homogenized ethnic economy in Hackney, it also represents for them an act of resistance. Indeed, because of its ubiquity and the mundaneness of its nature, it is all the more powerful as a source of identity demarcation that depends on distinctions drawn from culturally contiguous "others." Thus, even those who are not overtly political in their behavior or ideology can and do use the vehicle of food to express their distinction from and disdain for "Turkish" food culture in this respect.

Conclusion

Migratory experience by its nature leads to the interaction of differing cultural systems. This includes foodways, which, given that migrants are active agents in their own experiences and not simply victims of a transnational process beyond their comprehension or control, need not lead exclusively to trauma but can equally encourage "positive identity fusions and transformations" (Harbottle 2000: 10). While I have tried to challenge the prevailing emphasis on migrants and refugees as passive, "struggling" with their position of being "caught" between cultures, we must also be careful not to overlook the possibility that such processes—both positive fusions and negative experiences—can coexist, even or perhaps especially at the individual level. For Kurds embedded within the local ethnic economy of Hackney, and often identified from an external perspective as "Turkish speakers" or even as Turks, there is a considerable degree of ambivalence toward the many factors which shape their identifications: the sense of responsibility to and pride in maintaining a distinct ethnic identity, the desire to integrate socially into the local culture, and the very pragmatic economic interests met by the essentializing expressions of ethnicity in the commercial arena. Indeed it is perhaps because of this ambivalence, and the relative fluidity of ethnic identifications at the local level (particularly within the economy and public space), as familiar social relations are juxtaposed, reconfigured, and rendered unfamiliar in the multicultural metropolis, that foodways become more salient as signifiers of historical experience, ethnic integrity, and purity. As Van den Berghe (1984: 393–94) reminds us:

> "[I]t is only through a process of alienation that ethnic food in itself becomes ethnic food for itself. Ethnic food is most likely to become a fully self-conscious product in the context of the multiethnic city."

In the case of food culture in Hackney's "Little Turkey," the consciousness of ethnic identity, and its expression through food, is informed by wider economic, political, and social discourses at local and transnational levels.

Notes

1. I use the phrases "food culture" and "foodways" loosely and interchangeably, much in the same way that Van den Bergh refers to the "culinary complex" consisting of "the inventory of foods, the repertoire of recipes and the rituals of commensalisms" in a given culture (1984: 392). To this list I would also add the traditional ways of producing particular foodstuffs, and the social relations that structure its production, distribution, and consumption.

2. Kurds in the diaspora, and indeed in the sending context as well, are made up of a diversity cross cut by social class, gender, generation, region of origin, and even language and religion, the only unifying feature of which is their desire to be identified as Kurds regardless of the internal diversity of their population. Therefore I use monolithically conceived labels such as "The Kurdish diaspora" or "Kurdish Refugees" not without reservation and fully aware of the limitations this imposes, but I have culled my framework from my informants themselves, according to self-ascription of ethnic identity. It is this last sense to which I refer when using such terminology throughout the paper.

3. See John Eade (1997) on "strategic essentialism," or the ways in which pragmatic goals shared by members of diverse and often competing interest groups can override antagonisms which may exist between them, particularly in the UK policy context (cf. Baumann 1999).

4. This phenomenon can, of course, also be attributed to the criminalization of certain expressions of Kurdish culture within regimes in the sending countries, in contrast to the freedom to express oneself culturally and politically that one finds in the United Kingdom.

5. The Kurdish population, which numbers some thirty million in the Middle Eastern lands of their origin, are dispersed across southeastern Turkey, northern Syria and Iraq, and western Iran. The border regions where these states converge are viewed by Kurdish activists (and many in the diaspora) as the historical Kurdish homeland, and referred to as "Kurdistan." Though this geographic label is recognized at the regional level within Iran and Iraq, for example, and as a culture area by many historians, "Kurdistan" is not an internationally recognized geopolitical entity.

6. These are open charcoal grills often said to be an innovation imported directly from the Kurdish area in the southeast of Turkey by the many refugees who came to Hackney in the 1990s, who took up employment in the Turkish and Cypriot owned catering businesses in the area.

References

Appadurai, A. 1981. "Gastro-politics in Hindu South Asia." *American Ethnologist* 8: 494–511.

Baumann, G. 1999. *The multicultural riddle: rethinking national, ethnic, and religious identities*. New York: Routledge.

Bowman, G. 2003. "Constitutive violence in the nationalist imaginary: antagonism and defensive solidarity in "Palestine" and "former Yugoslavia."" *Social Anthropology (Journal of the European Association of Social Anthropologists)* 11: 37–58.

Bell, D. and G. Valentine 1997. *Consuming geographies: We are where we eat.* London: Routledge.

Caglar, A. 1995. "McKebap: Doner kebap and the social positioning struggle of German Turks." In *Marketing in a multicultural world: Ethnicity, nationalism and cultural identity,* ed. G.J. Bamossy and J.A. Costa. London: Sage Publishing, 209–230.

Cohen, R. 1999. *Global diasporas: An introduction.* London: UCL Press.

Douglas, M. 1997. "Deciphering a meal." In *Food and culture: A reader,* ed. C. Counihan and P. Van Esterik. London: Routledge, 36–54.

Eade, J. 1997. "Reconstructing places: Changing images of docklands and spitalfields." In *Living the global city: Globalization as social process,* ed. J. Eade. London: Routledge, 127–145.

Hackney Council 1999a. *Ethnic diversity in the making of Hackney: The Kurdish population.* London: Hackney Borough Council.

———. 1999b. *Planning for the Turkish/Kurdish community in Hackney: A profile and needs study.* London: Hackney Borough Council, Division of Planning.

Hall, S. 1997. "The local and the global: Globalization and ethnicity." In *Culture, globalization and the world system: Contemporary conditions for the representations of identity,* ed. A.D. King. Minneapolis: University of Minnesota Press, 19–39.

Harbottle, L. 2000. *Food for health, food for wealth: Ethnic and gender identities in British Iranian communities.* Oxford: Bergahn Books.

Howes, D. 1996. "Commodities and cultural borders." In *Cross-cultural consumption: Global markets, local realities,* ed. David Howes. London: Routledge, 1–18.

James, A. 1996. "Cooking the books: global or local identities in contemporary British food culture." In *Cross-cultural consumption: Global markets, local realities,* ed. D. Howes. London: Routledge, 77–92.

Keeler, S., forthcoming. "There has been enough crying and now I want to dance! Belonging, cosmopolitanism and resistance in London's Kurdish diasporic spaces." Ph.D. Dissertation. Canterbury: University of Kent.

Kloosterman, R. and J. Rath. 2001. "Immigrant entrepreneurs in advanced economies: Mixed embeddedness further explored." *Journal of Ethnic and Migration Studies* 27: 189–202.

Mandel, R. 1996. "A place of their own: Contesting spaces and defining places in Berlin's migrant community." In *Making Muslim space in North America and Europe,* ed. B. D. Metcalf. Berkley: University of California Press, 147–166.

Parker, D. 1994. "Encounters across the counter: Young Chinese people in Britain." *New Community* 20: 621–634.

Pecoud, A. 2000. "Entrepreneurship and identity among German Turks in Berlin." *Intergraph: Online Journal of Dialogic Anthropology* 1(3). Retrieved 17 May 2004 from http// www.intergraphjournal.com/enhanced/contents1_3.htm

Staub, S. 1981. "The near Eastern restaurant: A study of the spatial manifestations of the folklore of ethnicity." *New York Folklore* 31: 113–127.

Sutton, D.E. 2001. *Remembrance of repasts: An anthropology of food and memory.* Oxford: Berg Publishing.

Van den Berghe, P. 1984. "Ethnic cuisine: Culture in nature." *Ethnic and Racial Studies* 7: 387–397.

Wahlbeck, O. 1999. *Kurdish diasporas: A comparative study of Kurdish refugee communities.* London: MacMillan Press.

Warde, Alan (1997) "Novelty and tradition." In *Consumption, food and taste: Culinary antinomies and commodity culture.* London: Sage Publishing, 57–77.

Zubaida, S. 1994. "National, communal and global dimensions in Mid Eastern food cultures." In *Culinary cultures of the Middle East,* ed. S. Zubaida and R. Tapper. London: Taurus, 33–45.

Chapter 9

A Strange Drug in a Strange Land

Neil Carrier

Introduction

In the last couple of decades a plant stimulant has traveled into lands not traditionally associated with its consumption and in the process has become highly controversial, causing panic not only for worries over its impact on health, but for far wider reasons as well. *Khat* is the substance in question, and unlike other plant stimulants such as coffee, tea, and tobacco, which also met with considerable opposition when they first entered the West, khat arrived along with representatives of the indigenous groups who consumed it in its lands of origin. This is a crucial factor in its reception, and the main focus of this chapter. After first introducing the substance and its effects, I describe the growth in its international trade, linking this to the spread of khat-chewing diaspora communities. I focus in particular on the Somali diaspora, emphasizing khat's social and cultural importance for the life of Somalis in the West. Of course, not all Somalis approve of the substance, and I bring out its role as a focus of tension between men and women and emphasize the social conditions that underlie the problematic use of khat in the United Kingdom. I then detail its reception in the West and its glib conflation by the media and others with substances that have very different cultures of consumption. Khat has become absorbed into the "war on drugs," and even into the more recent "war on terrorism," creating much moral panic in the process. I argue that despite such moral panic, khat's social and cultural importance, and the tangible link it offers to far away homelands, means that its consumption in the West will continue for some time to come.[1]

Khat and Its Effects: A Brief Description

Khat is the name most commonly used in reference to the stimulant stems and leaves of *Catha edulis* (Forssk.). Other common names for the substance include *miraa* (Kenyan), *qat* (transliterated from Arabic), and *chat* (Ethiopian). Consumers purchase the leaves and stems in bundles and chew over several hours. The highest concentrations of consumers are found in East Africa (especially Kenya, Somalia, Ethiopia, and Djibouti) and the Middle East (especially Yemen). But consumption is also known in Madagascar and South Africa, as well as in countries hosting Somali, Ethiopian, and Yemeni diaspora communities (Australia, Canada, Denmark, Italy, the Netherlands, New Zealand, Sweden, United Kingdom, the United States, and so on). Khat is harvested wild almost everywhere it is known to grow, although commercial cultivation is concentrated in Yemen, Ethiopia, and Kenya. The substance is subject to legal restrictions in several countries (see below) including Tanzania, Eritrea, some European countries, New Zealand, and the United States and Canada. Following a recent report on possible legislation against khat by the Advisory Council on the Misuse of Drugs, the Home Secretary announced in January 2006 that khat will remain legal in the United Kingdom.[2]

Catha edulis favors an altitude band between 5000–8000 feet (c.1500–2450 meters). Wild trees can grow as high as 80 feet, although the farmed variety is kept at around 20 feet with constant pruning. The harvested commodity varies from region to region as to what is considered chewable and how it is presented. Thus, in Yemen often just leaves are chewed, whereas in Kenya the stems are also chewed. Stems are a mixture of green and purple hues depending on variety; small leaves are normally quite dark, becoming greener as they mature. They can taste bitter, although the better the khat, the sweeter the taste.

Research on the chemical actions of khat's constituent alkaloids has been undertaken for over a century and continues to reveal new effects, although none seem especially strong. The principal alkaloid is called *cathinone,* now known to be more powerful than *cathine* (d-norpseudoephedrine) which was once thought the main active ingredient (see Kennedy 1987: 181). Cathinone releases dopamine in the brain and affects the central nervous system in a manner similar to amphetamine, increasing "heart rate, locomotor activity and oxygen consumption" (Weir 1985: 46). Cathinone has been calculated as "about half as potent as amphetamine" (Zaghloul et al. 2003: 80), hence the common description of its effects as rather like a mild form of amphetamine. Also, just as chewing coca leaves is different from taking pure cocaine, so chewing khat has a gentler effect than would taking isolated cathinone. Cathinone degrades rapidly post-harvest, and so the potency of khat is affected by the length of time that elapses between harvest and consumption. Trade networks delivering khat therefore must operate highly efficiently (Carrier 2005a).

Cathinone and cathine are not the only active constituents. A further group of alkaloids—the *cathedulins*—were isolated in 1979 (Crombie 1980), and research has now identified more than sixty cathedulin variants within khat samples. The psychoactivity prompted by cathedulins appears limited, although recent research suggests that they do play some role in the effects of *miraa* as they, too, release dopamine (*The Guardian,* 5 February 2004). Once again, although this can sound dramatic and potentially dangerous, it must be realized that many other substances and human activities (including sex) also release dopamine. Other khat constituents include tannin and ascorbic acid. The range of these constituents suggests that there might be a significant synergistic action: as Weir suggests (1985: 46), "the effects of cathinone could be modified or counteracted by the action of the other chemical constituents of qat."

The medical side effects of khat have been researched extensively over the past fifty years. Many side effects appear linked more to indirect effects rather than to the direct effects of its chemical action: some chewers do not sleep for long periods while chewing, and sleep deprivation obviously can cause problems. Kennedy (1987) surveys the literature on khat and health. He mentions the research of Halbach, "the most noted and cited medical authority on qat" (1987: 214). Halbach "asserts unequivocally that certain ailments are "common" among qat chewers: gastric problems such as stomatitis, esophagitis, gastritis, constipation, malnutrition, and cirrhosis of the liver." Cases of anorexia, sexual problems, and anemia have also been associated with khat consumption. The most frequently mentioned side effect is insomnia. Kennedy's research team also conducted surveys in Yemen on khat and health. The data collected led him to conclude that "the argument that qat is responsible for the health problems of Yemen is exaggerated, but it also shows that they are not without foundation" (Kennedy 1987: 231). Recent Yemeni research suggests a link with khat chewing and cardiovascular problems, as some chewers suffered heart attacks at times of the day other than early morning, the time most patients are reportedly admitted with "acute myocardial infarction" worldwide (Al-Motarreb et al. 2002: 407). The suggestion is that "khat chewing may be a risk factor for ischemic heart disease" (ibid). Such research is at an early stage and might suggest that those with heart conditions would be better advised not to chew. However, khat would hardly be alone in being relatively harmless *ceteris paribus,* but a risk factor to those with heart conditions. Kennedy states that Yemenis themselves "are nearly unanimous in the opinion that qat is *not* an important threat to their health" (Kennedy 1987: 213).

Most frightening to those unacquainted with the substance is its association with forms of psychosis. Such a link is often mentioned in the literature (for example, Critchlow 1987) and is widely repeated in nonmedical accounts of khat. However, there is in fact no clear empirical evidence to establish the existence of the condition sometimes termed "khat-psychosis." Reports on individual cases are mentioned—notably from Ethiopia (Alem and Shibre 1997), and other examples from among the Somali immigrant population in the United Kingdom—with those who wish to prohibit khat being only too ready to generalize. In the United Kingdom much talk of khat as a cause of psychosis is linked to the Somali community. There are no reports at all of khat-related psychosis among

Yemeni, Ethiopian, or Kenyan consumers in the United Kingdom. This suggests that social issues are behind cases of psychosis among Somali consumers. High rates of unemployment and social exclusion might very well play a part, as might also possible post-traumatic syndromes linked with incidents in their country of origin. Halbach (1972) compared khat and amphetamine in regard to toxic psychosis (a condition arising from an "excessive intake of amphetamines") and noted the rare occurrence of psychotic reactions to the former:

> "Whereas . . . toxic psychosis is characteristic of an excessive intake of amphetamines, it seems to occur much less frequently in areas where khat is consumed, even if consumption is heavy. In fact, personal inquiries in Harrar and Dire-Daua [sic] [Ethiopia], Djibouti, Somalia, Kenya, Aden, and the Yemen in 1959 gave the impression that toxic psychosis was very seldom, if ever, diagnosed in these areas" (Halbach 1972: 26–27).

A recent report on consumers in Yemen backs up such remarks. It concludes: "Overall, it appears that khat use is not necessarily linked to psychological morbidity; any association that is found may reflect an interaction with other environmental factors" (Numan 2004: 64).

Research into khat is an ongoing process, so knowledge of its pharmacologico-medical effects is not definitive. However, from the literature—and from my own experience—it seems fair to state that khat is not an overwhelming substance, and therefore there is much scope for consumers to mold the khat experience creatively in the way they use and talk about khat. Various names are used for khat's effect. In Kenya, the word applied is *handas* (Carrier 2003), while the Somali word *mirqaan* is common throughout the diaspora, and *kayf* is used in Yemen (Weir 1985; Kennedy 1987). In Yemen and Ethiopia Khat consumption is associated with conversation, contemplation, and relaxation in the company of friends. However, khat's effect is very much dependent upon the context in which it is consumed: it stimulates farmers and other workers to persevere with arduous work, and yet it can also relax and render chewers loquacious in recreational settings. Khat is associated with indolence by many, despite often being consumed in work contexts. Associations with indolence derive especially from labor time lost to khat sessions in Yemen and Djibouti, where many stop work after lunch to chew.

Ethiopian khat is produced in many regions, but most intensively in Hararge province, which has an age-old association with the substance. From Hararge, khat is exported throughout Ethiopia and beyond: Djibouti, Somalia, Somaliland, and Europe. Khat in Kenya is grown most intensively in the Nyambene Hills (northeast of Mount Kenya) by Tigania and Igembe, subgroups of the Meru. It is the major cash crop of the region, bringing far better returns for farmers than coffee or tea. For the Tigania and Igembe, khat is also of great cultural importance, being a marker of identity and used ceremonially on such occasions as marriage negotiations. The network in Kenya, mainly run by Meru, delivers khat efficiently to an internal market, and a mainly Somali-run network delivers large quantities internationally to Somalia, Europe, and North America.

Khat and Migration

Migration has helped spread khat consumption from very early times as the practice became popular throughout East Africa, the Middle East, and also Madagascar. However, it is its more recent spread to Europe (see Nencini et al. 1989; Halliday 1992; Griffiths et al 1997; Ahmed and Salib 1998; Nabuzoka and Badhadhe 2000; Patel et al. 2005), North America (Browne 1991), and even Australia (Stevenson et al. 1996) and New Zealand that I focus upon. While khat consumption is important for many Yemenis, Ethiopians, and Kenyans now living in the West, my own experience of consumption in the United Kingdom, Europe, and North America has been mainly with Somalis, and so I concentrate on them.[3]

Somalis began settling in the United Kingdom over a century ago. As was the case with Yemenis, many Somalis, especially those from the north (now Somaliland), joined the British Merchant Navy and found themselves spending time in British ports. Communities gradually developed in Cardiff, Manchester, Liverpool, and London. In writings on migrant populations in Britain, "a portrait emerges of Somali seafarers who, by the 1940s and 1950s, were numerically the second largest Muslim group after the Yemenis in a number of British ports" (El-Sohl 1991: 539). Few married locally, maintaining families at home, and led a rather segregated existence, suffering some racial abuse (ibid. 540).

The decline in the Merchant Navy caused great problems for the community, but still it grew with a steady flow of migrants. From 1981 onward, however, numbers grew more rapidly due to conflict in Somalia. Still the influx was predominantly from the north, the former British protectorate. The regime of Barre began its most violent period in 1989 with all out civil war, and numbers increased yet more. Not all were from the north either; a large proportion now came from the south. Somalis spread in vast numbers around the world—Ontario and Minnesota absorbed a large part of those fleeing—but the diaspora also spread to most European countries, and even to Australia and New Zealand (Pérouse de Montclos 2002).

This dispersal had a huge effect on the khat trade. As cathinone degrades quickly, and as chewers prefer the taste and texture of fresh khat, the trade was once limited geographically by lack of speedy transportation. The growth in infrastructure over the course of the twentieth century meant that fresh khat could reach further afield. The advent of air travel between Europe and East Africa (and the Arab Peninsula) then provided the option of sending khat to Europe. Exactly when this opportunity was first seized is difficult to say, but small quantities are likely to have begun reaching the United Kingdom in the 1950s. According to El-Solh (1991: 549), up to the 1970s it was Yemenis who imported khat into the United Kingdom, mainly feeding Yemeni demand. In the last decades of the twentieth century, however, khat in the United Kingdom has become more associated with Somalis, and nowadays Kenyan Somalis send khat to fellow Somalis in Britain and elsewhere in the diaspora.

In the United Kingdom, the two most common varieties of khat sold are *kangeta*—a long-stemmed Kenyan variety named after a town in the Nyambene Hills—and *harari,* the Ethiopian variety (Gebissa 2004). The former retails at the standard price of £3 per bundle (known as a *mijin* in Somali), and the latter for £5. Occasionally one can also buy the shorter-stemmed Kenyan variety of *giza* (again at £3 per *mijin*), which many chewers in the United Kingdom regard as more potent. UK consumers often describe the Ethiopian variety as less potent, yet some Somalis prefer *harari,* especially older ones from the north of their homeland: Hargeisa and other towns in what is now the self-declared state of Somaliland obtain their supplies of khat from nearby Ethiopia; hence this is the variety to which many UK Somalis are accustomed. Younger Somali chewers—whether from the north or south—usually prefer *kangeta* or *giza.*

Somalis retail khat at establishments called *mafrish,* private houses where rooms are set aside for chewing. Khat is also sold in grocery stores run by Somalis, Ethiopians, and Yemenis: near Kings Cross in London, one Asian manager of a branch of a well-known UK grocery chain stocks Ethiopian khat in the fresh vegetables section. He told me that most khat buyers were Ethiopians and Somalis, although the occasional curious European had also been known to invest in a bundle or two. Because of their perishability, all varieties must be traded efficiently, but it is interesting to note how much more formalized trade in the Ethiopian variety appears compared with the Kenyan: Ethiopian khat retailers and wholesalers in London are even listed on a pocket-sized directory of Ethiopian businesses. This reflects the more formalized nature of much khat production and trade in Ethiopia compared with what seems a more haphazard—albeit still efficient—chain of individuals operating the international trade of Kenyan khat.

In the United Kingdom the Kenyan khat network encompasses not only retailers in London, but also many in cities throughout the country. Kenyan khat comes four times a week, and much of it, having arrived at Heathrow from Nairobi and been cleared through customs, is delivered by van to a depot in Southall. There it is collected by distributors who speed it on to retailers waiting at their *mafrish.* Khat arrives in boxes containing forty *mijin,* and distributors charge retailers about £80 per box. Providing a retailer sells all the bundles at £3, he or she makes a profit of £40. Khat is sent up to northern cities by van: deliveries are made in Birmingham, then Manchester, and finally Liverpool. Until recently deliveries were at regular times as khat was transported on scheduled flights. However, Kenya Airways and British Airways stopped accepting shipments of khat—apparently because of irregularities in payment—and for a while khat arrived in the United Kingdom on charter flights. Deliveries were delayed by a day at times, if there was not enough space on a flight, and placed in cold storage at the airport in Nairobi until space was found. The situation is now resolved, more or less, and shipments are routed on scheduled KLM flights.

As regards making money from khat, retailers in the United Kingdom are unlikely to become rich. One Somali retailing khat in Manchester is supplied with two or three boxes (always three on the weekend and holidays) four times a week. He has cut out middlemen in London and set up a deal direct with a Nairobi exporter, but has extra costs in transport up from London: I reckon him to make about £60 per box providing all is sold, giving him earnings of around £600 a week. However, he does have significant costs: rent of the lower floor of the house (he lives with his family in another house while others occupy the first floor of the *mafrish*), rental of, and subscription to, satellite television, payment for a friend who helps run the *mafrish,* and small expenses such as the cost of the tea and plastic cups provided free to customers. He also rarely sells all bundles, and many customers obtain khat on credit, meaning that business can be rather precarious. All told, he does reasonably well out of his trade but is hardly making a fortune. Similarly, a friend who runs a *mafrish* in London has a moderate income from retailing the few boxes he is supplied with four times a week but finds he has to work hard for this income, often spending much of the night cleaning up the establishment after customers have departed.

Better money is made by those established as exporters in Nairobi or as distributors in the United Kingdom as their turnover is often far higher. Some exporters based in Nairobi send hundreds of boxes to Europe: money from this trade is often plowed back into building up the Nairobi suburb of Eastleigh—where many Somalis live—or into vehicles used in transporting khat from the Nyambene Hills to Nairobi. Not even Nairobi exporters are guaranteed good money, however: it is not unknown for traders in the United Kingdom to cut off contact with suppliers in Kenya when money is owed for shipments received. Large-scale exporters can cope with such losses, whereas smaller-scale exporters—those sending only twenty boxes perhaps—cannot. Of course, much money can be made from successful smuggling runs to the United States and Canada: a bundle of khat in the United States retails for around $30. However, seizures of shipments and costs of hiring couriers eat into profits, and khat traders in the United States and Canada are unlikely to match stereotypes of wealthy drug barons.

Not all khat consumption in the United Kingdom takes place at a *mafrish* but most does. As Weir (1985) describes with regard to Yemen, a *mafrish* provides a social center. One goes to relax after work, one goes to make contacts to help find work, or to catch up on the latest news in the local community and from further afield. Khat provides a focal point around which social groups gather. Chewing sessions I have attended at a *mafrish* in Manchester are quite typical of those throughout the United Kingdom. It is located in a terrace house in an area with a Somali population of a few thousand. There are four other such establishments in the district. Two rooms of a rented house are set aside, one room for those who come for conversation and another for those who prefer listening to music. Furniture is scarce, and customers sit on cushions laid around the sides. One trader vacuums the rooms before the khat arrives and prepares flasks of tea. He also stocks up with disposable cups and bottled water. As customers fill up the establishment from mid-afternoon onwards—almost all in the 25–50 year age range, the noise level rises. Some customers buy khat and leave to chew at home, while most linger for at least an hour or two. Discussions revolve around the latest goings on in Somalia, job opportunities in the area, the quality of the khat for sale, and soccer. The satellite television is usually turned on, often tuned to a Somali channel, and sometimes videos of weddings and other gatherings in Somalia are shown. Once, a Somali musician played on a keyboard while all the gathered chewers joined in singing popular Somali songs. *Mirqaan*—the effect of khat—is itself redolent of sessions back in the Horn of Africa, a redolence reinforced by the link home provided by Somali television channels, videos, music, and, of course, friends. While chewing, although in the northwest of England, one almost feels as if one were in Brava, the southern coastal town where many of Manchester's Somalis originate. Sessions I have attended in London, where most participants are from Somaliland, also provide these tangible links to Africa.[4]

Somali women do chew khat, Patel et al. (2005: 24) state that 16 percent of their sample of 278 Somali women in the United Kingdom had chewed recently—although never at the male *mafrish,* compared with 51 percent of a sample of 324 Somali men. Some establishments in London are said to have separate rooms for women chewers, but usually women chew with friends at home. Khat is associated far more with men than women, and, in fact, much of the opposition to chewing within the Somali community comes from women's groups.[5] While there is criticism among women in Africa of khat use men, the gendered nature of criticism of khat consumption seems more pronounced in the Somali diaspora. Women were often the first to arrive and adjust in the new countries, while their men came later and relied on the knowledge their wives had of the social systems, and it is suggested that there has been something of a role reversal between the genders: unemployed men rely on bread winning women. McGowan (1999: 93) paraphrases the president of the Somali Youth Association of Toronto thus:

> "[He] maintains that the problems with marriages can be traced to qat chewing. The husband arrives in Canada, accustomed to being the main breadwinner, cannot find a job, loses hope, and chews qat, which exacerbates his problem because he is not employed, but neither is he contributing to the work of the home. He does not help with

the running of the home or with the children's homework, which is all the more problematic in cases where the mother is illiterate. After some months of this situation, the
wife is distressed and wants him out of the house."

The man can also deplete household money through his expenditure on khat. In
a survey of young Somalis (N=94) in Sheffield, Nabuzoka and Badhadhe (2000:
16) found that "[m]ost of the respondents (43.6 percent) reported spending less
than £25 on Khat or other drugs each week; 37.2 percent spent between £25 and
£50 a week and 19.1 percent spent over £50 a week. The majority (58.5 percent)
indicated that spending on khat caused them financial problems." This is in the
United Kingdom where a bundle costs the moderate sum of £3.[6] Where khat is
illegal—in the United States and Canada—those who chew (and there are
plenty) pay a much higher price.

Clearly, there is some problematic use of khat among Somali diaspora communities in situations where adjusting to the host country and finding work is
difficult, and such usage is a source of tension between men and women, leading to strong calls for a UK khat ban from Somalis themselves (49 percent of
Somalis questioned by Patel et al. [2005: 48] were in favor of a ban, with a far
higher proportion of men opposed to a ban as compared to women). There is
also a suggestion that consumption by Somalis in the diaspora is higher than in
Africa. Research conducted on behalf of the Home Office in 1994 to 1995 revealed that the majority of men and women out of 207 London-based Somalis
interviewed had tried khat (Griffiths et al. 1997: 281). Most of those interviewed
chewed with friends, and khat "was generally used two or three times a week,
often during the evenings or weekend" (ibid. 282). Seventy-six percent of the
sample "reported using more or much more now than they did in Somalia"
(ibid). The researchers hypothesize that a factor in this increase in use is that khat
has been "taken from its original sociocultural context and transplanted to a foreign setting in London where previous roles and restraints may be less effective."
However, such a hypothesis appears to be based on shaky premises: Griffiths in
the Home Office report of 1998 states that khat "problems do not appear evident
where its use is acculturated, as with the Mehru [sic] and Isiolo tribal people of
Kenya, where only the elders are permitted to chew it" (Griffiths 1998: 13). This
is incorrect, and my own experience researching khat in Isiolo and Meru—where
khat is considered *poa* ("cool" in Swahili) by many youth—shows its consumption to be common also among the young (see Carrier 2005b).

Employment appears a factor in decisions about when and for how long a
consumer chews, both in Kenya and in the United Kingdom (although from
Patel et al.'s figures (2005: 24–25), it does not appear linked in any clear way to
whether one chews or abstains in general): those with jobs are more likely to resist
chewing until their free time. For some who are unemployed, khat chewing fills
the time and alleviates boredom. The Home Office research in the 1990s revealed

that only 17 percent of those interviewed had work (Griffiths et al. 1997: 281), leaving many with time on their hands and few pursuits besides chewing, hence the perceived increase in use. Khat and employment are not mutually exclusive—khat is often used functionally to keep alert while working—and most Manchester and London chewers that I know hold down jobs. Some only chew when not at work: one Somali in London told me how chewing during the weekend helps relax him after working all week as a bus driver. If a Somali man sees an opportunity for work, khat is unlikely to stop him grasping it.

Griffiths et al.'s report also emphasized the strong feeling among Somalis that "the use of qat helps to maintain cultural identity," a statement with which 66 percent of the sample agreed (ibid. 282). However, in an alien environment with few available jobs it is hardly surprising that some Somalis—especially those who chewed khat back in Africa—would seek solace in the company of friends and the warm embrace of *mirqaan* (although its use should not be generalized as escapist).[7] Such cultural importance is clearly not restricted to Somalis in the United Kingdom. In the United States, Canada, Sweden, and elsewhere, despite its illegality khat reinforces cultural identity and acts as a social adhesive for many within displaced communities. In the case of Australia, Stevenson et al. report that already khat "is proving to be important as an identity marker. Through their use of khat, East Africans are able to assert their desire to preserve distinct identities within a culturally diverse community" (1996: 80).

Khat's Reception in the West: Moral Panic

In khat use among Somalis of the diaspora, there is much continuity with the consumption in East Africa: as Goldsmith states (1997: 478), "a Somali *khat* gathering in London or Toronto is indistinguishable from one in Eastleigh [a suburb of Nairobi with a large Somali population]." For many Somalis, Ethiopians, and Yemenis of the diaspora, khat and khat sessions are something very familiar indeed. However, for the wider public in countries like the United Kingdom and the United States, khat is something decidedly unfamiliar, and as the substance gains more exposure through media coverage, it comes to be commonly, and glibly, portrayed through more familiar prisms: the "war on drugs," immigration, and, more recently, the "war on terror."

Rather than focusing on the social and cultural aspects of khat consumption, the Western media often choose to emphasize the idea of khat as the latest "addictive drug," an idea also promulgated by many government agencies. Khat consumption is conflated with consumption of other—much stronger—stimulants like cocaine and ecstasy. A recent example is an article in *The Guardian* (5 February 2004) with a picture of a khat bundle on the cover of its science supplement and the headline: "This has the same effect as cocaine and ecstasy. And it's legal." Scotland's *Sunday Mail* even dubbed khat "legal crack" (15 August 2004).

This link to the "war on drugs" is reflected in its varying legal status through-out the world. Prohibition of khat is not a new phenomenon as colonial author-ities in Somaliland, Djibouti, and Kenya attempted unsuccessfully to reduce consumption through legal restrictions, and independent Somalia banned the substance, again with little success in reducing consumption. However, after years of international concern and conferences on khat,[8] cathinone, and cathine, the main constituents of khat were added to the list of scheduled substances in the UN Convention on Psychotropic Drugs of 1971, obliging countries signed up for the convention to put them under control.

This move led to a recent wave of legal restrictions against khat. One signa-tory to the convention was the United States, and to fulfill its international ob-ligations it declared cathine a Schedule IV substance in 1988 (Federal Register vol. 53, no. 95) and cathinone a Schedule I substance in February 1993 (Federal Register vol. 58, no. 9). To be added to Schedule I a substance must be consid-ered to have a high potential for abuse and no accepted medical use in the United States. It was deemed that cathinone fulfilled these criteria. In order to avoid confusion, the entries to the Federal Register made it clear that by scheduling the chemicals, khat was also scheduled: "When khat contains cathinone, khat is a Schedule I substance. . . . When khat does not contain cathinone, but does con-tain cathine, khat is a Schedule IV substance."[9]

How countries interpret their obligations under the UN Convention appears to be flexible: other countries—still including the United Kingdom—distin-guish between the alkaloids and the substance in plant form, allowing khat to re-main legal. However, by conflating khat and its alkaloids, a substance viewed by many as more akin to espresso than anything stronger was placed on a legal par with LSD by February 1993, and customs officers nicknamed 'khat busters' strive to keep the United States khat-free.[10] These changes to khat's legal status occurred as thousands of Somali refugees began arriving in the United States and trade in khat was expanding. Perception of khat in the United States has been strongly colored by events of the early 1990s in Somalia: US troops encountered Somali militiamen who, it was claimed, were "high" on khat. Media attention fo-cused on khat consumption, presenting Somali militia-men as drug crazed and trigger happy. Goldsmith (1997: 475) describes well the hysteria that built up around khat at the time:

> "*Khat* was exploited by the Western media to add a sensational element to the tragic events in Somalia, where it was equated with violence. A news item in *Africa Confi-dential* at the peak of Somalia's civil war reported that "for the gunmen and soldiers in what is a violent society, chewing gives them a focused edge." The US Assistant Sec-retary of State for Africa alluded to it around the same time in terms of "teenage *khat*-chewing Rambos getting pumped up for early evening raids" on the MacNeil-Lehrer news hour."

The emphasis on a link between khat and violence is dramatically explicit in an article on the "supporters of the US Border Patrol" website.[11] There, a picture of a *mooryaan*—a derogatory term for young militiamen of Mogadishu—holding an AK47 is captioned: "Khat eater waiting for a new target." A disturbing video clip on the same site shows the mutilated bodies of American troops killed in Mogadishu. The following text accompanies the clip:

> "Almost all of the barbarians who attacked the UN peacekeepers and the barbarians who hacked them into chunks and the barbarians who attacked our troops and the barbarians who then—by the thousands—attacked our rescue forces all chewed a narcotic called khat."

Canada has followed the United States in restricting khat, and in both countries seizures are increasing year by year, probably as much a result of increased vigilance as increased shipments. Over twenty-two tons of khat was seized in Canada in 2003 according to a Royal Canadian Mounted Police website.[12] Smuggling khat into the two countries has, of course, encouraged the creation of smuggling networks, and couriers—usually British nationals—are sometimes caught trying to get suitcases full of the substance through customs. Other countries to make the substance illegal include New Zealand—where some khat is imported for the small East African community, Sweden, and Germany. Australia, meanwhile, has developed a permit system allowing limited importation of khat.

The United Kingdom is a signatory to the 1971 UN Convention, and cathine and cathinone have both been declared Class C substances, putting them on a par with the recently downgraded cannabis. Khat in leaf form remains legal after the recent review by the Home Office. In the run up to the Home Office's recent decision, stoked up by the media hype of "legal crack," there have been many calls for the United Kingdom to ban the substance, and there is pressure from the United States for the United Kingdom to restrict the substance and so reduce the amount of khat passing through London en route to the United States. Members of Parliament have raised questions in parliament asking what measures the government will take to control khat. One Liberal Democrat Member of Parliament used condemnation of khat to gain leverage in arguing for the decriminalization of cannabis, asking rhetorically why this harmful "hallucinogenic" substance is legal when the supposedly less harmful cannabis is not (Hansard 15 June 1998: column 16). The Member of Parliament misleadingly reported that khat users "become stoned and unable to do anything," hardly accurate considering its functional use. A Member of Parliament for Newport (where khat is used) did argue against a ban, however: "Is not the lesson clear—that the prohibition of khat would not decrease but increase its use? It would increase the amount of khat crime, and would drive a wedge between the Somali and Yemeni populations and the police" (ibid). More recently, a Member of Parliament for

Ealing (London) —spurred on by concern for his large Somali constituency—
referred to khat in a parliamentary debate as "corrosive, vicious and pernicious"
(Hansard 18 January 2005: column 774). He also declared that chewers "stay up
all night, come home in the morning, beat up the wife and try to sleep through
the day" (ibid).

Concern about khat cuts across the British political spectrum. On the left
there is worry that khat is causing all sorts of social problems for minority communities, while on the right there is indignation at immigrant communities spending benefits on a "drug." In fact, the moral panic over khat sometimes appears
to camouflage what is really concern about immigration. This appears the case
not just in the United Kingdom, but in most countries hosting khat-chewing
minorities. The US Border Patrol website mentioned above makes this concern
explicit. It continues:

> "The number of violent aliens and illegal aliens in America at this moment from that
> part of the world can be calculated by the amount of this drug being smuggled into the
> country. The narcotic is an acquired taste and certainly not for an American palate."

In the United Kingdom, the far-right British National Party are reported to have
become interested in khat consumption in London, using it as another rhetori-
cal tool with which to attack immigration, while a member of the Canadian
"white nationalist" group "Stormfront" has attacked khat on their website as "Islam's dirty little secret."[13]

The panic surrounding khat in the West has recently risen further in the wake
of the "war on terror." Some suggest that khat is funding Islamist terrorist
groups. However, little evidence for this link to khat is provided, and what is
offered relies on conjecture and cultural and religious stereotypes. The line of
argument proceeds thus: US authorities are unclear where profits from khat end
up; khat consumers are often from Yemen (a Muslim country strongly linked to
terrorism by the United States) or Somalia (a Muslim country regarded as a
"failed state" and as a safe haven for terrorists); therefore money made from khat
must be going to fund terrorism. One person worried about the link is Harvey
Kushner, described on a CBS report as the "director of the criminal justice department at Long Island University."[14] On the same report, he is quoted thus:

> "You don't have to make a quantum leap to link drug smuggling from the Middle East,
> to Middle Eastern communities and . . . the great possibility of that funding terrorist
> conspiracies, both here and abroad . . . It's certainly known worldwide that Osama bin
> Laden was dealing with the [heroin] trade in Afghanistan . . . coming into this country.
> It's a great cash flow. It's hard to trace, and it's quick money to support terrorist activities."

A stereotype of people of the Middle East being terrorists seems to be the main
evidence used by Kushner (2004) in spreading moral panic about khat, and a

whole chapter on khat in his book on Islamic terrorist networks within the United States offers little in the way of proof for his claims. There is good evidence showing how first Siyad Barre, and more recently Southern Somali "warlords," have used khat as a "political weapon" (see Goldsmith 1997: 476). However, regarding a link of khat to international terrorism, the evidence seems tenuous. In my own experience researching the international khat trade, most profits from Kenyan khat end up being sent to help relatives in Somalia, Somaliland and Kenya (part of the huge sums sent as remittances to Somalia by the Somali diaspora each year[15]).

Furthermore, adding to the moral panic is the fear that khat consumption may spread beyond migrant communities and become popular more widely in Western countries. On the CBS report, Kushner suggests there may be an epidemic of khat consumption in Western countries similar to that of crack cocaine. This appears unlikely. After all, the niche of stimulant drugs is already filled with readily available amphetamine and ecstasy. The idea that khat consumption may spread widely originates in the conflation of khat with drugs more familiar to Westerners, ignoring the very different cultural and social aspects of khat consumption compared with typical drug consumption in the West: khat is a substance whose relatively mild effect of enhancing sociability is enjoyed over the course of a few hours and requires some effort for such effects to be felt; on the other hand, popular recreational drugs in the West tend to be those taken in pill form or snorted, thus providing an instantaneous, intense experience. Western tourists who try khat in Africa often expect quite a strong sensation given the hype that currently surrounds the substance, and its more subtle effect can lead to disappointment.

One has to learn how to be a khat chewer, just as Becker (1953) reported one has to learn how to become a marijuana smoker: one needs time and effort to overcome the bitterness and disgust at the manner of consumption and learn to appreciate its effects. To put in such effort requires a social link to chewers, and consumption among Westerners seems limited to those who have spent time in East Africa and Yemen, or who have become friendly with East Africans or Yemeni migrants in the West. Those who try khat without such social links can end up with an experience similar to that of Paul Ross, who tried the substance as part of a report for the BBC: "take it from me," he said, "khat tastes disgusting"; "My blood pressure has shot up, and I feel a bit trembly, like I've had too much coffee."[16] There are social and cultural aspects to all drug use, but Shelagh Weir rightly emphasized the greater importance of such aspects for substances as mild as khat (Weir 1985: 53):

> "[I]f the pleasurable and addictive properties of a drug are known to be powerful, considerable weight can be placed on medical, physical explanations for its consumption. But in the case of a weak substance such as qat, we must lean more heavily on social explanations for its popularity and expense."

The translation of khat into the discourse of the "war on drugs," and its conse-
quent conflation with stronger substances, means that such subtlety is hard to find
in Western commentary on the substance and has also led to much unnecessary
anxiety about a new drugs epidemic.[17]

Of course, reactions to khat in the West are not always negative. There are
those who are not deterred but are attracted to khat because of the current hype:
the sale of dried khat on many websites selling "legal highs" testifies to that.
Some such websites also sell khat seeds and plants for those wishing to cultivate
their own supply. However, khat is also sold by more "respectable" outlets, less
for its stimulating properties and more for its charm as an exotic houseplant.
Khat has not only been translated into rhetoric of the war on drugs, it has also
been admitted into the logic of garden centers, and one such center near Bristol
sells young khat plants for £5. It is popular among New Zealand gardeners too,
and even some US garden centers sell it despite haziness as to the legal status of
the plants.[18] It is interesting to note that the source of a chemical listed as Schedule I in the United States, and the source of a commodity some fear is linked to
the shady world of terrorism, graces many a suburban garden.

Conclusion

The spread of khat and khat consumers to the West has raised concerns for the
welfare of migrant communities among whom it is used. However, when what
is needed is a calm appraisal of how best to minimize any problems that arise, we
have instead much moral panic. Anguished remarks about rising consumption,
the idea that khat profits help sponsor terrorism, and the conflation of khat with
very different psychoactive substances have created an environment where khat
is demonized. Indeed, debate about khat might be said to be suffering from "Reefer
Madness Effect," after the film *Reefer Madness* that—comically in retrospect—
sought to demonize cannabis. Often evident in media discussions of the substance is the assumption that people indulge in khat because they are in the thrall
of its chemical constituents. Clearly khat's stimulant effects are important but
not to the degree that one can posit some kind of simplistic explanation for consumption of a pharmacological-determinist variety. Weir refuted two decades
ago the common view of outsiders that Yemenis spent time and money on the
substance because they are "addicted" to it, emphasizing convincingly instead the
social and cultural importance of khat sessions in Yemeni life as an explanation
for the value placed on the substance (Weir 1985). For some Somalis, Ethiopians, and Yemenis in the West, khat sessions maintain this significance. Indeed,
this significance becomes more pronounced in the context of migration: the familiar taste and effect, the convivial atmosphere of the *mafrish*, the stories told,
the advice given, the shared reminiscences of home all increase in meaning when

home is so far away. Despite the moral panic, khat consumption in the West is not going to disappear any time soon for these very reasons.

Acknowledgements

My interest in khat began with my Ph.D. research into the farming, trade, and production of khat in Kenya, for which I was kindly sponsored first by a grant from the Carnegie Foundation for Scottish Universities, and later by an Economic and Social Research Council studentship. Further research was undertaken in Kenya, Europe, and North America during a research assistantship on a project (*The Khat Nexus*) sponsored jointly by the ESRC and the Arts and Humanities Research Council. Special thanks to Nicholas Mwambia, Abdikadir Araru, Michael Carrier, Douglas Webster, Pauline Whitehead, Paul Baxter, Roy Dilley, Paul Goldsmith, David Anderson, Susan Beckerleg, and Axel Klein.

Notes

1. As well as a comprehensive survey of the literature on khat, this chapter is based on long-term anthropological fieldwork on khat in Kenya and the United Kingdom (beginning in 1999), and also on research trips to the United States. My main methodology has been participant observation combined with key informant interviews. In both Kenya and the United Kingdom, khat consumers and others concerned with the substance have treated my presence and questions with great tolerance despite its controversial status. In the United Kingdom I made some contacts amongst Somali traders and consumers on my own initiative, but have also been introduced to several others through UK-based Somali and Kenyan friends, visiting many chewing sessions in the process.
2. The ACMD's recommendations are published on the Home Office website: http://www .drugs.gov.uk/publication-search/acmd/khat-report-2005/
3. Axel Klein once humorously remarked at a meeting on khat how Somalis had "hijacked" the issue. All reports in the media nowadays appear related to Somali consumption alone. Debate on khat does disproportionately focus on Somalis, and I fear that in concentrating on them once more I am not helping to redress the balance. For an earlier look at khat consumption amongst Yemenis in Britain, see Halliday (1992: 122 ff.).
4. Compare khat's link back to home when used in the diaspora with the way young Somalis talk of their hopes and dreams of success in the west at khat sessions held during their premigration wait (Rousseau et al. 1998).
5. Much opposition to khat amongst Somalis is also religious in nature as those embracing a more conservative Islam often see khat as *haram* (forbidden).
6. Khat in the United Kingdom does seem much better value than that other social drug, alcohol.
7. Clearly many chew because they positively like chewing with friends, rather than as a means of seeking escape from an uncomfortable reality.
8. See Weir (1985: chapter four) for a good account of the attitudes of international agencies and other outsiders to what is usually dubbed the "qat problem."
9. There is much haziness, however. The direct mention of miraa/khat was only made in the Federal Register, and only cathine and cathinone were added by name to the actual schedules. This has opened the way for those caught with the substance to mount the defense that they were not given due warning of miraa's illegality, because how could a layman know that the chemicals listed in the schedules were the constituents of miraa.

10. The US Customs Service has made a novelty badge for their "khat-busting" officers. It features a fierce looking bulldog, two bundles of khat with no-entry signs superimposed, and the motif: "US Customs Service Khat Busters."
11. (Accessed 26th March 2005): http://www.borderpatrol.com/borderframe901.htm
12. (Accessed 31st March, 2005): http://www.rcmp.ca/crimint/drugs_2003_e.htm#khat
13. See: http://www.stormfront.org/forum/showthread.php?t=211785 (accessed 2 October 2005)
14. See: http://cbs11tv.com/investigations/local_story_061103132.html (accessed 2 June 2005)
15. Pérouse de Montclos has calculated that remittances sent back to Somalia by Somalis in the diaspora add up to over $140 million (United States) each year (2002).
16. From the BBC news program "Inside out," http://www.bbc.co.uk/print/insideout.
17. Fortunately, in the case of the United Kingdom, the Advisory Council on the Misuse of Drugs was not caught up in the hype, and avoided moral panic and demonization in their recent report, recommending that educational measures, rather than a ban, are the way forward in alleviating problems associated with khat consumption.
18. From reading the Federal Register entry on cathinone, one would assume the plants to be illegal, but representatives of the Drug Enforcement Administration whom I met in Washington D.C. seemed to think it was perfectly legal to grow khat as an ornamental plant.

References

Ahmed, A.G., and E. Salib. 1998. "The khat users: A study of khat chewing in Liverpool's Somali men." *Medicine, Science and the Law* 38: 165–169.

Alem, A., and T. Shibre. 1997. "Khat induced psychosis and its medico-legal implication: A case report." *Ethiopian Medical Journal* 35: 137–141.

Al-Motarreb, A., K. Baker, and K.J. Broadley. 2002. "Khat: Pharmacological and medical aspects and its social use in Yemen." *Phytotherapy Research* 16: 403–413.

Becker, H.S. 1953. "Becoming a marijuana user." *American Journal of Sociology* 59: 235–242.

Browne, D.L. 1991. "Qat use in New York City." NIDA Research Monograph 105: 464–465.

Carrier, N. 2003. "The social life of miraa: The farming, trade and consumption of a plant stimulant in Kenya." Ph.D. Dissertation. St Andrews: University of St Andrews.

———. 2005a. "The need for speed: Contrasting timeframes in the social life of Kenyan miraa." *Africa* 75: 539–558.

———. 2005b. "Miraa is cool: The cultural importance of miraa (khat) for Tigania and Igembe youth in Kenya." *Journal of African Cultural Studies* 17: 201–218.

Critchlow, S. 1987. "Khat-induced paranoid psychosis." *British Journal of Psychiatry* 150: 247–249.

Crombie, L. 1980. "The cathedulin alkaloids." *Bulletin on Narcotics* 32: 37–50.

Kennedy, J. 1987. *The flower of paradise.* Dordrecht: Reidel.

El-Sohl, C.F. 1991. "Somalis in London's East End: A community striving for recognition." *New Community* 17: 539–552.

Gebissa, E. 2004. *Leaf of Allah: Khat & agricultural transformation in Hararge, Ethiopia 1875–1991.* Oxford: James Currey.

Griffiths, P. 1998. *Qat use in London: A study of qat use among a sample of Somalis living in London.* Drugs Prevention Initiative, Paper 26. London: Home Office.

Griffiths, P., M. Gossop, S. Wickenden, J. Dunworth, K. Harris, and C. Lloyd. 1997. "A transcultural pattern of drug use: Qat (khat) in the United Kingdom." *British Journal of Psychiatry* 170: 281–284.

Goldsmith, P. 1997. "The Somali impact on Kenya, 1990–1993: The view from outside the camps." In *Mending rips in the sky: Options for the Somali communities in the 21st century,* eds. H.M. Adam and R. Ford Lawrenceville. NJ: The Red Sea Press.

Halbach, H. 1972."Medical aspects of chewing khat leaves." *Bulletin of the World Health Organization* 47: 12–19.

Halliday, F. 1992. *Arabs in exile.* London: IB Tauris.

Kushner, H. 2004. *Holy war on the home front: The secret Islamic terror network in the United States.* New York: Sentinel.

McGowan, B. 1999. *Muslims in the diaspora: The Somali communities of London and Toronto.* Toronto: University Press.

Nabuzoka, D., and F.A. Badhadhe. 2000. "Use and perceptions of khat among young Somalis in a UK City." *Addiction Research* 8: 5–26.

Nencini, P., M.C. Grassi, A.A. Botan, A.F. Asseyr, and E. Paroli. 1989. "Khat chewing spread to the Somali community in Rome." *Drug and Alcohol Dependence* 23: 255–258.

Numan, N. 2004. "Exploration of adverse psychological symptoms in Yemeni khat users by the Symptoms Checklist-90 (SCL-90)." *Addiction* 99: 61–65.

Patel, S.L., S. Wright, and A. Gammampila. 2005. *Khat use among Somalis in four English cities.* Home Office Online Report 47/05. Retrieved 16th October 2005 from http://www .homeoffice.gov.uk/rds/pdfs05/rdsolr4705.pdf

Pérouse de Montclos, M-A. 2002. "A refugee diaspora: When the Somali go west." In *New African diasporas,* ed. K. Koser. London: Routledge.

Rousseau, C., T.M. Said, M.-J. Gagné, and G. Bibeau. 1998. "Between myth and madness: The premigration dream of leaving among young Somali refugees." *Culture, Medicine and Psychiatry* 22: 385–411.

Stevenson, M., J. Fitzgerald, and C. Banwell. 1996. "Chewing as a social act: cultural displacement and *khat* consumption in the East African communities of Melbourne." *Drug and Alcohol Review* 15: 73–82.

Weir, S. 1985. *Qat in Yemen: Consumption and social change.* London: British Museum Press.

Zaghloul, A., A. Abdalla, H. El-Gammal, and H. Moselhy. 2003. "The Consequences of Khat use: A review of literature." *European Journal of Psychiatry* 17: 77–86.

Chapter 10

Traditional Health Care and Food and Medicinal Plant Use among Historic Albanian Migrants and Italians in Lucania, Southern Italy

Cassandra L. Quave and Andrea Pieroni

Introduction

In this chapter, we explore the scientific questions related to the issue of traditional health care and food practices in a community founded by a historical ethnic group of Albanians who migrated to southern Italy during the fifteenth century and among an autochthonous south Italian community. In doing this, we employ the use of food and medicinal plants as a lens for better understanding the ethnomedical practices, including the perception and use of *medicinal foods,* which we recorded in that area during four years of fieldwork. The specific aim of our reflections is a cross-cultural comparison of traditional medical practices, botanical remedies, and foods in two communities: Ginestra/Zhurë, inhabited by Arbëreshë Albanians, and Castelmezzano, inhabited by autochthonous south Italians.

The questions we will address here are:

- Do two communities who share a similar flora and environment, but which are of different cultural or ethnic origin, utilize the same environmental resources in their food and medical practices?
- Does the culture with a longer history of occupation in that environment maintain a greater base of knowledge regarding the use of the flora for its dietetic and medicinal applications?

Here, we present our findings of the present day use of wild and semi-cultivated plants in the traditional medical and food practices of these two communities for comparison. In addition, we also reflect on the sociocultural issues of acculturation, and the impact that such phenomena have on local food and medical traditions.

Ethnographic Background

In this study, the traditional use of medicinal and food plants in two communities in the Basilicata region (also named Lucania) of southern Italy were compared (figure 10.1).

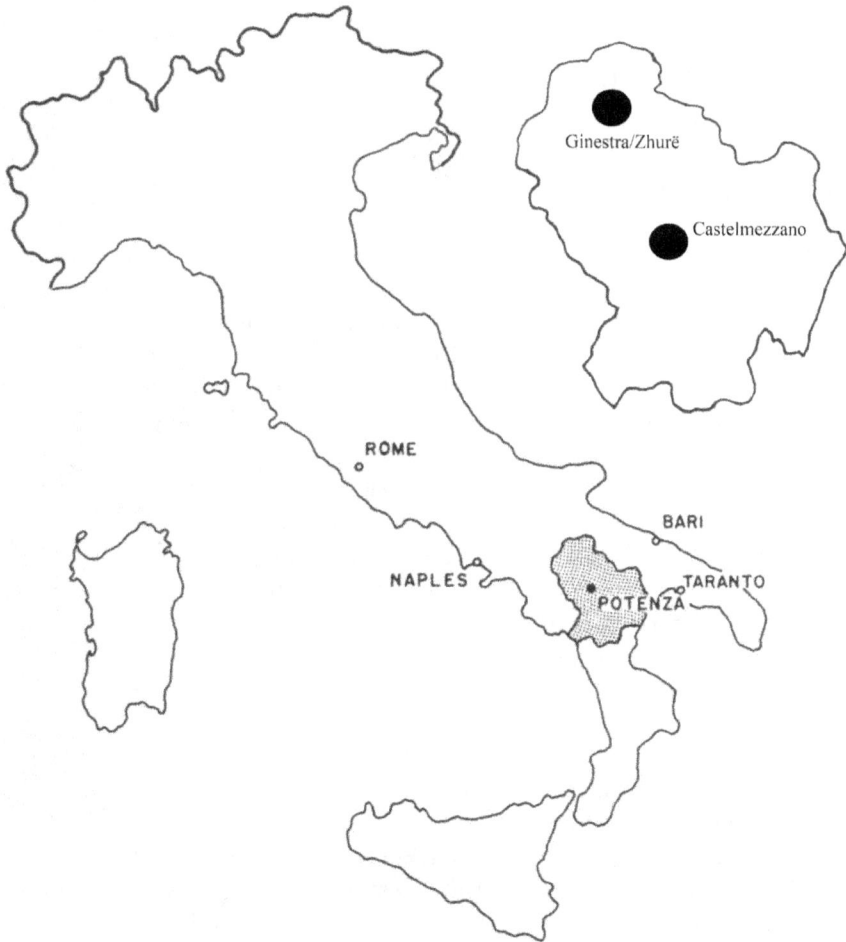

Figure 10.1. *Location of the selected communities*

The Italian National Statistical Institute (ISTAT 2000) reports that Basilicata is the Italian province with the lowest percentage of urban population (17 percent, calculated in 1997–1999), the highest life expectancy (75.7 years, calculated in 1991–1995), and the lowest utilization of allopathic medical services (23.9 percent among men and 32.5 percent among women, calculated in 1997).

The two communities selected for this study share similar socioeconomic and demographic characteristics (table 10.1), but different ethnic origins: Ginestra/Zhurë, a historic Arbëreshë Albanian community located in the Vulture region in northern Lucania; and Castelmezzano, an autochthonous Italian community located in the Dolomiti Lucane area of central Basilicata. The two communities are separated by a distance of 60 kilometers, and there is no regular exchange between the two populations at present. Moreover, it is unlikely that they have ever been in contact in the past.

2.1 GINESTRA/ZHURË: THE ARBËRESHË IN LUCANIA

Ginestra/Zhurë (Zhurë in Arbëresh) has approximately 700 inhabitants and is located in the northern part of Basilicata in a territory characterized by the dormant volcano, Monte Vulture. The majority of members of this community are descendants of Arbëreshë Albanians, who arrived in the Monte Vulture region from Albania during the second half of the fifteenth century (Dessart 1982). This historic mass immigration of Albanians to southern Italy from the fifteenth to eighteenth centuries was instigated by the Ottoman Turk invasion of their country. Years before the exodus of this populace, the Albanian general Skanderberg had come to the aid of King Alfonso I of Naples in his battle with the King of Sicily. Consequently, Albanian soldiers had set up military outposts scattered throughout southern Italy. It was upon Skanderberg's death in A.D. 1468 that the last resistance to the Turkish invasion fell apart. Using Skanderberg's military connections in southern Italy, his son led the first Albanian refugees to the region. Today, approximately 80,000 descendants of these original migrant populations live in various Arbëreshë communities across southern Italy (Grimes 2000). A small percentage of these descendants remain bilingual in Italian and Arbëreshë. For example, in the community of Ginestra/Zhurë, it can be estimated that only 10–15 percent of the population is fluent in Arbëreshë Albanian (Pieroni et al. 2002b).

The Arbëreshë Albanian language belongs to the Tosk Albanian subgroup of Albanians, which represents the only surviving language from the ancient Paleo-Balkan group (Illyrian, Messapic, and Thracian) of the Indo-European family (Grimes 2000). In the Redbook of Endangered Languages (UNESCO), Arbëreshë Albanian has been classified as an "endangered language" (Salminen 1999), and in December 1999, together with eleven other non-Italian speaking groups, the Arbëreshë obtained official recognition as an "historical ethnic minority" from the Italian Parliament. This should ensure a future for the integration of their language in local schools and should also give the Arbëreshë the legal right

Table 10.1. Summary of the Ethnographic, Socioeconomic, and Medical Frameworks of Ginestra/Zhurë and Castelmezzano

	Ginestra/Zhurë	Castelmezzano
Eco-geographic characteristics		
Altitude	564 m	750 m
Climate	mild winter	cold winter
Natural landscape	countryside	mountains
Ethno-demographic data		
Population (1998)	730	840
Change of population (1991–1998)	−6.3%	−9.5%
% population over 65 (1998)	25.6%	26.5%
Ethnic group	Albanians (Arbëreshë)	South Italians
Historical immigration flows from Albania	1470–1478	none
Current Albanian speakers (bilingual)	10–15%	none
Emigration flows during the period 1950–1980 (to northern Italy and/or Central Europe)	+++	+++
Immigration flows of foreign newcomers in the last 10 years	+ (from Albania and Poland)	+ (from Ukraine)
Economic features		
Commercially important crops	olive trees vine durum wheat	durum wheat
Animal breeding (sheep, goats, *Podolica* cows)	+ (sh and go)	++ (sh, go, and co)
Labor in nearby factories	+++	+
Labor in nearby services	+	++
Household incomes dependent on public pension payments to the oldest members of the households	+	++
Socio-pharmaceutical and medical frameworks		
Presence of stationary General Practitioner	yes	yes
Presence of a General Practitioner night service	yes	yes
Presence of a community pharmacy	yes (part-time)	yes (full-time)
Closest hospital	4 km far	32 km far
Specific traditional healers using medicinal plants	none	none
Character of traditional phytotherapy (and zootherapy)	household PHC	household PHC

to use their idiom in official acts of administration. In addition, this government recognition should also be helpful in measures for sustaining cultural initiatives concerning the defense of their heritage.

The Arbëreshë living in the territory of Monte Vulture are quite isolated from most other Arbëreshë communities, which are concentrated in Calabria and Sicily, as well as in a few other ethnic isles in southern Basilicata, Apulia, Campania, Molise, and Abruzzo. The Arbëreshë of Vulture are concentrated in Ginestra/Zhurë and the nearby communities of Barile/Barilli (ca. 3,400 inhabitants) and Maschito/Masqiti (ca. 1,900 inhabitants). The local economy was sustained originally by small-scale pastoralism and agriculture. At present, however, the cultivation of olive trees (*Olea europaea*), a local variety of grape (*Vitis vinifera* var. *aglianico*), and durum wheat (*Triticum durum*), as well as a car factory in the nearby center of Melfi (open for the last ten years), represent the primary economic assets for the community.

In Ginestra/Zhurë, a distinct cultural gap exists between generations, and today only the oldest subset of the community is able to actively speak Arbëreshë Albanian. The majority of the mid-aged population (35–55 years) can recall some words and basic customs of their Arbëreshë history but do not incorporate these facets of traditional life into their daily activities. The impact of "modernization," or transition into the mainstream Italian culture, is most apparent in the youngest subset of the population. This group, for the most part, has abandoned the traditional agro-pastoralist way of life as a primary source of income and is sustained instead by factory and service labor (Pieroni and Quave 2005).

2.2 Castelmezzano: Autochthonous Italians of the Lucania Dolomites

Although there are several Italian villages in the Vulture area, their close proximity to Arbëreshë communities has fostered a flux of people and traditional knowledge with the Arbëreshë. In addition, while many of these communities, such as Melfi and Rionero, are acculturated today to southern Italian models, they share a history of Albanian occupation. In an effort to control for such confounding variables as informational flux between study sites and historical ethnic hybridization, an isolated community in the Dolomiti Lucane was selected for intercultural comparison in this study.

Castelmezzano has approximately 840 autochthonous Italian inhabitants and is located in central Lucania at a distance of ca. 60 kilometers from Ginestra/Zhurë in a mountainous terrain bordering the Basento River Valley. The history of Castelmezzano has been characterized by Norman (eleventh century), Swabian (thirteenth century), and Spanish Bourbon (fifteenth century) domination. The economy is primarily based on small-scale agriculture, including the management of sheep, goats, and the Podolica breed of cattle for making cheese. However, like Ginestra/Zhurë, the younger subset of the population relies on service and industrial labor, while the older generations are involved in an agro-pastoralist economy.

A scheme of the geographic, ethnic, demographic, socioeconomic, and socio-medical characteristics of Castelmezzano and Ginestra/Zhurë is reported in table 10.1.

Field Methods

Fieldwork was conducted in Ginestra/Zhurë during the periods of April–June 2000 and March–July 2001, and during three other weeks in August and November 2000; in Castelmezzano, fieldwork was conducted during the periods of March–June 2002, October 2002, March–June 2003, and two weeks of September 2003. Information regarding the use of traditional medicines and the preparation of traditional foods was gathered using a variety of standard ethnographic techniques, including participant observation, focus groups, and semi-structured interviews with approximately 240 persons selected using snowball techniques.

Traditional knowledge regarding plant use was assessed using standard ethno-biological and cognitive anthropological analyses for a better understanding of folk-taxonomical hierarchies and systems, and for studying the most frequently cited plants by free-listing, triad tests, pile-sorts, and through construction of a consensus index (Berlin et al. 1966; Berlin 1992; D'Andrade 1995; Alexiades and Sheldon 1996; Cotton 1996). Based on the responses to these interviews, a consensus level of 66 percent of informants reporting a plant's use(s) was employed in the construction of our summary tables. These consensus-based tables provide a basic means for comparison of the botanical practices of the two study communities.

Field methods employed followed strictly the ethical guidelines set forth by the American Anthropological Association, Italian Association for Ethnological Studies (AISEA), and the International Society of Ethnobiology. Prior Informed Consent (PIC) was requested and obtained verbally before each interview, and specific permission for the use of any recording devices (audio or visual) for documentation of ethnobotanical practices was obtained before use.

For each plant or plant product quoted by an informant, a botanical specimen was collected, its identification was confirmed by the same informant, and taxonomic identification followed the standard botanical work "Flora d'Italia" (Pignatti 2002). Voucher specimens of all non-domesticated plants are deposited at the Herbarium of the Laboratory of Pharmacognosy at the School of Life Sciences of the University of Bradford (Herbarium Code: BRAD).

Results and Discussion

Traditional Medicine in Present Day Lucania

It has become evident from our field studies that traditional medicine (TM) in the two chosen areas is in a state of rapid decline. Most of the plant remedies recorded are not used at present. In Ginestra/Zhurë, only 40 percent of the quoted uses were also directly observed during field research, while an even lower proportion of 31 percent was observed in Castelmezzano. Complementary medicine in both centers demonstrates a household character and is intended mainly as a mode of primary health care (PHC), which is perceived to have a preventive action or ability to heal minor illnesses.

No specific traditional healers, reputed to deal solely with the use of natural remedies (plants), can be found anymore in the two communities. Yet these healers still exist in the collective memory of the population. For example, in Castelmezzano, there is still a clear remembrance of a female healer ("Zia Teresa 'a Lia," alias Teresa Vertino), who died one year before our field study began, and who reportedly healed many illnesses with the use of herbs or mixtures of them. In addition, many people still remember the most famous "healer" of southern Italy (well described by the Italian anthropologist De Martino in 1959: the "Mago Ferramosca" (alias Giuseppe Calvello from Pietrapertosa) who died in 1968 in Castelmezzano, and from whom Zia Teresa learned the practice of herbal healing.

These two cases, however, are isolated examples: both in Ginestra/Zhurë and in Castelmezzano traditional phytotherapy was and is managed normally by women within the household. However, the role of these women as medical caregivers is threatened nowadays since new generations have lost most of the traditional knowledge concerning plant and folk medicines (Pieroni 2003).

On the other hand, small commercial pharmacies in Ginestra/Zhurë and Castelmezzano provide a wide variety of pharmaceuticals from today's mainstream European market, most of which are bought by locals only via prescriptions from a physician. Self-medication with over-the-counter pharmaceuticals is still very limited. These local pharmacies also provide a few new modern phytotherapeuticals and herbal remedies in the form of "nutraceuticals," or herbal supplements. These commercial phytotherapeutical products, however, represent only a very small portion of Complementary and Alternative Medicines (CAMs) used in the community when compared with the traditional phytotherapeutical means (decoctions) administered in households. These commercialized products tend to become important primarily for new couples, young women, and other locals who are not connected with familiar networks, in which their older relatives would normally provide traditional herbal drugs.

In Castelmezzano, other valued commercial plant-derived nutraceuticals and beverages are sold in the local pub to young people in the form of beverages containing gingko, ginseng, and green tea extracts, normally traded by northern Italian companies. Consciousness of the commonality in plant derivatives between commercial products and traditional remedies among these young consumers is very low. These products are consumed instead in large part due to popular culture, which is strongly reinforced by extensive television advertising directed at this age group, and in some small part, due to the belief that such beverages are "good for you."

This scenario is one example of the complex cultural transition phase, which both communities are faced with at present. This transition is quickly leading to a loss of the last remnants of these old agro-pastoral societies, and acculturation to mainstream metropolitan Italian customary trajectories.

Folk Ritual Medical Practices

Although it is evident from our field studies that traditional medicine is in a state of rapid decline in these communities, the population is still dependent on some of its forms (Pieroni and Quave 2005).

The Arbëreshë community of Ginestra/Zhurë displays a greater reliance on ritual-based folk healing, which involves the use of a healer who functions as a medium between the mythic, or perceived universe, and the patient's state of disease. The approach to treating illness within this medical framework involves the use of distinct oral formulas and light massage in a ceremonial format. We have documented twenty-one folk illnesses in Ginestra/Zhurë that are treated through such ritual means. Although the incorporation of medicinal plants into these therapeutic sessions is often important, the invocation of holy entities plays a central role. In part, healers gain authority by drawing on religious symbols and ideologies. The details of this facet of Arbëreshë traditional medicine have been documented in previous publications (Quave and Pieroni 2002; Quave and Pieroni 2005). The practice of ritual healing in the Italian community of Castel-mezzano was much less evident: only one folk illness, *malocchio* (evil eye), was recognized and treated through ritual means (table 10.2).

The frequent occurrence of evil eye and other psychosomatic illnesses in Ginestra/Zhurë could also be an indicator of high social stress in the community. An assessment of such illnesses, which are strongly correlated with sociopsychological components of causality, could explain how the complex cultural dynamics to social stress—motivated by acculturation processes in this case—are expressed in community wide psychopathologies. The community of Ginestra/Zhurë has a long history of exposure to extraordinary social stressors. Its first inhabitants were Albanian refugees/shepherds, who escaped the political turmoil of their homeland (due to the invasion of the expanding Ottoman empire) in the second half of the fifteenth century and moved to a sort of "no-man's-land," without any familiarity with the tradition of dominant classes/landowners, as was typical of the South Italian socioeconomic framework of the time. It is not by chance then, that in Ginestra/Zhurë still today, the discourse of climbing social status is tremendously central and difficult to explain if one considers the real income of most of the households. Less decisive in this context of social status, however, is

Table 10.2. Illnesses Managed by Magic Rituals in Ginestra/Zhurë and Castelmezzano

Illness	Symptoms	Ritual healing practices recorded in the Albanian (Al) or in the South-Italian community (It)
Acqua dalla bocca ("water in the mouth")	dry, cracked corners of mouth from excessive drooling of saliva	Al
Acqua nel pipi ("water in the penis")	inflammation of the penis	Al
Cigli alla testa (migraine)	sharp pin-like pain runs from the front to back down on the top of head	Al
Fuoco di Sant'Antonio ("Saint Anthony's fire")	dermatitis: pronounced, red, round inflammations with fluid on the skin (eruptions are more pronounced than in *Fuoco morto*)	Al
Fuoco morto ("dead-fire illness")	dermatitis: pronounced, red, round inflammations with fluid on the skin	Al
La serra (fallen fontanelle)	fallen fontanelle	Al
La zilla (head lice)	Head lice; scabies	Al
Mal d'arco ("rainbow-illness")	jaundice, hepatitis symptoms	Al
Mal di denti (toothache)	toothache	Al
Mal di gola (sore throats)	red inflamed throat	Al
Mal di pancia (abdominal pains)	abdominal pain and gas; constipation	Al
Mal di testa (headache)	posterior pain at the base of skull	Al
Mal vint ("wind-illness")	dermatitis: small, round, red inflammations of the skin	Al
Malocchio (evil eye)	frontal headache with pain behind the eyes	Al and It
Nervi accavallati ("crossed nerves")	nerve/muscular pain	Al
Occhi secchi ("dry eyes")	dry, red, inflamed eyes	Al
Orecchioni (mumps)	enlarged lymph nodes of the neck	Al
Pelo alla menna ("Breast hair illness")	mastitis: red, inflamed breast with fever, unable to give milk	Al
Risibola (erysepalas)	region of isolated dark, hardened, withdrawn skin	Al
Sangue dal naso (nosebleed)	nosebleed	Al
Vermi ("worms")	helminthiasis, weakness	Al

the concept of "liminals" (those who are at the borders of the wide "middle rural class," both near the lower and within the upper classes), as proposed in an analysis of the same phenomenon by Galt on Pantelleria Isle in Sicily at the end of the 1960s (Galt 1982).

Shifts in medical dominance in Italy, noted most prominently in the changing paradigms of social medicine in the country over the past three decades, may be another factor influencing communities like Ginestra/Zhurë, which are already experiencing more social stressors, to cling more tightly to traditional medicine than other autochthonous Italian populations. Furthermore, as suggested in a study on traditional healers in South America, social factors underlying these traditional medical practices, such as a familiar historical background in traditional medicine, play a crucial role in the transmission and survival of medical traditional medicine in communities (Vandebroek et al. 2004).

The persistence of ritual-based healing amongst the Arbëreshë may be indicative of the effect that complex sociocultural changes—such as those associated with acculturation to mainstream Italian life—can have on an ethnic community. The frequent occurrence of evil eye and other psychosomatic illnesses in this community certainly influences their social and medical frameworks. Thus, in a certain sense, culture mediates the practice of traditional medicine here—in a way, which excludes, to some degree, many of the medicinal products that are common to Italian communities positioned in similar ecological landscapes.

Medicinal and Wild/Non-Cultivated Food Plants Uses

We have compiled a list of plants used for medicine (table 10.3) and of the wild plants used for food (table 10.4) by the Italians and Albanians in the study communities. Knowledge pertaining to the use of medicinal plants is clearly more prominent in the Italian community (see figure 10.2), while the traditional knowledge regarding wild food plants is slightly higher among Albanians.

MEDICINAL PLANTS

While the practice of ritual healing in Ginestra/Zhurë is still commonplace, some illnesses are recognized as being outside of the realm of ritual or spiritual healing. These are often approached through the application of phytotherapeutic means. Yet, the Arbëreshë report a much lower number of medicinal plants unique to their pharmacopoeia—representing only 9 percent of the 101 medicinal plants reported (figure 10.2). The autochthonous Italians, on the other hand, maintain knowledge of 50 medicinal plants (table 10.3) unique to their pharmacopoeia, representing 54 percent of all medicinal plants reported (figure 10.2).

Although the folk pharmacopoeia of Italians in Castelmezzano is considerably much larger than that of the ethnic Albanians in Ginestra/Zhurë, they do share some basic similarities. Seven of the ten most quoted medicinal species, as reported

Table 10.3. Botanicals Used as Medicines in the Studied Communities

Family	Species	Part(s) Used	Medicinal Uses	Albanians	Italians
Adiantaceae	*Adiantum capillus-veneris* L.	ap	enhance uterine contractions during childbirth	+	
	Ceterach officinarum DC.	ap	to eliminate renal calculus; shoulder pains	+	+
Apiaceae	*Conium maculatum* L.	wp	anti-warts		+
	Petroselinum crispum (Mill.) Nyman ex A.W. Hill	ap	abortive; to treat insect bites		+
Araceae	*Arum italicum* Mill.	sa	anti-warts		+
Aspleniaceae	*Asplenium trichomanes* L.	ap	enhance uterine contractions during childbirth	+	
Asteraceae	*Achillea millefolium* L.	ap	hemostatic, diuretic		+
	Anthemis altissima L.	ap	digestive		+
	Cichorium intybus L.	wh	depurative		+
	Cynara cardunculus L.	ap	anti-rheumatic		+
	Cynara cardunculus ssp. *scolymus* (L.) Hayek	le,fl	digestive; liver depurative	+	+
	Erigeron acer Bivona	ro	against toothache, bruises, and arthritis		+
	Lactuca sativa L.	le	against gingival abscess, toothache, and sore throat		+
	Matricaria recutita L.	ft	anti-hypertensive; anti-inflammatory for eyes; against sore throat and bronchitis; intestinal depurative; digestive; sedative	+	+
	Santolina chamaecyparissus L.	ap	anti-tussive		+
	Senecio vulgaris L.	ap	skin anti-inflammatory		+
	Silybum marianum (L.) Gaertn.	ap	laxative		+
	Sonchus asper (L.) Hill & *S. oleraceus* L.	le	anti-gastritis; anti-afta	+	+
	Tussilago farfara L.	le, ro	diuretic	+	+
Boraginaceae	*Borago officinalis* L.	ap	postpartum depurative; galactagogue; against sore throat	+	+
Brassicaceae	*Armoracia rusticana* P. Gaertn., B. Mey. & Scherb.	ro	anti-rheumatic		+
	Brassica oleracea L.	le	against mastitis and shoulder pains		+
	Diplotaxis tenuifolia (L.) DC.	le	against muscular pains (esp. in the shoulders)		+

Caprifoliaceae	*Sambucus ebulus* L.	ap	anti-rheumatic; diaphoretic	+	+
	Sambucus nigra L.	le, fl	against sore throat	+	+
Crassulaceae	*Sedum rupestre* L.	ap	diuretic		+
	Sedum telephium L.	le	anti-warts		+
	Umbilicus rupestris (Salisb.) Dandy	le	suppurative; against carbuncles and skin inflammation	+	+
Cucurbitaceae	*Ecballium elaterium* (L.) A. Rich.	fr	against toothache; antiseptic and vulnerary	+	+
Euphorbiaceae	*Euphorbia cyparissias* L.	le	anti-warts		+
	Euphorbia helioscopia L.	le	male aphrodisiac (penile vasodilator)		+
	Mercurialis annua L.	ap	laxative	+	
Fabaceae	*Glycyrrhiza glabra* L.	ro	against sore throat and anti-tussive	+	+
	Lupinus albus L.	se	anti-diabetes		+
	Robinia pseudoacacia L.	fr	anti-bronchitis		+
	Spartium junceum L.	sa	anti-warts		+
Gentianaceae	*Centaurium erythraea* Rafn.	ap	anti-fever		+
Hypericaceae	*Hypericum hircinum* L.	ap	against bronchitis		+
Hypolepidaceae	*Pteridium aquilinum* (L.) Kuhn	rh	against non-specific pains	+	
Lamiaceae	*Ballota nigra* L.	le	diuretic; hemostatic	+	+
	Marrubium incanum Desr. & *M. vulgare* L.	ap	diuretic; digestive; anti-malarial; against cysts; panacea	+	+
	Mentha spicata L.	le	against stomachache		+
	Ocimum basilicum L.	le	against headache		+
	Origanum heracleoticum L.	ft	anti-tussive; against toothache	+	+
	Rosmarinus officinalis L.	le	against sore throat and stomachaches		+
	Salvia argentea L.	le	hemostatic		+
	Salvia officinalis L.	le	against sore throat and headaches	+	+
	Teucrium chamaedrys L.	ap	anti-malarial		+
Lauraceae	*Laurus nobilis* L.	le	digestive; anti-stress	+	+
Liliaceae s.l.	*Allium cepa* L.	bu	heal purulent skin abscesses (caused by thorns), galactagogue, anti-bruises	+	+
	Allium sativum L.	bu	anti-hypertensive, heal insect bites, vermifuge, anti-wart, skin anti-inflammatory	+	+
	Asparagus acutifolius L.	sh	diuretic		+
	Leopoldia comosa (L.) Parl.	bu	anti-fever	+	
	Ruscus aculeatus L.	sh	liver depurative		+

(continued)

Table 10.3. Continued

Family	Species	Part(s) Used	Medicinal Uses	Albanians	Italians
Malvaceae	*Malva sylvestris* L.	ap, ft	mild laxative; against menstrual pains, sore throat and bronchitis; intestinal depurative	+	+
Moraceae	*Ficus carica* L.	fr,sa	against sore throat and bronchitis; intestinal depurative; heal insect bites; anti-warts	+	+
	Morus alba L. & *M. nigra* L.	le, st	against sore throat & bronchitis		+
Oleaceae	*Fraxinus excelsior* L.	le	against gastritis		+
	Olea europaea L.	ap	hepatoprotective; against stomachache	+	+
Papaveraceae	*Papaver rhoeas* L.	fl	mild sedative for children	+	
	Papaver somniferum L.	fr, se	tranquilizer; sedative; against toothaches	+	+
Plantaginaceae	*Plantago lanceolata* L. & *P. major* L.	le	suppurative		+
Poaceae	*Agropyron repens* L.	rh	diuretic	+	+
	Arundo donax L.	cm	hemostatic	+	+
	Avena sativa L.	se	reconstituent for small children against sore throat	+	+
	Hordeum vulgare L.	se	to heal sore throat and bronchitis; reconstituent for small children and elderly	+	+
	Triticum aestivum L. & *T. durum* Desf.	se	anti-tussive; against sore throat	+	+
	Zea mays L.	se, sti	antiseptic; diuretic; reconstituent	+	+
Primulaceae	*Cyclamen hederifolium* Aiton.	tu	anti-warts		+
Ranunculaceae	*Clematis vitalba* L.	fr	heal mouth inflammations		+
Rhamnaceae	*Ziziphus jujuba* Mill.	fr	against sore throat and cough	+	
Rosaceae	*Agrimonia eupatoria* L.	ap	prevent feet from sweating		+
	Crataegus monogyna Jacq.	fl	tranquilizer; enhances circulation		+
	Malus domestica Borkh.	fr	anti-tussive	+	+
	Potentilla reptans L.	ap	anti-hypertensive; anti-malarial; anti-rheumatic		+
	Prunus domestica L.	fr	laxative		+
	Prunus dulcis (Miller) D.A. Webb.	ep, se	against intestinal pains (in children); against sore throat	+	+

	Prunus spinosa L.	fr	hepatoprotector		+
	Pyrus communis L.	fr	depurative; mild laxative	+	+
	Rosa canina L.	le, fl, fr	against stomachache; antidepressant; diuretic; heal insect bites; against the evil eye	+	+
	Rubus ulmifolius Schott.	le	diuretic; against carbuncles and purulent skin abscesses	+	+
	Sorbus domestica L.	fr	anti-diarrhea		+
Rubiaceae	*Galium album* Mill. & *G. verum* L.	le	heal wounds and gingival inflammations		+
Rutaceae	*Citrus limon* (L.) Burm. f.	fr	anti-diarrhoeal		+
	Citrus sinensis (L.) Osbeck	ep	against sore throat and cough		+
	Ruta graveolens L.	ap	digestive; anti-helmintic; against muscular pains	+	+
Scrophulariaceae	*Linaria vulgaris* Mill.	ap	against stomachache		+
	Scrophularia canina L.	ap	anti-rheumatic; against muscular pains		+
	Verbascum thapsus L.	le	anti-tussive	+	
	Veronica beccabunga L.	ap	diuretic		+
Solanaceae	*Capsicum annuum* L.	fr	anti-hypertensive; anti-rheumatic; anti-fever; against evil eye	+	+
	Lycopersicon esculentum Mill.	fr, pe	diuretic; against sore throat and bronchitis	+	+
	Solanum nigrum L.	fr	against toothache		+
	Solanum tuberosum L.	tu	burn treatment; emollient for eyes		+
Tiliaceae	*Tilia cordata* Mill.	fl	heal body tremors		+
Ulmaceae	*Ulmus minor* Mill.	ga	anti-bruises; relieve muscular pain	+	+
Urticaceae	*Parietaria judaica* L.	ap	diuretic; against intestinal pains; postpartum depurative	+	+
	Urtica dioica L.	le	digestive		+
Vitaceae	*Vitis vinifera* L.	sh,sa,fr	against insect bites; galactagogue; against sore throat and eye inflammation; partum enhancer; anti-tussive; anti-gastritis; anti-fever; anti-rheumatic; anti-diarrhoeal	+	+

Part(s) used: ap: aerial parts; bu: bulbs; cm: cambium membrane; ep: fruit epicarps; fl: flowers; fr: fruits; ft: flowering tops; ga: galls; le: leaves; ls: leaf stalks; pe: fruit peduncles; pf: pseudo-fruits; re: flower receptacles; rh: rhizomes; ro: root; sa: sap; se: seeds; sh: shoots; st: stems; sti: stigma; tu: tubers; uf: unripe fruits; wh: whorls; wp: whole plant.

Table 10.4. Wild Botanicals Used as Foods in the Studied Communities

Family	Species	Part(s) Used	Culinary Uses	Albanians	Italians
Amaranthaceae	*Amaranthus retroflexus* L.	le	cooked	+	+
Apiaceae	*Apium nodiflorum* (L.) Lag.	ap	raw and cooked	+	
	Foeniculum vulgare ssp. *piperitum* Mill.	ap, fr	raw, cooked, and condiment	+	+
	Tordylium apulum L.	wh	cooked, condiment	+	
Asteraceae	*Carlina acaulis* L.	re	cooked		+
	Centaurea calcitrapa L.	wh	cooked	+	
	Chondrilla juncea L.	wh,sh	raw and cooked	+	
	Cichorium intybus L.	wh	raw and cooked	+	+
	Crepis vesicaria L.	wh	cooked	+	+
	Cynara cardunculus ssp. *cardunculus* L.	st, re	cooked		+
	Lactuca serriola L.	ap	raw and cooked		+
	Leontodon spp.	wh	raw and cooked		+
	Picris echioides L.	wh	cooked	+	+
	Reichardia picroides (L.) Roth.	wh	raw and cooked	+	+
	Scolymus hispanicus L.	ls	cooked	+	
	Silybum marianum (L.) Gaertn.	st, ro	cooked		+
	Sonchus asper (L.) Hill. & *S. oleraceus* L.	wh	raw and cooked	+	+
	Taraxacum officinale F.H. Wigg.	wh	cooked	+	+
Boraginaceae	*Borago officinalis* L.	le	cooked	+	+
Brassicaceae	*Capsella bursa-pastoris* (L.) Medik.	wh	cooked	+	
	Diplotaxis tenuifolia (L.) DC.	le	raw	+	
	Nasturtium officinale R. Br.	le	raw and cooked	+	
	Sinapis arvensis L.	ap	cooked	+	+
	Sisymbrium officinale (L.) Scop.	wh	cooked	+	
Cannabaceae	*Humulus lupulus* L.	sh	cooked	+	
Caryophyllaceae	*Stellaria media* (L.) Vill.	ap	raw and cooked	+	
Chenopodiaceae	*Beta vulgaris* ssp. *maritima* L.	bu	cooked	+	+
	Chenopodium album L.	le	cooked	+	+
Dioscoreaceae	*Tamus communis* L.	sh	cooked	+	
Lamiaceae	*Origanum heracleoticum* L.	ft	condiment	+	+
Liliaceae s.l.	*Allium ampeloprasum* L.	bu	cooked and condiment	+	+
	Asparagus acutifolius L.	sh	cooked	+	+

	Bellevalia romana (L.) Reichb.	bu	cooked		+
	Leopoldia comosa (L.) Parl.	bu	cooked	+	+
	Muscari atlanticum Boiss. & Reuter & *M. botryoides* (L.) Mill.	bu	cooked	+	+
	Ruscus aculeatus L.	sh	cooked		+
Papaveraceae	*Papaver rhoeas* L.	wh, le	cooked	+	+
Portulacaceae	*Portulaca oleracea* L.	ap	raw	+	+
Ranunculaceae	*Clematis vitalba* L.	sh	cooked	+	+
Scrophulariaceae	*Veronica beccabunga* L.	ap	raw		+
Solanaceae	*Lycium europaeum* L.	sh	cooked	+	
Urticaceae	*Urtica dioica* L.	le	cooked	+	
Valerianaceae	*Valerianella carinata* Loisel.	wh	raw	+	

Part(s) used: ap: aerial parts; bu: bulbs; fl: flowers; fr: fruits; ft: flowering tops; le: leaves; ls: leaf stalks; pf: pseudo-fruits; re: flower receptacles; ro: root/tuber; sh: shoots; uf: unripe fruits; wh: whorls.

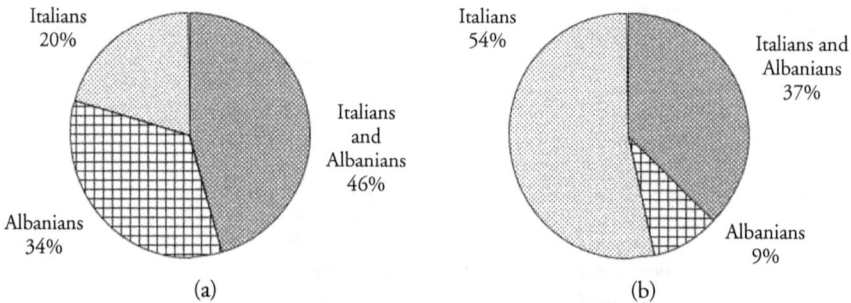

Figure 10.2. *Cultural distribution of traditional knowledge regarding wild botanicals as (a) foods and (b) medicines by percent of taxa reported*

by >66 percent of the study participants, are the same for both communities (table 10.5). The therapeutic targets for these phytopharmaceuticals, however, are distributed differently among the two populations (Pieroni and Quave 2005).

NON-CULTIVATED FOOD PLANTS

Since food and medicine in traditional practices are strongly embedded and ingestion of specific food plants is remarkably related to a precise perceived health benefit (Johns 1990; Etkin and Ross 1994; Pieroni and Price 2006; Pieroni and Quave 2006), traditions of gathering and preparing wild botanicals in the cuisine also represents an important aspect of domestic health care. Of our two study communities, the practice of gathering and preparing wild botanicals today is most popular in Ginestra/Zhurë. The Arbëreshë make a clear distinction

Table 10.5. Ten Most Quoted Medicinal Plants Used in Ginestra/Zhurë and Castelmezzano, as Reported by >66 Percent of Informants

Family	Species	English Name	Medicinal Uses	Albanians	Italians
Asteraceae	*Matricaria recutita* (L.) Rauschert	Chamomile	anti-hypertensive; anti-inflammatory for eyes; against sore throat and bronchitis; intestinal depurative; digestive; sedative	+	+
Boraginaceae	*Borago officinalis* L.	Borage	postpartum depurative; galactagogue; against sore throat	+	+
Euphorbiaceae	*Mercurialis annua* L.	Mercury	laxative	+	
Hypericaceae	*Hypericum hircinum* L.	Stinking tutsan	against bronchitis		+
Lamiaceae	*Marrubium vulgare* L. and *M. incanum* Desr.	Black horehound	diuretic; digestive; anti-malarial; against cysts; panacea	+	+
Lauraceae	*Laurus nobilis* L.	Bay leaves	digestive; anti-stress	+	+
Malvaceae	*Malva sylvestris* L.	Mallow	mild laxative; against menstrual pains, sore throat and bronchitis; intestinal depurative	+	+
Moraceae	*Ficus carica* L.	Fig	against sore throat and bronchitis; intestinal depurative; heal insect bites; anti-warts	+	+
Poaceae	*Agropyron repens* L.	Couch grass	diuretic	+	+
Rhamnaceae	*Ziziphus jujuba* Mill.	Jujube	against sore throat and cough		+

between edible weedy greens, or *liakra,* and non-edible grasses and herbs, or *bara.* The term *liakra* itself has an Albanian origin—*lakër* is "cabbage" in modern Albanian. However, from recent preliminary observations (Pieroni 2006) carried out in northern Albania, we found that a specific term for wild leafy vegetables does not exist in the modern Albanian language.

While today the term *liakra* is restricted to the Arbëreshë lexicon, the collection of wild edibles is not unique to this culture. In the south Italian dialect, these wild greens are referred to as *foglie,* or leaves. In Castelmezzano, as well as the rest of south Italy, wild greens are generally recognized as the "poor man's" food. As demonstrated in table 10.2, roughly half of the food plants reported is used both by Italians and ethnic Albanians in a similar way. In fact, when comparing traditional Arbëreshë cuisine with that of their Italian neighbors, only minor differences are notable. This commonality of knowledge regarding wild edibles may have been fostered by the gradual acculturation of Arbëreshë communities to an Italian way of life. Specifically, major changes in the local economy beginning in the 1960s spurred shifts in labor practices outward from the community, leading to regular exchange with Italians. The subsequent installation of Italian speaking government officials and schools in the Albanian communities reinforced this exchange.

In the past, when the south Italian economy was dominated by agro-pastoralism, the collection of wild edibles took place during fieldwork. Often, they were eaten raw as a snack by workers in the field, or brought home, washed at the communal fountain, then cooked in a traditional terracotta pot. In the poorest subset of the community, these wild greens were eaten raw with bread, but without olive oil or salt. Nowadays, only a few wild edibles are consumed in their raw form. Instead, they are commonly prepared by both Italians and ethnic Albanians by light boiling or frying in olive oil with garlic, salt, and sometimes, a few hot chili peppers. They are usually then eaten with pasta or bread.

Formerly, both in Ginestra/Zhurë and in Castelmezzano, women were the ones who taught their daughters how to identify the edible plants that had to be gathered while mothers and daughters worked in the field. This was a kind of experiential learning process, where daughters observed gestures renewed daily during the spring and summer seasons. Today this mechanism of passing on cultural knowledge has broken down. Young women no longer go into the fields but instead work at home or in factories in the surrounding areas, and do not gather wild edibles.

Moreover, nutritional patterns have changed in the study areas, and young couples are now accustomed to eating meat every day. For most of them, *liakra* and *foglie* represent only a kind of sporadic "exotic" variation that is primarily provided by an elderly female (usually the mother or mother-in-law), and which is often already prepared.

It has become clear from our field observations that the loss of Traditional Environmental Knowledge (TEK) and of the Arbëreshë language is not completely affected in Ginestra: a few middle-aged interviewees are able to remember Arbëreshë names of plants but cannot identify them or explain their traditional use. Upon analysis of the same recorded ethnobotanical competence by means of a curvilinear regression model, we found that among the men the percentage of correctly identified plants reaches a maximum for those aged between 50 and 60 years and decreases strangely for those older than 60. This trend might be explained by the fact that most of the male population over age 65 generally consists of people who emigrated for work, usually leaving their wives and children on their own in the village, and usually returned to the community several years later.

In Ginestra, the emigration that had its major peak at the end of 1960s and the beginning of the 1970s has certainly contributed to TEK loss: this generation of men did not work closely with the natural environment on a regular basis.

Moreover, when these same men returned home, they began to work in factories rather than in agriculture. They also seem to have played a certain role in the positive internalisation of the acculturation process and in the adoption of mainstream Italian/European cultural models. There are those who generally still tend to reject Arbëresh cultural practices—and not by chance. Contrary to this group of men, elderly women try actively to maintain most of the original expression of their unique culture through continued involvement in gathering weedy greens and preparing traditional meals.

Another factor that has certainly played a role in this matter is represented by the emigrated families of the middle-generation (those who left southern Italy during the 1980s to move to northern Italy), who normally come back to visit their parents or relatives in the summer. Among these people, the rejection of traditional culture is very strong, perhaps in part because of the negative images portrayed by the media concerning the recent immigrant flows from Albania. A man from this group tried to convince his interviewer that traditional Albanian culture is something that has be to hidden, because "Albanians are, after all, like *gypsies*"—a statement which is in strong contradiction to his own ethnic heritage. Similar power relations between old and new Albanian migrants have been observed in Sicily (Derhemi 2003).

On the other hand, all of the Arbëresh villages visited had had contacts with Albanians from Albania and Kosovo over the past decades, with cultural exchange and mutual visits sustained by local municipalities, and that involved many people at a time when informal cultural exchanges with "communist" Albania were very rare all over the world. Paradoxically, after 1991 and the political upheaval in Albania, these contacts ended and no interest remained in Ginestra/Zhurë to continue this kind of tradition.

Although the spread of Italian cuisine in Arbëreshë communities is evident, neither the knowledge nor application of Arbëreshë culinary tradition is noticeable among Italians. Some dishes that are unique to the Arbëreshë cuisine include *luljëkuq ma fazuljë* (corn poppy leaves [*Papaver rhoeas*] and beans) and *bathë e çikour* (mashed fava beans [*Vicia faba*] and wild chicory [*Cichorium intybus*]). Other characteristic Arbëreshë dishes are prepared only for special holiday feasts. For example, on Christmas Eve, anchovies or dried fish are prepared with the boiled and fried shoots of the semi-cultivated plant *çim de rrapë* (*Brassica rapa* ssp. *rapa*) or of the wild mustard (*sënap, Sinapis* spp.). For Easter, a special pie (*verdhët*) made of eggs, lamb, ricotta cheese, and the boiled leaf stalks of Spanish salsify (*Scolymus hispanicus*) are prepared.

"Traveling Plants"

The migration of Albanians to southern Italy in the fifteenth century does not seem to have resulted in the introduction of Albanian "traveling plants" to their new environment. Instead, the population has undergone a long process of cultural adaptation since then, and what we can observe today in Ginestra/Zhurë reflects the indirect consequences of that long process. Five centuries are probably too long a time span to be able to still trace original traveling plants brought by Albanians to southern Italy. Yet, other newcomers arriving in recent years have brought plants with them and are probably renewing the same long process that ethnic Albanians underwent centuries ago.

Newcomers Today

Both in Ginestra/Zhurë and Castelmezzano a few newcomers from Eastern Europe have arrived over the past few years. Their numbers are still too small to evaluate the nature of possible further cultural changes, but in the domain of the folk medicines they brought with them to both villages, new practices have spread. For example, "new" Albanians have brought *caj malhit* (*Sideritis* spp., Lamiaceae, unknown by the Arbëreshë and not native in southern Italy) from their country to Ginestra/Zhurë, which they use in decoctions for the relief of sore throats and cough. Ukrainian women in Castelmezzano have introduced the use of an alcoholic chili macerate with metamizole (classic example of a syncretism between medicinal plants and pure chemicals) for the external treatment of rheumatism, and the functional use of pickled tomatoes (with dill, bay, and horseradish leaves), consumed to recover from a state of drunkenness.

Changes in the cultural and ethnomedical systems of Lucania continue, and only future studies in the following decades will tell us if any of these new (migrant-based) pharmaceutical uses will become integrated into the folk medical heritage of the autochthonous population.

Conclusions

We have found in this chapter that two communities who share similar sociodemographic, economic, and geographic characters, but different cultural identities, will not necessarily utilize the products of their environment in the same way. The cultural and ethnic makeup of communities is critical to the formation of specific social and medical frameworks. These frameworks, in turn, play a central role in community behavior and tradition in relation to the local flora.

It is difficult to determine in our case study whether the Italian community's longer history of occupation in the environment enhanced their knowledge base regarding useful applications of plants for foods and medicines. While the autochthonous Italians are familiar with a much larger botanical pharmacopoeia than the Arbëreshë, they also follow a different medical framework.

It is probable that the lower number of medicinal plants known to the Arbëreshë is due to their heavy reliance on a ritualistic or spiritually-based system of complementary medicine. In this system, plants serve an underlying role as spiritual objects and are not necessarily associated with being the source of the cure. With regard to wild food botanicals, the Arbëreshë more frequently integrate *liakra,* or wild greens, into their daily diet. While their cuisine has certainly grown due to an informational flux with neighboring Italian communities, this exchange of culinary knowledge has probably not been a mutual phenomenon: we have not been able to trace clearly Albanian food plants and food plant uses in the surrounding South Italian culinary traditions.

Nowadays in Ginestra/Zhurë, Arbëreshë Albanians have been acculturated to the mainstream South Italian dominant culture, but their strong preference for consuming *liakra* and the practice of religious-magic healing serves as proof of their "different" origin, and speaks more about the strategy that this community put into place during this century in order to maintain their traditions and defend their cultural identity.

The impact of acculturation on the Arbëreshë has had an erosive effect on the passage of traditional ethnobotanical and ethnomedical knowledge to younger generations. Likewise, for the Italians of Castelmezzano, the enticement of "modern" life has pulled the younger generations away from an agro-pastoralist economy, a consequence of which is the rapid disintegration of daily interactions with the environment.

In conclusion, comparative ethnobiological studies on migratory populations are complex but can offer much insight into the way that culture mediates interaction with the environment. The case illustrated here demonstrates how intricate these processes are, and how "traveling plants" and related practices of use are sometimes much more hidden that we would think at first.

Acknowledgements

Special thanks to the communities of Ginestra/Zhurë and Castelmezzano, whose people generously agreed to share their knowledge regarding local folk-medical practices and traditional foods. Special thanks also to Giuseppe Pepice and Massimo Summa, Mayor and Vice-Mayor of Ginestra/Zhurë, and Nicola Valluzzi, Mayor of Castelmezzano, for their logistical assistance.

Funding for this 3-year research project was provided by a number of sources, to which we are deeply grateful. These include: Mars Nutritional Research Council (Postdoctoral fellowship awarded to A. Pieroni in 2000–2002); National Institute of Health NCCAM (Predoctoral training fellowship awarded to C. Quave, #1-T32-AT01060-01); National Kappa Alpha Theta Foundation research grant (assigned to C. Quave), Foundation for Science and Disability grant (assigned to C. Quave); EU Grant of the European Union, assigned to the Wageningen University, along with seven other universities and research centers (RUBIA, coordinated by A. Pieroni, #ICA3-2002-10023); and an EU Grant assigned to the School of Pharmacy of the University of London, along with six other research centers (Local Food Nutraceutical, coordinated by M. Heinrich, #QLK1-CT-2001-00173).

References

Alexiades, N.M., and J. W. Sheldon, eds. 1996. *Selected guidelines for ethnobotanical research: A field manual.* New York: New York Botanical Garden.

Berlin, B. 1992. *Ethnobiological classification: Principles of categorization of plants and animals in traditional societies.* Princeton: Princeton University Press.

Berlin, B., D.E. Breedlove, and P.H. Raven. 1966. "Folk taxonomies and biological classification." *Science* 154: 273–275.

Cotton, C.M. 1996. *Ethnobotany. Principles and applications.* Chichester: Wiley.

D'Andrade, R. 1995. *The development of cognitive anthropology.* Cambridge: Cambridge University Press.

Derhemi, E. 2003. "New Albanian immigrants in the old Albanian diaspora: Piana degli Albanesi." *Journal of Ethnic and Migration Studies* 29: 1015–1032.

Dessart, F. 1982. "The Albanian ethnic groups in the world: a historical and cultural essay on the Albanian colonies in Italy." *East European Quarterly* 4: 469–484.

Etkin, N. L., and P. Ross. 1994. "Pharmacologic implications of "wild" plants in Hausa diet." In *Eating on the wild side: The pharmacologic, ecologic, and social implications of using non-cultigens*, ed. N.L. Etkin. Tucson: University of Arizona Press, 85–101.

Galt, A. 1982. "The evil eye as synthetic image and its meanings on the island of Pantelleria, Italy," *American Ethnologist* 9: 664–681.

Grimes, B. F. 2000. *Ethnologue-CD ROM.* Dallas: Summer Institute of Linguistics.

Johns, T. 1990. *With bitter herbs they shall eat it.* Tucson: University of Arizona Press.

Pieroni, A., C. Quave, S. Nebel, and M. Heinrich. 2002. "Ethnopharmacy of the ethnic Albanians (Arbëreshë) of northern Basilicata, Italy." *Fitoterapia* 73: 217–241.

Pieroni, A. 2003. "Wild food plants and Arbëresh women in Lucania, Southern Italy." In *Women and plants. Gender relations in biodiversity management and conservation.* ed. P.L. Howard. London: Zed Press, 66–82.

———. 2006. "Ethnobiology, traditional medicine and post-communism in upper Kelmend, Northern Albania. " In *The dynamics of bio-cultural diversity in the New Europe: people, health and minor plant resource pools*, eds. M. Pardo de Santayana, A. Pieroni and R. Puri. Oxford: Berghahn Press.

Pieroni, A., and L. Price, eds. 2006. *Eating and healing: Traditional food as medicine.* Binghamton: Haworth Press.

Pieroni, A., and C.L. Quave. 2005. "Traditional pharmacopoeias and medicines among Albanians and Italians in southern Italy: A comparison." *Journal of Ethnopharmacology* 101: 258–270.

Pieroni, A., and C.L. Quave. 2006. "Functional foods or food-medicines? On the consumption of wild plants among Albanians and Southern Italians in Lucania." In *Eating and healing: traditional food as medicine*, eds. A. Pieroni and L.L. Price. Binghamton: Haworth Press, 101–129.

Pignatti, S. 2002. *Flora d'Italia.* Bologna: Edizioni Edigricole.

Quave, C., and A. Pieroni. 2002. "Magical healing. Traditional folk-medical practices of the Vulture area of southern Italy." In *Handbuch der ethnotherapien/handbook of ethnotherapies*, eds. C. Gottschalk-Batschkus and J.C. Green.Munchen: ETHNOMED Institut fur Ethnomedizin, 97–118.

————. 2005. "Folk illness and healing in Arbëreshë Albanian and Italian communities of Lucania, southern Italy." *Journal of Folklore Research* 42: 57–97.

Salminen, P. 1999. "UNESCO red book report on endangered languages: Europe." Retrieved 2 January 2006 from http://www.helsinki.fi/~tasalmin/europe_report.html

Vandebroek, I., P. Van Damme, L. Van Puyvelde, S. Arrazola, and N. De Kimpe. 2004. "A comparison of traditional healers' medicinal plant knowledge in the Bolivian Andes and Amazon." *Social Science & Medicine* 59: 837–849.

Chapter 11

Plant Knowledge as Indicator of Historical Cultural Contacts

Tanning in the Atlantic Fringe[1]

Ingvar Svanberg

Although I began to study ethnographic aspects of husbandry, traditional crafts, and exploitation of locally available biological resources among a historical sedentary Sami minority in Sweden in the mid 1970s (Svanberg 1986), it took some time before I decided to specialize in ethnobiological research. I actually continued with ethnographic fieldwork among Yörük herdsmen, Turkish peasants, and various marginal groups in Anatolia in the late 1970s and subsequently, in the mid 1980s, with Kazak nomads and other ethnic groups in Central Asia. In the beginning I focused on sustenance and utilization of the environment (Svanberg 1987; 1988; 1996). However, this research gradually took me away from ecological and local economic studies towards minority and immigrant research. For many years therefore I concentrated mainly on the relationship between social structures and the adaptation response of minority cultures in new political and social settings. I became a scholar in international migrations and ethnic relations rather than an ethnobiologist.

I continued to do fieldwork, however, and I noticed that many immigrants in Sweden were involved in gathering plants for various purposes. I observed, for instance, how immigrant women from the Anatolian countryside continued to gather wild sorrel, *Rumex acetosa* L., in the suburban settings of Stockholm and used it, mixed with yoghurt, as food in the same way as in their former home villages. I also saw Koreans in Sweden harvesting the spring shoots of bracken ferns, which they utilized as vegetables. Immigrant Thai women took advantage of the opportunity of gathering bilberries, *Vaccinium myrtillus* L., cowberries, *Vaccinium vitis-idaea* L., and mushrooms and sold them at the weekly markets in Swedish cities. I also noticed that my Czech and Polish colleagues (refugees from the Cold

War) at Uppsala University were much more skilled at differentiating between various kinds of wild edible mushrooms that are readily available in the forests around the city than most Swedes were (Svanberg and Szabó 1993: 151).

Berries, mushrooms, and common plants are, thanks to the legal right of access to private land, freely available resources for everyone to use in Sweden. Many immigrants utilize them, although these immigrants generally live in urban environments. Market trade is actually thriving thanks to the recent newcomers, who also sell many homemade food products. Immigrants have also revived the interest in allotment gardening, and many "new" vegetables are thus becoming available, also in the local markets. For instance, it was through Italian immigrant workers in my hometown Karlskoga that I first came in contact with zucchini, *Cucurbita pepo* L., and chicory salad, *Cichorium intybus* L., in the mid 1960s. Nowadays the supplies of "exotic" vegetables are huge. Also, marginal phenomena such as breeding singing canaries and training carrier pigeons have survived in Sweden thanks to immigrants from southeastern Europe and southwestern Asia. These examples indicate that contemporary immigrants are affecting the food habits, material culture, and management of public space in Swedish society in many ways.

Encounter with the Faroese Landscape

This introduction is given just to exemplify how immigrants in contemporary Scandinavia utilize wild and domestic species. However, my return to what I now call ethnobiology did not begin with these observations, although they certainly kept my interest alive. Instead it came in the mid–1990s, when for other reasons, I was more or less stranded on the Faroe Islands and was tired of writing papers about ethnic conflicts, minority policies, and immigrant communities.

I found myself living in a household with an elderly couple who were supporting themselves in many respects with a rather traditional economy. Most food came from locally available resources: cod (*Gadus morhua*), haddock (*Melanogrammus aeglefinus*), and other fish; sheep (*Ovis aries*), pilot whale (*Globicephala melaena*), common guillemot (*Uria aalge*), puffin (*Fratercula arctica*), and fulmar (*Fulmarus glacialis*). The couple also belonged to a generation who had in earlier days been utilizing other local biological resources, including mollusc shells, various birds, plants, algae, and so on. This increased my awareness of and insight into the importance of the traditional knowledge in societies with a low level of technological development, also in the North.

On the Faroe Islands I got to know a landscape that reflected the utilization of local resources. My hosts taught me about a landscape which, in their eyes, still was able to support them with many necessities. This was true, although they hardly used it anymore, with the exception of their dependency on fish from the marine biota, and hay production and sheep grazing on the terrestrial biota, which included not only traditional ecological knowledge but also sophisticated rules for mutual help, sharing, and social networking. Fowling and the occasional catch of pilot whale shoals were also part of their economy (Joensen 1976; 1982).

Many older Faroese men still had a detailed knowledge about where the cod and other fish were to be found, as well as about the state of currents and other important aspects about the sea. Contemporary and former bird cliffs and caves were revealed to me, and I got to know archaic techniques of hunting, preparing, and storing various foodstuffs.

The landscape itself, with its toponyms and lore, was a mirror of the traditional way of life and dependence on the locally available resources. Faroese sheep farmers and bird hunters, who move around in the birds' cliffs as well as in the mountains, still have a detailed set of toponyms for mountain cliffs, clefts, shelves, and other formations which, beside shape, also are distinguished through their vegetation or through the various bird species breeding there. There are, for instance, certain names for the stone clefts where the mountain angelica (*Angelica archangelica* L.) grows and for the slopes where the puffin with its droppings creates certain conditions for a vigorous form of red fescue *Festuca rubra* var. *fraterculae* Rasm.—known locally as *lundasina*, "puffin grass." This grass was sought after for thatching roofs on older houses. In the same way, a Faroese bird hunter with a very sophisticated taxonomy distinguishes various age groups and behavior of the various bird species, which can be captured on the bird cliffs (Rasmussen 1927; 1928; Skarði 1957).

Birds, fish, and sheep were preserved in ways that were known from prehistorical times. They taught me how shellfish were traditionally used and prepared, showed me where various plants could be gathered and how a primitive sponge, *Isodictya palmate* (Ellis and Solander, 1786), only a few decades ago, was used for wiping slates in the village school. Very old food customs still prevailed (Arge 1995). Although traditional plant knowledge was restricted to a few species, it included taxa used for various purposes: fodder, emergency food, handicrafts, tobacco substitutes, medicine, dyes, toys, and other purposes. British botanist James Nicol (1854: 379) concluded in the mid nineteenth century that the inhabitants of the Faroe Islands made use of few native plants. My own study proved that many more species have been traditionally utilized than what was recognized earlier (Svanberg 1998a).

I actually decided to devote the rest of my academic career to the relationship or—in the words of William Balée (1994: 13)—the "activity context," between human beings and other biological organisms in traditional societies, especially in preindustrial northwestern Europe. Since then I have devoted my time to studying and developing ethnobiology as a separate discipline in Sweden. It was possible for me, through interviews and occasional observations, to study the traditional use of many animal and plant taxa in the Faroes, since much knowledge had survived (Svanberg 1997; 1998b; 1998d; 2001b; 2003). As late as the 1960s and early 1970s, for instance, some children used homemade skin shoes, made of lambskin, and tanned with tormentil, *Potentilla erecta* (L.) Räusch., rhizomes (Olsen and Svanberg 1999: 5).

These kinds of shoes were known as early as the Viking age, and the technique to tan the skin with the rhizomes of tormentil is recorded in the late sixteenth century (Dahl 1951). However, it became clear, through literature research, that this use was known also in the island communities outside Scotland, once populated by Norse settlers. This study aims at describing and explaining the unique use of tormentil as an agent for tanning on the Islands of the northwestern Atlantic. I will emphasize the contextual aspects of its use by examining the dimension of distribution, the social context, and the time aspects of this particular function of the tormentil.

Tormentil

Tormentil is a famous medicinal plant, both in the scholarly medicine and in European folk therapy, and its qualities were well known back in ancient times (Allen and Hatfield 2004: 144; Olsen and Svanberg 2005). It has also been used locally as a dye plant among the Sami and settlers in northern Scandinavia and along the Baltic coast (Olsen 2004). However, the practice of using rhizomes of tormentil for tanning skin seems to be restricted to the Faroe Islands, Shetland Islands, Orkney Islands, and the Hebrides, and also perhaps in Ireland; i.e., areas once settled by Norse Vikings, thus indicating a relationship that will be dealt with here. Interestingly, we have no indication of this use from Norway or other parts of Scandinavia (cf. Høeg 1974: 523–528; Brøndegaard 1979: 161–163; Svanberg 1998c: 75–76).

The whole plant of tormentil is very astringent, especially the red, woody rhizome. This rhizome is from 2.5 to 5 cm long and thick as a finger, tapering to the end, with one to three short branches near the larger end. Externally it is brown or blackish, and internally it is a light brownish red. It has a peculiar aromatic scent and a strongly astringent taste. It contains 17 to 30 percent tannin (Cardon and Chatenet 1990: 291). The use of tormentil for producing tannin is still a well known method among Faroe islanders, and it is possible to find people who can actually prepare skin with the use of tormentil rhizome. It is also often mentioned in the local literature (Bjørk 1972).

The Scope of Ethnobiology

Although many scholars try to market ethnobiology as a type of research that can find new economically important products, the main academic interest in the subject is that the discipline in its broadest sense represents a unified approach to understanding the interaction between human beings and the environment in low-technological and preindustrial societies. A few years ago, Daniel Clément published a treatise on the history of ethnobiology. As he emphasizes, early ethnobotany was largely the listing of plants, names, and uses (Clément 1998). Most researchers in the past did not regard what the people thought about plants as being important. Nowadays the researchers in the field would like to include conceptualizations of plants and vegetation in their studies. Especially the concept of traditional ecological knowledge has been influential on ethnobiological methods and theories (Davidson-Hunt 2000).

Ethnobiology will in the same way as most disciplines within the humanities create consistency, understanding, and supply a comprehensive view. Modern ethnobiology may be defined as the study of the total complex of relations that has developed between human beings and plants and animals (Balée and Brown 1996; Martin 2001). In my view, contemporary ethnobiology aims at studying the *biocultural domains* that develop in the relations between human beings and other biological organisms. Therefore all ethnobiological data must be contextualized culturally and socially. They must give an answer not only to *how*, but also to *when, where*, and by *whom* a certain taxon has been of importance within folk biology. To understand how a group of people imagine or explain things, we must get to know an *emic* perspective of their culture, i.e., the ethnobiologists have to put aside their own cultural biases (Svanberg 2002a; 2005b).

While economic botany ignores the cultural context and focuses instead on the useful qualities of a certain plant, ethnobiology stresses the activity context between plants and people. The research must be conducted in a defined local ecosystem, which provides resources to a specific human society, i.e., hunters, fishermen, or villagers in a certain setting. Such a system includes renewable resources, e.g., fodder, game, fish, fuel, and water; and non-renewable resources, such as stones, sand, and so on. The scope of ethnobiology also includes the study of human manipulation—e.g., through domestication, cultivation, coppicing, pollarding, gardening, etc.—of the biological resources in order to increase the production (Svanberg 2004a).

A renowned Canadian scholar claimed recently that "Europe is the most understudied region of the world concerning indigenous knowledge." This statement is both right and wrong. Thanks to, among others, Carl Linnaeus and the method of documentation that he put into practice, the Nordic countries have a long tradition of collecting information about the traditional use of plants, folk plant names, local lore connected with plants, and indigenous knowledge in a wider sense (Cox 1999; Svanberg 2001a; 2002b; 2005a). The Linnaeans were followed by linguists who gathered dialect samples and plant names, as well as by ethnologists and folklorists who documented not only the economic and practical use of plants in every aspect, but also folk beliefs, myths, stories, sayings, and so forth about the biological diversity that surrounded peasants, crofters, fishermen, and nomads in the Nordic countries (Svanberg 2002a).

Most current studies in ethnobiology are conducted in contemporary settings (Ford 1994; Martin 1995). However, the situation in the Nordic countries seldom allows us to do field work. Our data and sources are first and foremost historical; we therefore need to test new analysis models to approach these sources. When ethnobiology is currently being developed as a separate discipline in the Nordic countries, we therefore use this material to study the traditional plant knowledge on the basis of modern-day ethnobotanical theory and methods.

From close contact with the plants, scholars are able to relate local and specialized plant taxonomies. Ethnobiologists pay special attention to culturally relative cognitive and symbolic properties of the plants in a region or area. Ecological relationships within the plant community are central to these studies (Turner 1997).

However, the plant knowledge and plant names as such can sometimes reveal interesting historical facts that can increase our knowledge about the whereabouts of a local population. The tormentil actually has very similar names from a common Norse background on the Faroes and on the Islands outside Scotland.

Plants and Plant Knowledge as Cultural Heritage

Ethnobiological research in a historical context demands various methods and sources. In sources we include all kinds of observation data that the ethnobiologist can use for research. Landscape historian Mårten Sjöbeck (1927) asserts that the flora itself is an excellent historical source material for our knowledge about the relationship between the human being and her environment (cf. Svanberg 2002c). It is a source material whose potential has hardly been used at all in ethnobiological research, with the exception of a few local studies. Every individual plant species has, as Sjöbeck expresses, "something important to say about the cultivated ground, working life, tools, climate, and soil."

We can exemplify this with the occurrence of butterbur, *Petasites hybridus* (L.) P. Gaertn., B. Mey. & Scherb., in southern Sweden. In the literature of popular science it is a wide-spread opinion that the introduction of the butterbur in Sweden was connected with the Black Death in the fourteenth century. Also some local folk myths connect its occurrence with plague epidemics. Although it is mentioned in some late medieval herbals, there is no evidence for its introduction into the Swedish countryside until the seventeenth century (Svanberg 2005c). Actually, it seems not to have been as important a medicinal plant as sometimes assumed. By studying its present localities in the countryside of the southernmost province of Skåne, amateur botanist Gösta Ilien could, in an ethnobotanically exemplary way, document its importance as a cultivated plant and its local use for various purposes during the eighteenth and nineteenth centuries. It was used, among other things, to cure hogs with diarrhea. However, by using a synchronic perspective, he could also show that in the 1940s it was still regarded and harvested as a fodder plant among the local farmers (Ilien 1945). I have myself studied white horehound, *Marrubium vulgare* L., as a former cultivated plant used in medicine, which at times has spread to the local flora in some parts of Sweden (Svanberg 2002a). Another study deals with Good King Henry, *Chenopodium bonus-henricus* L., probably introduced once as a cultivated plant, which is still to be found in the Swedish landscape as a reminder of its former importance (Svanberg 2004b). These kinds of plants are sometimes referred to as botanical relicts (Brøndegaard 1955).

Many plants have been dispersed with the help of human beings. Botanists have recently integrated the concept of archaeophytes for synanthropous plants established in northern Europe before AD 1500 (Preston et al. 2003). Of course, these plants are all part of a biological cultural heritage, still reflecting the activities of human beings. Although seldom studied, many of them actually constitute important indicators of earlier cultural contacts, trade relations, and economic activities. The existence of specific apomictic forms of *Ranunculus auricomus* L. in certain places along the coastal areas of the Baltic Sea indicates such relations in premodern, probably Viking times (Julin 1977; Vasari 2002). Also, fossil botanical remains provide evidence of human activities in the landscape. The occurrence of the synanthropous ribwort plantain, *Plantago lanceolata* L., in pollen samples has for instance been taken as an indicator of grazing effect and *landnám* of Atlantic Islands (Jóhansen 1978; 1989).

Plant Knowledge as a Cultural History Source

However, it is not only plants that bear witness to human history. The modern folk traditions sometimes work as evidence of older contexts and can in such cases be used as source material. Anthropologist M. Anderson (2000) has for instance in a study of Sami child traditions shown how very old ecological knowledge has been in the trust of children, knowledge about plants and animals that the adult world cannot remember. My own favorite example is my own experience from a summer visit in the Italian town Verbania Intra, north of Lago Maggiore in 1966. The local children taught me how to catch wall lizards (*Podarcis muralis*) with the help of snares made of a grass straw. This was an ancient technique, reproduced by the Apollo statue Sauroctonus from the fourth century B.C. by the Greek sculptor Praxiteles. To gain an insight, with the help of children's contemporary traditions, into a custom that was known in the same geographical area 2000 years earlier was of course thrilling.

The study of popular plant names is another important part of ethnobiology since they reflect the human ability to observe, group, and categorize organisms (Rydén 2001). In this context we must also stress the importance of language as a source of knowledge for our understanding of earlier conditions. Vocabularies, plant names, and methods of categorizations are in themselves relevant sources when studying ethnobiological matters. Crotal lichen, *Ochrolechia tartarea* (L.) A. Massal., locally known as *korki*, had at one time played an important role as a dye plant to color wool and garments in various nuances of red in the Faroes. It is mentioned in written sources as early as 1669. However, the plant name itself points to early Celtic contacts, and according to some linguists it indicates Viking age relations with the Irish Celts (Matras 1958; Svanberg 1998a).

My study of the use of tormentil as a tanning agent in the Atlantic Islands includes analyses of local flora and ecological restrictions, the traditional way of using the plant, as well as the local plant names.

Tanning Agents

In order to make leather, skins and hides must be prepared by tanning. Vegetable tanning materials have always been used by human beings. Many plants contain tannins, which are the substances that work as agents in the process. The most common traditional sources of tannins in Scandinavia have been bark from trees, especially birch, *Betula pubescens* Ehr., Norway spruce, *Picea abies* (L.) H. Karst, oak, *Quercus robur* L., white willow, *Salix alba* L., sallow, *Salix caprea* L., and bearberry, *Arctostaphylos uva-ursi* (L.) Sprengel.

However, many plants contain a substantial amount of tannins. With the increasing trade in the nineteenth century many vegetable tanning materials became available on the world market. Several of them played a considerable role for the economic development in the colonies in many transoceanic areas (Lewington 1990: 60).

The traditional use of tormentil for tanning seems to have been restricted to the Hebrides, the Orkneys, Shetland, and the Faroe Islands. This practice does not exist on the Scandinavian Peninsula or in Scotland.

Both on the Faroe Islands and on the islands outside Scotland, other vegetable tanning material has been available as well, although it has always been imported. The Faroe islanders, for instance, used bark from timber that drifted with the currents (Malmros 1994). The locals called this bark *spónbark* (Dahl 1951: 9). Large quantities of vegetable tanning material were also imported, so called *útlendskt bark* ("foreign bark"), or *handilsbark* ("commercial bark")." In the late 1730s, around 6,660 kilograms of foreign bark was imported annually for tannin, while in 1910 the amount was 17,700 kilograms. The imported bark originated from north European trees in the eighteenth century, while *cutch* and *gambier* have been used in more recent times (Bjørk 1972: 112–113).

However, as late as in the 1870s, the Danish botanist Emil Rostrup writes that the tormentil rhizome is the most important substance used for tanning among the Faroese people (Rostrup 1870: 29).

The Faroese Context

The Faroe Islands were probably colonized by Norwegian Vikings in the ninth century, and they soon developed a local peasant culture in which sheep breeding was of great importance (Blehr 1964). Shepherding was one of the main economic activities in the preindustrial time. The inhabitants used almost the entire animal. The sheep produced meat, intestines, blood, and marrow for food, fat for energy, excrement for fertilization, wool for the thriving sock knitting and textile protoindustry, bone for glue and toys, horn for household tools, and skin for leather (Joensen 1979; Guttesen 2001; 2004).

Norwegian vicar Peder Claussøn Friis wrote the oldest reliable topographical description of the Faroe Islands (Jóhansen 1994). Claussøn Friis' information is based on interviews with a Faroese student, Jacob Oudensøn, whom he met in Copenhagen in 1592. The book was actually written in 1593, although it was not published until forty years later. Jacob had told the vicar that the tormentil plant "grows abundantly there, which they use for tanning skins or leather because they don't have bark of trees, and it becomes good leather when it is tanned in this way: a girl or a boy should be able to dig half a barrow of tormentil a day" (Claussøn Friis 1632: 149). Thomas Tarnovius (1950: 70) also mentioned the use of *Potentilla erecta* (L.) Räusch. as a tanning plant in his description of the Faroe Islands in 1669.

According to the maps compiled by Danish botanist K. Hansen (1966: 98), the tormentil is common all over the Faroese archipelago, with the exception of Mykines, from where there is no record. The plant grows on meadows and hill slopes.

The vicar Jørgen Landt gives a detailed description of the use of the rhizomes as a tanning agent on the Faroe Islands at the end of the eighteenth century:

> Not only skin but also hide the inhabitants know how to tan with the tormentil root; but this work actually belongs to the women, and this is their way of doing it. For every lambskin you take a quarter of a pound of tormentil root; this is washed and crushed with a stone, then mixed with water into a gruel, that is spread all on the hairy side of the skin, though the hair must have been beforehand removed; it is thereafter fourfold folded, and put away for a couple of days, but if the skin has been moistened eight days in advance to loosen the wool, then it must lay with the smear for eight days, though not all skins, especially those from fat sheep, that are easily tanned. If you want them quite pretty, it must therefore be repeated in the same way once or twice, and between each time washed in seawater and dried. When the crushed root is moistened in seawater, the skin becomes pale yellow, but if urine is used instead of sea water, they become darker (Landt 1800: 408).

The utilization of the resources of the outfield (*hagi*), where the farmers kept their sheep, was reserved for the Faroese landowners and tenants in each village.

The right to gather tormentil rhizomes, as well as collecting lichens for dyeing, was regulated by local land rights, which the local council decided upon. In some places the landowners of a village stated in by-laws how and who could use the outfield. The poor, however, could be assigned the right to collect plants if the landowners approved. In other places poor people had free access to gathering tormentil for their own use, in some places with the restriction that they were not allowed to sell it outside the village. The right to the products of the infield (*bøur*)—the cultivated and fenced land around a village—was solely for the landowner. In 1866 a law regulating these rights was implemented for the archipelago. This was in function until 1937 when the power of legislation was transferred to the local community council (Bjørk 1972: 121–125).

It was mainly a task for women and children to gather the rhizomes and there seems to have been a widespread idea about different qualities of the rhizomes, depending on the habitat of the plant. Jens Christian Svabo (1959: 284–285) makes the interesting observation that rhizomes harvested on the infield—*heimabark*—give the skin a bright yellow color, while those from the outfield give a red-yellow color, and those growing among the heather give the deepest red color (Patursson 1966: 50).

To gather the rhizomes the women used a digging-stick, called *barksneis*. The digging-stick has been regarded as a primitive tool, known from lower strata and marginal gathering communities in Eurasia, especially in Eastern and Central Europe. The rhizomes were usually gathered during fall or spring.

The Faroese author Maria Eide Petersen (1977: 35–36) describes in detail how tormentil was gathered and thereafter prepared. According to her, the rhizomes were thoroughly cleaned. At first they were soaked in water, and then rubbed against a rock in a brook. When cleaned, the rhizomes were pulverized. This was done in a pothole, the so-called *barkhella*, where they were ground with an oblong stone, *barksteinur*. The ground material was then boiled and cooled down, after which a sheepskin, which previously had been cleaned of hair and fat, was steeped in the solution.

Another way of preparing the skin is described by Johanna Maria Skylv Hansen (1968: 35). According to her, the ground rhizomes could also be placed on the skin, but this method was more time consuming. When considered ready the skin was taken out of the solution and dried, after which it was rubbed until it became soft and ready to use.

The tanned skin and hide were used for several products, but the most common use was for shoemaking. According to Thomas Tarnovius' (1950: 70) description from 1669, the Faroese men wore shoes of ox hide, seal hide, cowhide, and lambskin, while the women always wore shoes of lambskin. He also describes how the hide and skin were tanned with tormentil. As recently as the early twentieth century, women and children wore slippers, *rotuskógvar*, made from tanned sheepskin, while men used mostly tanned shoes, *húðarskógvar*, made from hide of cattle. Hide shoes had to be kept in salt. According to Sverri Dahl, it was customary in Miðvagur at the end of the nineteenth century to put them in *spiklaki,* that is, brine in which whale blubber had been kept. This was considered particularly effective to preserve the hide shoes. The men also had tanned sheepskin jackets and trousers when they were out fishing. Tanned skin and hides were also used for instance as bags, as sieves, as covers for milk pails, and aprons. The light skin slippers were used among men and women as late as the 1940s and 1950s (Dahl 1951; Stoklund 1971: 48–51).

Nowadays women and children sometimes use *rotuskógvar* when wearing the national costume. These shoes are also available as souvenirs.

Tormentil for Tanning in the British Isles

The tormentil is also a native and common plant on the Islands outside Scotland (Spence 1914: 20; Scott and Palmer 1987: 176; McCosh 2002: 331).

Thomas Monat and James Barclay (1793: 187) wrote from Unst in Shetland that rhizomes of tormentil were used for tanning hides. In the extinct Nordic language Norn of the Shetland Islands, tormentil was known as *bark,* which Faroese linguist Jakob Jakobsen mentions in his dictionary. According to him, Tormentil was formerly "commonly used for tanning of skin and hide (for sea clothes and boots)" on the Shetland Islands (Jakobsen 1928: 29). In the same way as the Faroe islanders had their *heimabark* and *hagabark,* the Shetlanders seem to have made a distinction between *earth-bark* and *hill bark.* They also used a digging-stick, locally known as *berkiklepp* among the inhabitants. (Matras 1958: 75; Bjørk 1972: 115).

Sir Walter Scott noted tormentil on the moors near Lerwick on his 1814 visit and referred to it as an "astringent root with a yellow flower, called Tormentil, used by the islanders in dressing leather in lieu of the oak bark" (Lockhart 1902: 408).

Samuel Hibbert wrote some years later that the "skins of the Shetland sheep are in requisition for the purpose of affording the fisherman a sort of *surtout,* that covers his common dress." The *Tormentilla erecta* has long been used in the process of tanning (Hibbert 1822: 443).

Fifty years later, Robert Cowie (1871: 163) describes how the sheepskins were tanned with tormentil "and made into waterproof clothes, for the men at the fishing, and the women, when engaged in their more severe labors in the field."

As recently as the 1930s, the Shetlanders used tormentil rhizome for tanning, although they also used alum, soda, and calcium according to a Swedish ethnologist (Jirlow 1931: 123).

The practice of using tormentil as a tanning agent is also known from the Hebrides. If Sir Walter Scott described its use on Shetland, none other than Samuel Johnson fifty years earlier had given a detailed description of hide shoes from

Skye, and he noted that the islanders tanned the leather with tormentil (Johnson 1985: 40). In his important book on the Scottish flora, which includes interesting ethnobotanical information, botanist John Lightfoot (1789: 272–273) wrote:

> The roots consist of thick tuberales, an inch or more in diameter, replete with a red juice, of an astringent quality. They are used in most of the western isles, and in the Orkneys, for tanning of leather; in which intention they are proved, by some late experiments, to be superior even to the bark of oak. They are first of all boiled in water, and the leather afterwards steeped in the cold liquor. In the Islands of Tirey and Col the inhabitants have destroyed so much ground by digging them up that lately they have been prohibited the use of them.

The local fishermen of the Hebrides used tormentil for tanning their nets (McNeill 1910: 118). From the mid nineteenth century we have a more detailed description of the use of tormentil from the island of Eigg, Inner Hebrides. The process is described as follows:

> The island shoes were altogether the production of Eigg, from the skin out of which they had been cut, with the lime that had prepared it for the tan, and the root by which the tan had been furnished, down to the last on which they had been molded. There are a few trees and, of course, no bark to spare; but the islanders find a substitute in the astringent lobiverous root of the *Tormentilla erecta*, which they dig out for the purpose among the heath at no inconsiderable expense of time and trouble. It takes a man nearly a day to gather roots enough for a single infusion. It was not unusual for the owner of a skin to give it to some neighbor to tan and that, the process finished, it was divided equally between them, the time and trouble bestowed on it by the one being deemed equivalent to the property held in it by the other (Miller 1870: 17).

A visitor to the westernmost island, St. Kilda, could in 1698 describe how the islanders' "leather is dressed with the roots of tormentil" (Martin 1934: 455).

Our knowledge of tormentil's use from the Orkneys is sparse. Lightfoot (1789: 272) and Lindley (1849: 184) briefly mention its use there. A botanical record from the mid nineteenth century describes how the people of Orkney Islands, in the same way as their contemporaries in the Hebrides, used the rhizomes to tan fish lines (Schübeler 1888: 507). Tormentil used to be highly esteemed for its roots, which were employed for tanning nets, writes Hugh Marwick (1929: 10). The plant was locally known as *hill-bark* and a strong decoction of the rhizomes was used as a substitute for bark in the tanning of skin, according to botanist Magnus Spence (1914: 103).

Surviving Customs

For historical reasons, the people of the Faroes and the Islands outside Scotland still have a common cultural heritage. They were all colonized by Norse settlers,

and they had a common Norse language for centuries. Most of the settlers originated from Norway, a fact which is still indicated in many cultural manifestations. It is, for instance, a well-known fact that the Shetland islanders still, or
until recently, used many plant, bird, and fish names of Norse origin (Jakobsen
1897: 44–48; Iversen 1951).

Plant customs, still surviving in the Faroes as well as in the Shetlands, probably originate from Norway as well. One interesting example is ribwort plantain
(*Plantago lanceolata* L.), known as *Johnsmas-flooer* on the Shetlands and *Jóansøkugras* on the Faroes. It is also known as *Jonsokgras* in southwestern and western Norway (Høeg 1941). The young islanders and the young Norwegians have
until today shared the same custom of using the plant as a divination plant (Scott
and Palmer 1987: 273; Darwin 1996: 137; Milliken and Bridgewater 2004:
136). Jørgen Landt describes the custom connected with the evening of St. John's
day in the Faroes:

> The stamens are plucked off, and the flower head is placed in the night time on the
> body or between the shirt and under the waistcoat, either in the armpits or on the
> breast, and the inhabitants believe that, by these means, they can predict not only
> whether the person to whom the plant is applied will live out the year, but also whether
> he will be fortunate in love, and many other events they are desirous of knowing. If
> the flower head, in the course of the night, shoots out fresh stamens, the person's wish
> will be fulfilled. Those, therefore, who make choice of a spike, the first flowers of
> which only have blown, are always certain of a favorable answer; for the next flowers
> in succession will, in the consequence of the heat, throw out their stamens in the
> course of the night (Landt 1800: 180).

Further examples can be found in Svanberg (1998a; 2005d; 2005e).

Discussion and Summary

Although tormentil contains a high percentage of tannin, its usage is restricted
to some Atlantic Islands where trees are lacking or scarce. The cultural knowledge is a result of the interplay among a given physical and biotic environment,
and the local people. This interplay is, as pointed out by Sarukhán (1985: 431),
basically ecological in its nature. When ecology restricted the amount of available tanning plants, as in the case of the Faroe Islands, the Hebrides, the Orkney
Islands, and the Shetland Islands, tormentil was a good substitute. The outfields
of the islands were also highly productive in tormentil, and the harvest could
therefore be good. In any case, the occurrence of tormentil has enabled the islanders to produce their own tanning substance in addition to the much more
expensive import products, which also were available on the local markets for
those who could afford them. However, it is probably not a plant knowledge that
was brought by the Vikings from Norway, as in the case with the custom con-

nected with *Plantago lanceolata* L. More likely, the use of *Potentilla erecta* (L.) Räusch. was a local invention on the Faroes, Shetland, or Orkney Islands, which was spread to the other Islands. The similar plant name, *børkuvísa* on the Faroes, *bark* on Shetland, and "hill–bark" on the Orkneys, indicates such diffusion, as well as the common technique of gathering and using it. Where this knowledge was actually developed is not possible for us to establish. However, the common names and the common usages seem to be endemic and must therefore have been spread between the Norse settlements after the establishment of the island populations. There were close contacts between the populations of the Faroes and the Islands outside Scotland during the Viking age and early medieval time (Stoklund 1991). Knowledge of the use of tormentil as a source for tannin might therefore have been transferred between the island communities.

With the development of practical botany, some attempts were actually made in the eighteenth century to promote the knowledge of using tormentil for tanning in other places, especially Ireland, as well as in Scandinavia and Germany. However, its promotion was never followed in practice (Evans and Laughlin 1971; Nelson and McCracken 1987; Olsen and Svanberg 1999: 22). Bark of oak and spruce has been readily available on the continent as well as in the Scandinavian Peninsula. The use of tree bark, usually available locally, must have been much more economic than the time-consuming gathering of tormentil rhizomes. There was no reason to accept tormentil as a substitute for tree bark.

For the island communities of the Faroes, Shetland, Orkney, and Hebrides, on the other hand, who did not have any trees in their environment; the tormentil rhizomes provided an inexpensive source of tannin, readily available for everyone. The rather marginal island communities on the Atlantic fringe, therefore, continued to use the tormentil for tannin purposes within the household economy until the end of the nineteenth and the early twentieth centuries.

Notes

1. I would like to thank Professor Mats Rydén (Uppsala) and museum teacher Osva Olsen (Uppsala) for comments on earlier drafts of this essay.

References

Allen, D.E., and G. Hatfield. 2004. *Medicinal plants in folk tradition: An ethnobotany of Britain and Ireland.* Cambridge: Timber Press.

Arge, S. 1995. "Mergsogin bein: En aldargamal matsiður." *Fróðskaparrit* 43: 56–66.

Anderson, M. 2000. "Sami children and traditional knowledge." In *Ecological knowledge of the North*, eds. I. Svanberg and H. Tunón. Uppsala: Swedish Science Press, 55–65.

Balée, W. 1994. *Footprints in the forest: Ka'apor ethnobotany—The historical ecology of plant utilization by an Amazonian people.* New York: Columbia University Press.

Balée, W., and J.C. Brown. 1996. "Ethnobotany." In *Encyclopedia of cultural anthropology*. Vol. 2, eds. D. Levinson and M. Ember. New York: Henry Holt and Company, 399–404.

Bjørk, E.A. 1972. "Børkuvísa (*Potentilla erecta*)." *Fróðskaparrit* 20: 99–128.

Blehr, O. 1964. "Ecological change and organizational continuity in the Faroe Islands." *Folk* 6: 29–33.

Brøndegaard, V.J. 1955. "Folkloristiske reliktplanter." *Naturens Verden* 39: 143–148.

———. 1979. *Folk og flora: Dansk etnobotanik*. København: Rosenkilde og Bagger.

Cardon, D., and G. du Chatenet 1990. *Guide des teintures naturelles: Plantes, lichens, champignons, mollusques et insectes*. Paris: Delachaus et Niestlé.

Claussøn Friis, P. 1632. *Norriges oc omliggene øers sandfærdige bescriffuelse*. Kiøbenhaffn: Melchior Marssan.

Clément, D. 1998. "The historical foundation of ethnobiology (1860–1899)." *Journal of Ethnobiology* 8: 161–187.

Cowie, R. 1871. *Shetland, descriptive and historical: Being a graduation thesis on the inhabitants of the Shetland Islands; and a topographical description of that country*. Aberdeen: Lewis Smith.

Cox, P.A. 1999. "The unfinished journey of Carl Linnaeus." *Plant Talk* 16: 33–36.

Dahl, S. 1951. "Føroyskur fótbúni." *Varðin* 29: 1–48.

Darwin, T. 1996. *The Scots herbal: The plant lore of Scotland*. Edinburgh: Mercat Press.

Davidson-Hunt, I. 2000. "Ecological ethnobotany: Stumbling toward new practices and paradigms." *MASA Journal* 16: 1–13.

Evans, E.E., and S.J. Laughlin. 1971. "A country tyrone tan yard."*Ulster Folklife* 17: 85–87.

Ford, R.I. 1994. *The nature and status of ethnobotany*. Ann Arbor: University of Michigan Press.

Guttesen, R. 2001. "Plant production on a Faeroese farm 1813–1892, related to climatic fluctuations." *Geografisk Tidsskrift* 101: 67–76.

———. 2004. "Animal production and climate variation in the Faroe Islands in the 19th century." *Geografisk Tidsskrift* 104: 81–91.

Hansen, K. 1966. *Vascular plants in the Faeroes: Horizontal and vertical distribution*. København: Dansk Botanisk Forening.

Hibbert, S. 1822. *Description of the Shetland Islands*. Edinburgh: A. Constantable & Co.

Høeg, O.A. 1941. "Jonsokgras, *Plantago lanceolata.*" *Det Kongelige Norske Videnskabers Selskab Forhandlinger* 13: 157–160.

———. 1974. *Planter og tradisjon. Floraen i levende tale og tradisjon i Norge 1925–1973*. Oslo: Universitetsforlaget.

Ilien, G. 1945. "Bidrag till Skånes flora. Förekomsten av *Petasites hybridus* i Skåne." *Botaniska Notiser* 1945: 231–233.

Iversen, R. 1951: "Noen shetlandske plantenavn." *Nysvenska studier* 31: 112–125.

Jakobsen, J. 1897. *Det norrøne sprog på Shetland*. København: Wilhelm Priors Hofboghandel.

———. 1928. *An etymological dictionary of the Norn language in Shetland*, vol. 1. Copenhagen: Vilhelm Prior.

Jirlow, R. 1931. "Drag ur färöiskt arbetsliv." *Rig* 15: 97–133.

Joensen, J.P. 1976. "Pilot whaling in the Faroe Islands." *Ethnologia Scandinavica* 1976: 25–41.

———. 1982. "Man and the physical environment." In *The physical environment of the Faroe Islands*, ed. G. K. Rutherford. The Hague: Dr W. Junk Publishers, 125–141.

Joensen, R. 1979. *Fåreavl på Færøerne*. København: C.A. Reitzels Boghandel.

Jóhansen, J. 1978. "The age of the introduction of *Plantago lanceolata* to the Shetland Islands." *Danmarks Geologiska Undersøgelse. Årbog* 1976: 45–48.

Jóhansen, J. 1989. "Jóansøkugras (*Plantago lanceolata*) og forsøgulig búseting í Føroyum." *Fróðskaparrit* 34–35: 68–75.

———. 1994. "Medicinal and other useful plants in the Faroe Islands before A.D. 1800." *Botanical Journal of Scotland* 46: 611–616.

Julin, E. 1977. "Some Bothnian subspecies in the *Ranunculus auricomus* complex: Origin and dispersal." *Botaniska Notiser* 130: 287–302.

Johnson, S. 1985. *A journey to the western Islands of Scotland*. Oxford: Clarendon Press.

Landt, J. 1800. *Forsøg til en beskrivelse over Færøerne*. København: Tikjøbs Forlag.

Lewington, A. 1990. *Plants for people*. London: Natural History Museum.

Lightfoot, J. 1789. *Flora Scotica: or, a systematic arrangement, in the Linnæan method, of the native plants of Scotland and the Hebrides* vol. 1. London: R. Faulder.

Lindley, J. 1849. *Medical and oeconomical botany*. London: Bradbury & Evans.

Lockhart, J.G. 1902. *Memoirs of the life of Sir Walter Scott, Bart*. Vol. 2. Boston & New York: Houghton, Mifflin and Company.

Malmros, C. 1994. "Exploitation of local, drifted and imported wood by the Vikings on the Faroe Islands." *Botanical Journal of Scotland* 46: 552–558.

Martin, G.J. 1995. *Ethnobotany: A methods manual*. London: Earthscan.

———. 2001. "Ethnobiology and ethnoecology." In *Encyclopedia of biodiversity*. Vol. 2, ed. S.A. Levin. San Diego: Academic Press, 609–621.

Martin, M. 1934. *A Voyage to St. Kilda*. Stirling: E. Mackay.

Marwick, H. 1929. *The Orkney Norn*. Oxford: Oxford University Press.

Matras, C. 1958. "Atlantssiðir-Atlantsorið." *Fróðskaparrit* 7: 73–101.

McCosh, D. J. 2002. "Potentilla erecta." In *An atlas of the vascular plants of Britain, Ireland, the Isle of Man and the Channel Islands*, eds. C.D Preston, D.A. Pearman and T.D. Dines. Oxford: Oxford University Press.

McNeill, M. 1910. *Colonsay: One of the Hebrides. Its plants: Their local names and uses*. Edinburgh: David Douglas.

Miller, H. 1870. *The cruise of the Betsy, or a summer holiday in the Hebrides*. Edinburgh: William P. Nimmo.

Milliken, W., and S. Bridgewater 2004. *Flora Celtica: Plants and people in Scotland*. Edinburgh: Royal Botanical Garden.

Monat, T., and J. Barclay. 1793. "Island and Parish of Unst in Shetland." In *The Statistical Account of Scotland*. Vol. 5, ed. J. Sinclair. Edinburgh: William Creech, 182–202.

Nelson, C.E., and E.M. McCracken. 1987. *The brightest jewel. A history of the National Botanic Gardens Glasnevin, Dublin*. Kilkenny: Boethius Press.

Nicol, J. 1854. *An historical and descriptive account of Iceland, Greenland, and the Faroe Islands; with illustrations of their natural history*. Edinburgh: Oliver & Boyd.

Olsen, O. 2004. ""Roten samlas till garfwerierna:" om blodrotens (*Potentilla erecta*) nytta." In *Växter i Linnés landskap*, eds. M. Manktelow and I. Svanberg. Uppsala: Swedish Science Press, 71–84.

Olsen, O., and I. Svanberg. 1998. *Tanning with tormentil (Pontentilla erecta): Ecological and ethnobotanical aspects*. Uppsala: Swedish Science Press.

————. 2005. "Blodrot *Potentilla erecta*." In *Människan och floran: etnobiologi i Sverige 2*, eds. M. Iwarsson, B. Pettersson and H. Tunón. Stockholm: Wahlström & Widstrand, 418–419.

Patursson, J. 1966. *Tættir úr Kirkjubøur søgu*. *Endurminningar*. Tórshavn: Felagiið Variðin.

Petersen, M.E. 1977. *Rekar*. Tórshavn: Bókaforlagiið Grønalííð.

Preston, C.D., D.A. Pearman, and A.R. Hall. 2003. "Archaeophytes in Britain." *Botanical Journal of the Linnaean Society* 145: 257–294.

Rasmussen, R. 1927. "Vegetationen i de færøske Fuglebjærge og deres nærmeste Omgivelser." *Botanisk Tidsskrift* 48: 46–70.

————. 1928. "Lundasina: eitt merkiligt tilbrigdi av grasaslagnum, *Festuca rubra*." *Varðin* 8: 45–51.

Rostrup, E. 1870. "Færøernes flora: en floristisk skitse." *Botanisk Tidsskrift* 4: 5–109.

Rydén, M. 2001. "Växternas namn." In *Människan och naturen: Etnobiologi i Sverige 1*, eds. B. Pettersson, I. Svanberg and H. Tunón. Stockholm: Wahlström & Widstrand, 45–54.

Sarukhán, J. 1985. "Ecological and social overview of ethnobotanical research." *Economic Botany* 39: 431–435.

Schübeler, F.C. 1888. *Viridarium norvegicum* vol. 2. Christiania: W. C. Fabritius & Sønner.

Scott, W. and R. Palmer. 1987. *The flowering plants and ferns of the Shetland Islands*. Lerwick: The Shetland Times.

Sjöbeck, M. 1927. "Bondskogar, deras vård och utnyttjande." *Skånska folkminnen* 1927: 36–62.

Skariði, J. 1957. "Eitt sindur um bjargarøkt í Skuvoy." *Varðin* 37: 99–122.

Skylv Hansen, J.M. 1967. *Gamlar gøtur* vol. 2. Tórshavn: Helge Justinussen Bókhandil.

Spence, M. 1914. *Flora Orcadensis. Containing the flowering plants arranged according to the natural orders*. Kirkwall: D. Spence.

Stoklund, B. 1971. "Skindsko og træsko." *Nationalmuseets arbejdsmark* 1971: 47–56.

Stoklund, B. 1991. "Fra center til periferi: Hovedlinjer i den nordatlantiske kulturudvikling fra middelalder til nyere tid." In *Nordatlantiske foredrag*, eds. J.P. Joensen, R. Johansen and J. Kløvstadd. Tórshavn: Noriðurlandahúsiið í Føroyum, 54–62.

Svabo, J.C. 1959. *Indberetninger fra en Rejse i Færøe 1782 og 1783*. Udgiven av N. Djurhuus, København: Selskabet til Udgivelse af færøske Kildeskrifter og Studier.

Svanberg, I. 1986. "A forest Lapp culture in Central Sweden in the 17th and 18th centuries." *Svenska Landsmål och Svenskt Folkliv* 109: 69–75.

————. 1987. "Turkic ethnobotany and ethnozoology as recorded by Johan Peter Falck." *Svenska Linnésällskapets Årsskrift* 1986–1987: 53–118.

————. 1988. "Marginal groups and itinerants." In *Ethnic groups in the Republic of Turkey*, ed. P.A. Andrews. Wiesbaden, 602–612.

————. 1996. "Ethnic categorizations and cultural diversity in Xinjiang: The Dolans along Yarkand River." *Central Asiatic Journal* 40: 260–282.

————. 1997. "Field horsetail (*Equisetum arvense*) as a food plant." *Fróðskaparrit*, 45: 45–55.

————. 1998a. "The use of wild plants in the Faroe Islands 1590–1990: A Contribution to Scandinavian Ethnobotany." *Svenska Linnésällskapets Årsskrift* 1996–1997: 81–130.

————. 1998b. "The use of rush (*Juncus*) and cottongrass (*Eriophorum*) as Wicks: An ethnobotanical background to a Faroese riddle." *Svenska Landsmål och Svenskt Folkliv* 323: 145–157.

————. 1998c. *Människor och växter: Svensk folklig botanik från "ag" till "örtbad."* Stockholm: Arena.

―――. 1998d. "The use of lichens for dyeing candles: Ethnobotanical documentation of a local Swedish practice." *Svenska Landsmål och Svenskt Folkliv* 324: 133–139.

―――. 2001a. "The Sami use of *Lactuca alpina* as a food plant." *Svenska Linnésällskapets Årsskrift* 2000–2001: 77–84.

―――. 2001b. "The snow bunting (*Plectrophenax nivalis*) as food in the northern circumpolar region." *Fróðskaparrit* 48: 29–40.

―――. 2002a. "Att se med allmogens ögon." In: *Ett växande vetande: Vetenskapsrådets temabok 2002*. Stockholm: The Swedish Research Science Council, 45–56.

―――. 2002b. "The trade in and use of *Gentiana purpurea* in Sweden." *Svenska Landsmål och Svenskt Folkliv* 326: 69–79.

―――. 2002c. "Kransborren (*Marrubium vulgare*) som odlingsväxt i Sverige." *Svenska Landsmål och Svenskt Folkliv* 327: 217–231.

―――. 2003. "Musslor och snäckor som föda och agn på Färöarna." *Gardar: Årsbok för Samfundet Sverige-Island* 33: 20–36.

―――. 2004a. "Ängsskäran (*Serratula tinctoria*) som färgväxt i 1700-talets Europa." *Svenska Linnésällskapets Årsskrift* 2002–2003: 49–67.

―――. 2004b. ""Växer på Uppsala gator": Lungrot (*Chenopodium bonus-henricus*) som levande kulturminne." In *Växter i Linnés landskap*, eds. M. Manktelow and I. Svanberg. Uppsala: Swedish Science Press, 107–128.

―――. 2005a. "Etnobiologien Linné." In *Varför reser Linné? Perspektiv på Iter Lapponicum*, ed. R. Jakobsson. Stockholm: Carlssons, 135–162.

―――. 2005b. "Källor till folklig växtkunskap." In *Människan och floran: Etnobiologi i Sverige 2*, eds. M. Iwarsson, B. Pettersson and H. Tunón. Stockholm: Wahlström & Widstrand, 23–34.

―――. 2005c. "Pestskråp." In *Människan och floran: Etnobiologi i Sverige 2*, eds. M. Iwarsson, B. Pettersson and H. Tunón. Stockholm: Wahlström & Widstrand, 447–448.

―――. 2005d. "En folklig lek med växter: Kämpar (*Plantago media*) i Linnés skånska resa." *Svenska Linnésällskapets Årsskrift* 2004–2005: 81–86.

―――. 2005e. "Jóansøkugras (*Plantago lanceolata*) på Färöarna: om växter som kulturhistoriska källor." *Gardar: Årsbok för Samfundet Sverige-Island* 35: 36–40.

Svanberg, I., and M. Szabó. 1993. *Etniskt liv och kulturell mångfald*. Stockholm: Nordiska museet.

Tarnovius, T. 1950. *Ferøers beskrifvelser*. Færoensia: Textus & Investigationes, II. København: Ejnar Munksgaard.

Turner, Nancy J. 1997. "Traditional ecological knowledge." In *The rain forests of home: Profile of a North American bioregion*, eds. P.K. Schoonmaker, B. von Hagen and E.C. Wolf. Covelo, CA: Island Press, 275–298.

Vasari, Y. 2002. "Botaniska bevis för västligare flyttare till Finland." In *När kom svenskarna till Finland?*, eds. A.-M. Ivars and L. Huldén. Helsingfors: Svenska Litteratursällskapet i Finland, 137–148.

Chapter 12

Procurement of Traditional Remedies and Transmission of Medicinal Knowledge among Sahrawi People Displaced in Southwestern Algerian Refugee Camps

Gabriele Volpato, Abdalahe Ahmadi Emhamed,

Saleh Mohamed Lamin Saleh,

Alessandro Broglia, and Sara di Lello

Introduction

Traditional medicinal systems worldwide are based on natural resources from the surrounding environment and on the ethnobiological knowledge needed to exploit those resources. Culture is seen as the filter between man and the surrounding environment; this implies that when the latter changes, traditional knowledge and practices come under pressure. When displacements occur because of war or other calamities, migrants and refugees strive to keep the connection between cultural identity, traditional resources, and their homeland (Brainard and Zaharlick 1989; Pieroni et al. 2005). However, their cultural identity comes under threat due to the loosing of ties with the place of origin, but is also threatened by the process of integration in the host culture (see Kim 2001).

In these contexts, the main forces that drive changes in the cultural domain of traditional medicinal knowledge are: (1) the adaptation of the original knowledge to the new (host) environment; and (2) the development of strategies to obtain the original remedies. Examples of the first process are the substitution of traditional remedies with locally available ones (traditional and/or modern), and the inclusion of remedies from the host culture into refugees' own pharmacopoeia (with the same use[s] or new ones elaborated according to their own belief systems) (Pieroni et al. 2005). Examples of the second process are strategies for the procurement of traditional remedies, such as their cultivation, gathering (in the new environment or through trips back to the source when these are possible), marketing, and the development of social networks that link migrants and refugees to relatives and friends in the place of origin.

About 165,000 (UNHCR —United Nations High Commissioner for Refugees—2005) Sahrawi people have been living in four refugee camps located on an isolated desert plateau in southwestern Algeria since 1975 (figure 12.1), when Morocco occupied the former Spanish colony of Western Sahara, and they were forced to flee. In a context in which refugees find themselves in a long-lasting and intractable state of limbo, with a high level of dependency from donors, and a peace process that seems to have irredeemably stagnated (Seddon 2000; Cozza 2003; Shelley 2004), Sahrawis have been fighting against identity loss both militarily and through social and cultural resistance. In the realm of traditional medicine, refugees established practices to obtain their traditional medicinal remedies through social networks within and outside the camps. These strategies permitted the continuation of traditional folk medicinal practices when this would not have been possible otherwise due to the lack of botanical resources in the area of the camps. Moreover, they represent a link between displaced people and their homeland (Western Sahara), and ultimately constitute a form of cultural resistance in the wastelands of the refugee camps.

In this chapter, we examine the process of adaptation of Sahrawi traditional medicine under the conditions of displacement, particularly focusing on the origin of remedies used in the camps and on the strategies developed by refugees to obtain them, as well as on the changes that occurred to traditional medicine in relation to knowledge transmission and cultural identity.

Methodologies

Sahrawi History and Life in the Refugee Camps

Sahrawis, literally "people from the desert," is the name given to the tribes of nomadic and pastoral people who traditionally inhabit a coastal area of northwestern Africa called Western Sahara. The origin of the Sahrawis traces back to the fusion between the Bani Hasan Arabic groups that migrated from Yemen from the eleventh to the thirteenth century B.C. and the Sahjaha nomadic Berber group that was living in Western Sahara. A slow process of integration of Arabic migrants with the autochthonous Berbers took place until the sixteenth century, from which arose a nomadic population of Sunnite Muslems, the Sahrawis. Sahrawi people were essentially nomadic, pasturing camels, goats, and sheep in the sandy low-lying plains of Western Sahara and relying for food on camel milk and meat, dates, sugar, and small amounts of cereals and legumes (OXFAM 1995; Cozza 2003). They moved in accordance with the seasons, their routes dictated by wells, watering holes, and rains. As an informant stated, they were "people moving towards every sign of rain."

During the 1960s, under Spanish colonization, the Sahrawis became increasingly sedentary. In 1975, following the occupation of Western Sahara by Mauritanian and Moroccan forces (Mauritania pulled out from Western Sahara in 1979), about 70,000 Sahrawis became refugees after fleeing the Moroccan army (Loewenberg 2005; Spiegel and Qassim 2003). As most of the men joined the resistance army immediately, it was the women, children, and old people who fled to the camps (Lippert 1992). Nowadays, Sahrawi people live in the refugee camps of southwestern Algeria (about 165,000), and in the part of Western Sahara under dispute (about 65,000), as well as residing as minority groups in south Morocco, Mauritania, the Canary Islands, and Algeria.

Camps are located in a desert plateau called Hamada near the Algerian town of Tindouf. The Sahrawi Arab Democratic Republic (RASD)/Frente Polisario (hereafter called "Polisario")—the political and military organization that represents the refugees—was granted administrative and governing autonomy over this area by the Algerian government. In the camps, refugees live in canvas tents and mud brick huts, with severe problems of water and food supply, and with car batteries as the main source of electricity. The European Union, some bilateral development cooperations, UN agencies, and several solidarity groups existing all around Europe make food, shelter, and other basic commodities available (Loewenberg 2005). Indeed, the general food ration covers only 68 percent of the estimated energy requirement of the population, and anemia caused by malnutrition is a common health problem among the general Sahrawi population, especially children and women (Branca 1997; UNHCR 2002). In spite of the efforts of the RASD and donors to build an efficient health system in the camps, the average life expectancy is around fifty, with extremely high child mortality rates mainly due to diarrhea and acute respiratory infections (Branca 1997; Mezzetti 1994).

In an attempt to cope with the situation and looking for an improvement of the quality of life in the camps, throughout the years refugees developed an informal economy with the marketing of many products (from clothing to personal hygiene products and food items to supplement the base diet provided by humanitarian assistance) from within and outside the camps (Bhatia 2001). The development and expansion of trading routes through the camps—from Sene-

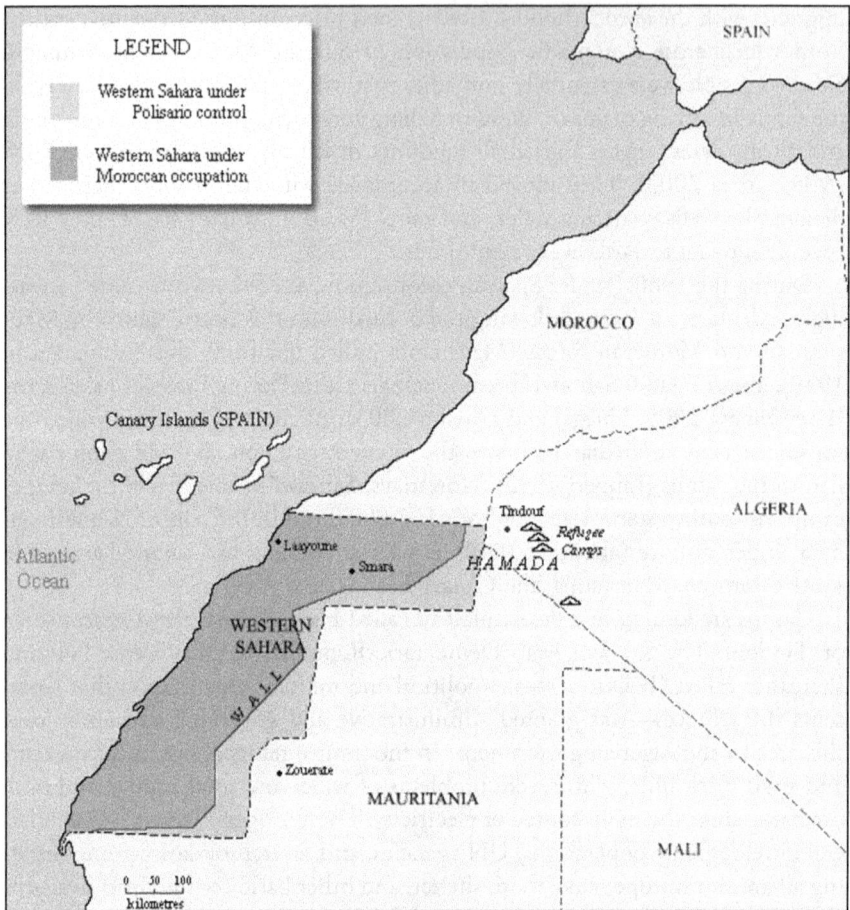

Figure 12.1. *Location of the study area and geographical locations reported in the text*

gal, Mali, Mauritania, Algeria, and Spain—brought about the emergence of small neighborhood shops and distinct commercial areas within the camps.

Besides the camps, the Polisario also has political control over the eastern part of the Western Sahara, which has been taken away from Moroccan control through a guerrilla war that lasted until the peace agreement of 1991 (Bhatia 2001). Consequently, Western Saharan territory is divided geographically and politically into two parts separated by 2,200 kilometers of earthen wall, constructed by Morocco in the late 1980s and protected by 150,000 soldiers and about one million landmines (San Martín 2004; Loewenberg 2005). The wall crosses into northwest Mauritania, physically separating the eastern portions of the territory under

Polisario control. These portions are the so-called "liberated territories" (approximately 20 percent of the Western Sahara), while the remaining "occupied territories" are under the administering authority of the Moroccan government.

FIELD STUDY

The investigation of Sahrawi traditional medicine in the Sahrawi refugee camps has been undertaken within a cooperation project funded by the European Union ("Salud animal en la tendopoli Sahrawi—Algeria," nr. ONG-PVD/2002/ 020-151) and carried out by two Italian NGOs (Africa 70 and SIVtro Veterinari Senza Frontiere Italia). Fieldwork was conducted over a period of fourteen weeks in October 2003, November–December 2004, and January–February 2006. Investigation methods for ethnobotanical and cognitive-anthropological analysis were structured interviews with a random-selected sample of households, semi-structured interviews with traditional healers and informants regarded as knowledgeable by local people, consensus analysis using free listings, a "walk in the woods" approach with knowledgeable informants, and a voucher specimens collection of the remedies cited (see Berlin 1992; D'Andrade 1995; Martin 1995; Alexiades and Sheldon 1996; Cotton 1996; Weller 1998; Cunningham 2001; Puri and Vogl 2004).

During the 2004 period, structured interviews were conducted with thirty-seven households (people living together in the same tent and mud brick huts) in the refugee camps of Awserd and Smara to investigate Sahrawi popular medicine. In each household, members were asked to identify the person in the household responsible for keeping and administering traditional remedies; the interview was then conducted with that person. Respondents had a mean age of 56 (ranging from 26 to 84), thirty-three of them being women and four men. Household members numbered five on average, ranging from two to ten. Informants were asked about the frequency of use of traditional remedies in the household, the remedies available within the household, their use, origin, and method of procuring, difficulties in the procuring of remedies, and the origin and transmission of their knowledge about traditional medicine. Interviews were conducted in Hassanya, the Arabic dialect with Berber substrate spoken by Sahrawis (see Cohen 1963; Quitout 1999; Ould Mohamed Baba 2004), and translated back into Spanish by local research assistants.

In every case, prior informed consent was obtained verbally before the interview was conducted and before a camera or audio-recorder was used. Participants were given an explanation of the methodology, aims, and possible outcomes of the study. Throughout the field study, the ethical guidelines adopted by the AAA (American Anthropological Association) (1998) were followed, and methodological and ethical advice according to Jacobsen and Landau (2003) were taken into consideration.

During the interviews, pharmacognostic (dried) specimens, and in a few cases fresh specimens, were collected and inventoried. Voucher specimens of the plants cited coming from the Hamada and the part of Western Sahara under Polisario control were collected in the field with knowledgeable informants. Plant identi- fication followed the Sahara and Western Sahara botanical standard treatises (Ozenda 1991; Lebrun 1998). Botanical names are written in full with author(s) and family name only the first time they appear in text or tables. Voucher spec- imens, digitally recorded interviews, and digital pictures of plants and remedies are available at the first author's address.

Sahrawi Traditional Medicine

Sahrawi traditional medicine finds its origin centuries ago, in Arabic and Islamic medicine and in the pharmacopoeia of Berber populations based on local re- sources from Western Sahara, but also from Senegal, Mali, and Mauritania, where nomads were moving about looking for pastures and water for their herds. At the same time, they were exchanging products with traders from Mauritania and Mali from the south and from Morocco and Algeria from the north, thus obtaining medicinal remedies not available in their nomadic areas (Guinea 1949; Monteil and Sauvage 1953; Caro Baroja 1955). In addition, remedies from the coast were marketed eastward and vice versa. With the partial process of the 1960s of becoming sedentary, the marketing of medicinal remedies in the main cities of Western Sahara, such as Laayoune and Smara, became another impor- tant source for obtaining these products.

Two types of traditional medicinal knowledge were distinguishable: a special- ized knowledge practiced by experts, and a popular medicine practiced by women within families. In relation to the former type, during war expeditions there was always a male expert carrying the knowledge needed to heal wounds, to adjust broken bones, and even to carry out surgical operations (Caro Baroja 1955; Mezzetti 1994). At the same time, female specialists were responsible for repro- ductive health and ophthalmologic problems, often through complex medicinal recipes. While specialized knowledge of women (based on medicinal plants and other remedies) has been preserved in the camps, specialized knowledge of men is quickly disappearing. In fact, the latter practices were used soon after the dis- placement took place to treat the effects of Moroccan bombings (Mezzetti 1994). However, at present they are contested by donors, development agencies, NGOs, and by the Polisario government as well, all of which have been trying to develop a modern biomedical system in the camps.

Popular medicinal knowledge is practiced and transmitted within families and deals mainly with common ailments such as digestive disorders, broncho- pulmonary afflictions, and *eghindi,* a folk term that includes allergies and intoxications. In this chapter, we will present and discuss data only about this second type of medicinal knowledge and related traditional remedies.

Informants reported 287 citations of remedies used, with a mean of almost eight remedies each, the number of remedies cited per household ranging from zero to nineteen. Of these remedies, 240 (84 percent) are of vegetal origin, 26 (9 percent) of mineral origin and 21 (7 percent) of animal origin. Vegetal remedies correspond to 68 products belonging to 57 species and 32 botanical families, Fabaceae and Lamiaceae (five species each) being the most cited.

In spite of the forced settlement of most of the population, the almost complete absence of vegetal life around the camps, and the fact that refugees have been living in the Hamada for thirty years, traditional medicine has been maintained within households. About 78 percent of the households reported having used traditional remedies in the week before the interview, and this figure rises to 95 percent when the time period considered is six months prior to interviews. The person who prepares and administers the remedies, and who holds the knowledge about them, is the oldest woman of the household in 89 percent of the cases, and the oldest man in the other 11 percent. Traditional remedies are kept dried and often triturated, enveloped in cloth pieces or in plastic bags, and form some of the few items present in the tents of the camps. This habit of conservation of remedies is characteristic of Saharan nomadic populations, who need to store the remedies as they encounter them in specific phytogeographic regions and during limited periods of the year (i.e., during the reproductive cycle of plants after the few and irregular rains), and not when the particular remedies are needed.

Table 12.1 lists the medicinal plants that informants had in the household at the time of the interview. Plants are reported in alphabetical order of scientific name, along with botanical families, voucher specimens when available, vernacular names as collected in the refugee camps, place(s) of procurement of the remedies (in order of relative importance within each specific product's row), and percentage of quotation. The two most frequently reported species are *Acacia tortilis* and *Ammodaucus leucotrichus;* the resin and the dried leaves of the first species are reported by 76 percent and 38 percent of the informants respectively, while the dried fruits of the latter are reported by 54 percent of the informants, the species being described by Sahrawi as their "traditional antibiotic." Other frequently mentioned species are characteristic of the flora of Western Sahara and include *Cleome amblyocarpa* and *Maerua crassifolia,* as well as species of southern Sahara-Sahelian areas such as *Adansonia digitata,* or are important Arabic remedies such as the seeds of *Trigonella foenum-graecum.*

In table 12.2, other biological remedies are presented, along with a description of the remedy, vernacular names, places of procurement and frequency of quotation. Red hematite, a mineral used as an antiseptic and cicatrizer, is reported

Table 12.1. Traditional Phytotherapeuticals Stored in Tents by Sahrawi People in South Algerian Refugee Camps

Botanical taxon and voucher specimen code(s)	Botanical family	Folk name(s) recorded in the camps	Specific products/ Plant part(s)	Provenience(s) of the plant material	% of quotation (n=37)
Acacia ehrenbergiana Hayne GV1015/GV1058	Fabaceae	Tamat	resin: el-alk tamat leaves: warga tamat	IWS-g IWS-g	2.7 2.7
Acacia senegal (L.) Willd. var. *senegal* GV1076	Fabaceae	Amour	seeds: sallaha	IWS-b/Mr-b	5.4
Acacia tortilis (Forssk.) Hayne subsp. *raddiana* (Savi) Brenan var. *raddiana* GV1010/GV1025/GV1061	Fabaceae	Talha	resin: el-alk leaves: warga talha resin (from lower part of trunk): abakak bark: dhbag	IWS-g/RC-b/Tn-b/Mr-b IWS-g/RC-g,b IWS-g IWS-g	75.7 37.8 2.7 2.7
Adansonia digitata L. GV1016	Bombacaceae	Teidoum	fruit pulp: tashmaht leaves: taghia teidoum	Mr-b/IWS-b/oWS-b Mr-b	24.3 8.1
Allium cepa L.	Alliaceae	Besla	bulbs	Tn-b	2.7
Allium sativum L.	Alliaceae	Thum	bulbs	Tn-b/RC-b	16.2
Ammodaucus leucotrichus Coss. & Durand GV1013/GV1033	Apiaceae	Kammuna, kammuna t'rag	fruits	IWS-g/oWS-b	54.1
Anastatica hierochuntica L. GV1027	Brassicaceae	Kamsha	leaves	IWS-g	8.1
Argania spinosa (L.) Skeels	Sapotaceae	Argan	seeds	oWS-b	2.7
Artemisia herba-alba Asso GV1042	Asteraceae	Shih	flowering tops	oWS-b/Tn-b/Ag-b	8.1
Asphodelus tenuifolius Cav. GV1078	Asphodelaceae	Tazia	seeds	IWS-g	2.7
Atriplex halimus L. GV1052	Chenopodiaceae	Legtaf	leaves	IWS-g/Tn-b	10.8

Species	Family	Local name	Part used	Code	%
Balanites aegyptiacus (L.) Delile GV1086	Balanitaceae	Teichat	fruits	IWS-g	5.4
Calligonum comosum L'Hérit	Polygonaceae	Shdida	leaves	IWS-g	8.1
Cassia italica (Mill.) Spreng. GV1020	Fabaceae	Fellajit	leaves	IWS-g	5.4
Caylusea hexagyna (Forssk.) M.L. Green GV1031	Resedaceae	Dhenban	aerial parts and flowers	IWS-g	2.7
Centaurea pungens Pomel GV1079	Asteraceae	Zreiga	leaves	IWS-g	5.4
Cotula cinerea Delile GV1083	Asteraceae	Gartufa, rebruba	flowering tops	IWS-g	5.4
Cinnamomum verum J. Presl GV1072	Lauraceae	El-qarfa	bark	Tn-b	2.7
Cleome amblyocarpa Barr. et Murb. GV1026	Capparaceae	Mkheinza	leaves	IWS-g/RC-g	32.4
Commiphora africana (A. Rich.) Engl. GV1017	Burseraceae	Dirs	resin: umm nass	Mr-b/Tn-b/oWS-b/ IWS-b	16.2
Coriandrum sativum L. GV1073	Apiaceae	Kasbour	seeds	Tn-b	2.7
Corrigiola telephiifolia Pourret GV1089	Caryophyllaceae	Tassergin't	roots	oWS-b	2.7
Chamomilla pubescens (Desf.) Alavi GV1090	Asteraceae	Lerbien	flowering tops	IWS-g/Ag-b	16.2
Cuminum cyminum L. GV1073	Apiaceae	Kammuna	seeds	Tn-b/Ag-b	5.4
Cymbopogon schoenanthus (L.) Spreng. GV1087	Poaceae	Idkhir	aerial parts	IWS-g	2.7
Echium horridum Batt. GV1091	Boraginaceae	Harsha	flowers	IWS-g	2.7
Eugenia caryophyllata Thunb.	Myrtaceae	Qronfel	flowering buds	Tn-b	2.7
Euphorbia calyptrata Coss. & Durieu var. *involucrata* GV1080	Euphorbiaceae	Rammadah	roots	IWS-g	2.7

(continued)

Table 12.1. Continued

Botanical taxon and voucher specimen code(s)	Botanical family	Folk name(s) recorded in the camps	Specific products/ Plant part(s)	Provenience(s) of the plant material	% of quotation (n=37)
Euphorbia officinarum L. subsp. *echinus* GV1001	Euphorbiaceae	Daghmus	inner stem / inner stem mixed with honey: daghmus honey	IWS-g/oWS-g / oWS-b	10.8 / 2.7
Ferula communis L.	Apiaceae	Leklekha	resin: fasoukh	Tn-b/oWS-b	10.8
Hammada scoparia (Pomel) Iljin. GV1009/GV1021	Chenopodiaceae	Remth	aerial parts	IWS-g/RC-g	10.8
Lavandula sp.[a] GV1022/GV1032	Lamiaceae	Lehzema	flowering tops	Tn-b/oWS-b	24.3
Lepidium sativum L. GV1060	Brassicaceae	Reshad	seeds: habb er-reshad, afatash	Tn-b/IWS-b	5.4
Lycium intricatum Boiss. GV1085	Solanaceae	Ghardeq	leaves / fruits: asako	IWS-g / IWS-g	5.4 / 2.7
Maerua crassifolia Forssk. GV1007/GV1019	Capparaceae	Atil	leaves: sadra el-hadra / stems: mesuak	IWS-g / IWS-g	21.6 / 2.7
Mesembryanthemum cryptanthum Hook. f.	Aizoaceae	Afzu	seeds	IWS-g	5.4
Myristica fragrans Houtt. GV1040	Myristicaceae	El-gouza	seeds	Tn-b	5.4
Nigella sativa L.	Ranunculaceae	Habba souda	seeds	Tn-b	13.5
Nucularia perrini Batt. GV1047	Chenopodiaceae	Askaf	stems	IWS-g	2.7
Panicum turgidum Forssk. GV1051	Poaceae	Umm rokba	roots	IWS-g	5.4
Peganum harmala L. GV1066	Zygophyllaceae	Harmal	seeds	Tn-b	5.4

Species	Family	Local name	Plant parts	Provenance	%
Piper nigrum L. GV1073	Piperaceae	Felfel	seeds	Tn-b	2.7
Rhus tripartita (Ucria) Grande GV1023/GV1064	Anacardiaceae	Shdari	leaves seeds: dhmah	IWS-g/Tn-b IWS-g	10.8 2.7
Rosmarinus officinalis L. GV1062	Lamiaceae	Iazir	aerial parts	Tn-b	8.1
Salsola vermiculata L. GV1093	Chenopodiaceae	Shifne	aerial parts	IWS-g	2.7
Salvia aegyptiaca L. GV1049	Lamiaceae	Azoukni, tezouknit	seeds: afatash flowering tops	IWS-g IWS-g	16.2 8.1
Tamarindus indica L. GV1004	Fabaceae	Aganat	seeds	Mr-b	2.7
Terfezia ovalispora Pat. GV1008	Terfeziaceae	Terfas	rhizome	IWS-g	5.4
Tetraclinis articulata (Vahl) Masters GV1003/GV1059	Cupressaceae	Ar'ar	leaves	RC-b	2.7
Thymus sp.[a] GV1028/GV1029	Lamiaceae	Azoukni	flowering tops	Tn-b	5.4
Trigonella foenum-graecum L. GV1018/GV1044	Fabaceae	Halba	seeds	Tn-b/IWS-b	21.6
Withania somnifera (L.) Dunal in DC. GV1081	Solanaceae	Sekran	seeds	Tn-b	5.4
Zingiber officinale Roscoe GV1039	Zingiberaceae	Skenjbir	rhizome	Tn-b	5.4
Ziziphus lotus (L.) Desf. GV1002	Rhamnaceae	Shdir	fruits: n'beg	IWS-g	2.7
ND[b]		Lebtena	leaves: taghia lebtena	IWS-g	21.6

Provenance(s) of the plant material. IWS: part of the Western Sahara under Polisario control; oWS: part of the Western Sahara under Moroccan occupation; RC: refugee camps and Hamada; Tn:Tindouf market; Tn:Tindouf market; Ag: other Algerian markets but Tindouf; Mr: markets and traders from Mauritania; g: gathered from the wild; b: bought in markets and shops, from traders or obtained for free from other people.

[a] Botanical identification at species level not possible, since only dry plant parts were been available
[b] Botanical identification not possible, since only triturated specimens of the plant could be obtained

Table 12.2. Other Biological Remedies Used by Sahrawi Refugees
in South Algerian Camps

Remedy	Folk name recorded in the camps	Provenance of material	% of quotation (n=37)
Red hematite (Fe_2O_3)	Hemera	lWS-g/oWS-g/Tn-b/Mr-b	35.1
Honey	Lasal	Sp-b/Tn-b	18.9
Piece of salt block	Shabba	Tn-b/Mr-b	16.2
Black galena (PbS)	Kehla	lWS-g/oWS-b	10.8
Skin and fat of spiny-tailed lizard (*Uromastix* spp.)	Dab	lWS-g	8.1
Unrefined sugar	Azucar el-har	oWS-b	8.1
Goat's fat	Adhin dsam	RC-pr	5.4
Ostrich (*Struthio camelus* L.) egg	Naama	lWS-b/Tn-b	5.4
Sea urchin (*Echinus* spp.)	Dghemissa	oWS-b/Ag-b	5.4
Chameleon skin	Buya	oWS-b/RC-b	5.4
Cooked camel skin	Aotye	RC-pr	2.7
Cream of goat's milk	Dsam leghem	RC-pr	2.7
Camel fat	Ludek	RC-b	2.7

Provenance(s) of the plant material. lWS: part of the Western Sahara under Polisario control; oWS: part of the Western Sahara under Moroccan occupation; RC: refugee camps and Hamada; Tn:Tindouf market; Ag: other Algerian markets but Tindouf; Mr: markets and traders from Mauritania; Sp: Spain. g: gathered from the wild; b: bought at markets and in shops, from traders and obtained for free from other people; pr: own production.

by 35 percent of the informants, while almost 20 percent of them cited honey as a medicinal remedy.

Origin of the Remedies Used and Procurement Strategies

In table 12.3, we present the places of provenance of the remedies used in the camps along with their frequencies of mention (the number of times that remedies available in the household were obtained from a specific place) and related percentages. In discussing this table, we will start from within the camps and continue our discussion in relation to other places more distant from the camps.

The area where the refugee camps are situated is a rocky plateau about 500 meters above sea level, the Hamada, characterized by rocks eroded by sun, wind, and scarce rainfall, and almost completely without vegetal life. With less than 50 mm of rain per year, the Hamada is classified as an absolute desert of Libyan or

Table 12.3. Origin of Traditional Remedies Used in Sahrawi Refugee Camps

Place of provenance	Percentage	Frequency of mention
Western Sahara under Polisario control	58.5	168
Tindouf market	20.1	58
Western Sahara under Moroccan occupation	6.7	20
Mauritania	5.6	16
Refugee camps' markets and shops	2.8	8
Algerian markets other than Tindouf	1.8	5
Spain	1.8	5
Own production in the camps	1.4	4
Hamada (surroundings of the camps)	1.4	4
Total	100	287

continental type (Ozenda 1991), considered unfit for human life. Less then 1.5 percent of the remedies used are gathered in the area surrounding the refugee camps; among them, the leaves of *Acacia tortilis* (cited by one informant who gathered them from one of the very few individuals growing in the camps), the leaves of *Cleome amblyocarpa* (mentioned by two informants and gathered some kilometers north of the camps), and the aerial parts of *Hammada scoparia* (reported by one informant). Plants readily available are usually not stored in tents but instead collected when needed. This is the case with the few medicinal plants growing around the camps, including *Pergularia tomentosa* L. (Asclepiadaceae), *Zygophyllum gaetulum* Emb. & Maire subsp. *gaetulum* (Zygophyllaceae), and *Hammada scoparia*. Their local availability is probably the reason for their absence or low scores in survey results. In addition, informants collected the aerial parts of *H. scoparia* around the camps only in one case out of four, since plant individuals growing in the Western Sahara were being regarded as "more powerful."

Almost 3 percent of the remedies cited were bought in the camps. They are sold in food item shops and come mainly from Western Sahara or Mauritania (e.g., resin and dried leaves of *A. tortilis*), and from Algeria (e.g., leaves of *Tetraclinis articulata* and chameleon skin, a highly regarded but difficult-to-procure remedy). Common food items used as remedies such as garlic and onion are also found in these shops. The selling of these products usually occurs as the result of personal initiatives of local traders, who buy well-known remedies from other traders, herders, or from other people coming back from trips outside the camps. Given the efforts made by refugees to procure traditional therapeuticals for home consumption, more medicinal products could be expected to be sold in the camp markets and shops. Nevertheless, trading in medicinal remedies within the camps has not been developed. Refugees prefer to obtain Western Saharan remedies mainly through family and other non-commercial social networks, or to buy remedies at Tindouf market, or directly from "traveling" traders. Reasons for this include the perception that remedies sold in the camps are of low medicinal quality, the high prices of remedies in camp markets and shops, and the irregularity in the supply of these remedies, all of which render other networks more reliable for procurement.

Given the climatic situation of the Hamada, no medicinal plants are cultivated in the camps, and the 1.4 percent of remedies from refugees' own production is of animal origin. Remedies obtained from wild and domesticated animals are traditionally relevant in Sahrawi medicine. Camels and goats are kept profusely around the camps in spite of the extreme difficulties people have in feeding them: it is estimated that families in the camps own approximately 45,000 sheep and goats and 500 camels (DNV-RASD 2005). Remedies from these animals are the only ones that refugees can produce by themselves in the camps, and are usually given free to friends and relatives when needed.

The Algerian town of Tindouf is situated about forty kilometers from the refugee camps. This city of about 50,000 inhabitants serves as the southern headquarters for the Algerian Armed Forces and is important in trading routes across the Sahara desert. Tindouf is the pole of the refugees' commercial networks, and informants buy about 20 percent of their remedies in markets and shops in Tindouf, and from traders in the city's streets. Typically, these remedies are plants cultivated in northern parts of Algeria or imported from other countries and characteristic of Arabic and Islamic medicine. Examples include the highly regarded seeds of *Trigonella foenum-graecum, Peganum harmala,* and *Nigella sativa,* the latter being cited in the Koran as a panacea (Siouti 1994), and remedies known worldwide like *Myristica fragrans, Cinnamomum verum,* and *Eugenia caryophyllata.* Among the non-vegetal therapeuticals bought in Tindouf, four informants mentioned pieces of salt blocks from local salt deposits (see McDougall 1990). Apparently, the medicinal remedies' market of Tindouf has been developing with the refugee camps, and traditional Sahrawi remedies coming from the part of Western Sahara under Polisario control or from Mauritania can be found on sale increasingly in the city. Nevertheless, refugees buy these kinds of remedies in Tindouf only if they have no other possibilities of obtaining them directly. As an example, out of twenty-eight informants reporting to possess *el-alk* (the resin of *A. tortilis*) in their tents, only one bought it in Tindouf, while twenty-three informants obtained it from the "liberated territories." Three people bought it in camp markets and shops and one from Mauritanian traders. Similarly, only one informant out of four bought the leaves of *Rhus oxycantha* in Tindouf, while the others obtained them from the "liberated territories." Reasons for this preference again lie in the cultural ties of Sahrawis with their homeland and the perception that remedies coming without intermediation from Western Sahara or Mauritania are of better quality and more powerful than from other locations.

The resin of *Commiphora africana*, for example, is bought preferentially from Mauritanian traders, informants stating that the resin sold in Tindouf is not good as a remedy. Another reason for this preference are the high prices of remedies in Tindouf; as it happens with remedies on sale in the camps, prices increase with increasing intermediation of different traders, and with increasing distance from their geographical area of production or gathering. Refugees have no source of income, and work for the military or political administration is typically unpaid. As a result, on the one hand they seek to diversify family income, with individual members separately engaged in commerce, military and administrative services, and small private activities, and on the other hand they try to establish non-commercial networks for obtaining items for living, including traditional remedies. Commercial networks are growing in the camps, and the source of the starting capital for these networks include pensions for former Spanish civil servants, and remittances from Sahrawis working abroad (mainly in Algeria, Mauritania, the Canary Islands, Spain) (Bhatia 2001). This process has expanded also the private ownership of cars and trucks, typically imported from Spain or Mauritania. Consequently, it also facilitated the movement of refugees among and between camps and Tindouf, increasing also the marketing of medicinal remedies and the supplying of these remedies in Tindouf.

Informants bought almost 2 percent of the remedies cited at other Algerian markets except Tindouf, mainly in Alger and Beshar. There, refugees obtain otherwise difficult-to-find remedies, like *Artemisia herba-alba* or crushed sea urchins. Those refugees who have the legal and economic possibility to travel across Algeria or to and from Europe, for example, buy remedies, which they bring back to the camps.

In spite of having been forced to become sedentary, Sahrawis reflect their nomadic culture by depending (in almost 60 percent of the reports) on products and species gathered from the wild in Western Sahara. Medicinal species are mostly gathered in the eastern stripe of Western Sahara under Polisario control. This area is characterized by a sub-oceanic desert climate where the lack of rains is partly replaced by a high hygrometric content in the atmosphere that allows more plant species to grow, and by permanently dispersed vegetation, at least in the southern part (Guinea 1949; Ozenda 1991). About 65 percent of the remedies that were cited (counting the number of remedies and not the number of citations) are procured exclusively or at least partly from these territories.

The control of this eastern stripe of Western Sahara is crucial for the conservation of Sahrawi traditional medicine in the refugee camps, and refugees' families have established social networks to obtain remedies traditionally gathered in this area. Out of thirty-seven households interviewed, thirty-two (about 86 percent) have some established network to obtain remedies from the "liberated territories." Figure 12.2 shows the networks through which these resources are gathered and brought to the camps as well as their respective percentages.

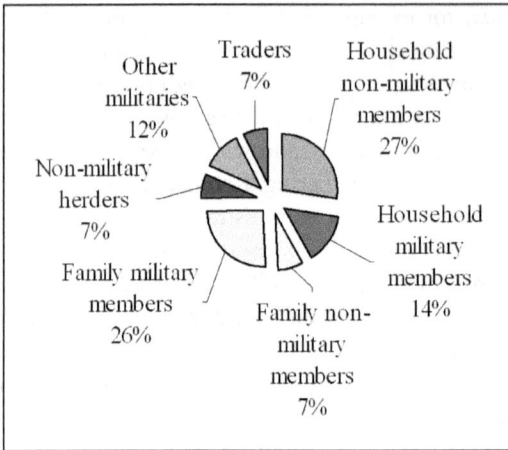

Figure 12.2. Social networks for procuring traditional remedies from the "liberated territories"

Household members bring more than 40 percent of the remedies to the camps. If we also include the members of the family, the figure rises to 74 percent, which shows how large parts of the networks are intrafamilial non-commercial networks. Family members who travel to the territories usually bring back to the camps a large variety and amount ("full bags") of remedies to be distributed to other family members who had asked for them. Forty percent of the family members who are in charge of bringing the remedies to the camps are combatants of the Polisario, a figure that rises to 52 percent of the total networks if we include also the combatants not belonging to the family. In fact, members of the Sahrawi population in this area are mainly combatants cantoned with their units in order to control the territory and pasture the about 27,000 camels belonging to the Polisario (DNV-RASD 2005). The number of soldiers is estimated to be between three and six thousand; at any given time, one third of them are on permission, which allows them to leave their posts at the front in order to return to their families in the camps (Bhatia 2001). These soldiers gather the plants and other remedies during favorable seasons and store them until they go back to the camps on permission. Other than the militaries, in the territories live some Sahrawi nomads who use the area as grazing land for camels and goats. A large majority of them are people and families from the camps who travel with their herds to the territories during the cold season (from September–October and February–March) and live according to a traditional lifestyle. During favorable cold seasons, plants grow and are gathered to be used *in loco* as well as to be brought back to the camps for their own consumption or that of relatives and friends.

In some cases, these non-commercial and intrafamilial networks for procuring traditional remedies are embedded into non-familial commercial activities like the production and marketing of coal from the "liberated territories" to the camps, or, during favorable years, the gathering and marketing in Tindouf of the desert truffle (*Terfezia ovalispora*).

About 7 percent of the remedies coming from the "liberated territories" are obtained through commercial relations with traders who come from Mauritania and sell to families and combatants some specific products that are then brought to the camps. In table 12.3, we can see that more than 5 percent of the remedies reported in the camps have been bought from traders in Mauritania (e.g., in Zouerate); these traders are an important source of specific products that are otherwise difficult to obtain. These products are mainly those characteristic of Sahelian areas, like the dried fruit pulp and the dried and powdered leaves of *Adansonia digitata,* and the resin of *Commiphora africana.* On some occasions, traders coming from Mauritania arrive in the camps to sell remedies and other stuffs (including clothes, personal products, and handicrafts).

As reported in table 12.3, almost 7 percent of the remedies come from the territories of the Western Sahara under Moroccan occupation, on the other side of the *berm*—the Moroccan defensive wall. The western portion of the territory has a population of 200,000 Moroccan soldiers, a similar number of Moroccan settlers from the post–1975 period, and more than 65,000 indigenous Sahrawi (1997 figures) (Bhatia 2001). Most of these Sahrawis have been separated from their families for three decades, but have kept family ties in spite of the physical separation. Products are sent to the camps from relatives or are brought back to the camps by those refugees who have had the permission recently to travel to Western Sahara for visiting the family. Nevertheless, procurement of traditional remedies in the "occupied territories" is still difficult for most refugees, and usually takes place only for specific products that can not be procured in other ways. These products are usually bought at the markets of Laayoune or Smara cities, and include plants characteristic of the oceanic climate of Morocco and Western Sahara, as well as endemic species like *Argania spinosa* and *Corrigiola telephiifolia,* and products of oceanic origin like sea urchin. Sweet remedies are highly regarded in Sahrawi culture, especially for treatment of *eghindi,* a very common folk illness. Unrefined sugar arrives in the camps from the occupied Western Sahara, probably coming from the Canary Islands. The habit of using this sugar goes back to the commercial relations that Sahrawi people living in the coast traditionally had with the people of these islands (see Caro Baroja 1955).

Beyond *el-alk* (acacia resin) and unrefined sugar, another most highly regarded sweet remedy is honey, which can sometimes be found on sale in Tindouf. Indeed, the main place of origin of the honey consumed in the camps (in six out of seven citations) is Spain, and honey is completely responsible for the almost 2 percent of remedies of Spanish provenance in table 12.3. Honey arrives in the camps mainly through Sahrawi children who go to Spain during summer holidays under the auspices of NGOs and solidarity groups; in some cases these children bring back pots of honey to the camps.

Difficulties in the Procurement of Remedies

We have shown up to now that traditional Sahrawi medicine is still practiced by refugees after thirty years of living displaced in the desert. We have identified the remedies used, and investigated the variety of solutions that have been established by refugees to procure the remedies they depend upon. The following research questions consequently arise: are there traditional remedies that are difficult to procure? If so, which are these remedies? We investigated these issues through free-listing ("please tell me which are the remedies that you would like to have but that you have been unable to procure") The results are reported in table 12.4 for remedies of plant origin, and in table 12.5 for other biological remedies of animal or mineral origin, cited by at least two informants. On average, each in-

Table 12.4. Medicinal Plants Reported as Difficult to Procure in Sahrawi Refugee Camps

Botanical taxon and family	Folk name(s) recorded in the camps	Specific products/ Plant part(s)	% of quotation (n=37)
Salvia aegyptiaca L. (Lamiaceae)	Azoukni, tezouknit	seeds: afatash	13.5
Artemisia herba-alba Asso (Asteraceae)	Shih	flowering tops	10.8
Balanites aegyptiacus (L.) Delile (Balanitaceae)	Teichat	fruits	10.8
Euphorbia officinarum L. subsp. *echinus* (Euphorbiaceae)	Daghmus	daghmus honey inner stem	8.1 5.4
Acacia tortilis (Forssk.) Hayne subsp. *raddiana* (Savi) Brenan var. *raddiana* (Fabaceae)	Talha	leaves: warga talha resin: el-alk	5.4 5.4
Lycium intricatum Boiss. (Solanaceae)	Ghardeq	leaves fruits: asako	5.4 5.4
Adansonia digitata L. (Bombacaceae)	Teidoum	fruit pulp: tashmaht	5.4
Ammodaucus leucotrichus Coss. & Durand (Apiaceae)	Kammuna, kammuna t'rag	fruits	5.4
Commiphora africana (A. Rich.) Engl. (Burseraceae)	Dirs	resin: umm nass	5.4
Maerua crassifolia Forssk. (Capparaceae)	Atil	leaves: sadra el-hadra	5.4
Mesembryanthemum cryptanthum Hook. f. (Aizoaceae)	Afzu	seeds	5.4
ND	Lebtena	leaves: taghia lebtena	5.4

Table 12.5. Other Biological Remedies Reported as Difficult to Procure in
Sahrawi Refugee Camps

Remedy	Folk name recorded in the camps	% of quotation (n=37)
Ambra grisea, whale's fat	Enebra	45.9
Honey	Lasal	21.6
Chameleon egg	Buya	16.2
Ostrich (*Struthio camelus* L.) fat	Naama	10.8
Yellow and hard clay	Unkel	10.8
Gazelle (*Gazella* spp.) meat	Dama	5.4
Unrefined sugar	Azucar el-har	5.4

formant reported about three products, ranging from zero to eight. We found no relation between the number of remedies available in the household at the time of interview and the number of remedies reported as difficult to procure by each informant.

As far as plant remedies are concerned, the list shows all the products that are most frequently used by Sahrawis as reported in table 12.1—i.e., the leaves and resin of *A. tortilis,* the fruits of *A. leucotrichus,* and the leaves of *Maerua crassifolia*—usually gathered in the "liberated territories," as well as the dried fruit pulp of *A. digitata* and the resin of *C. africana,* usually bought from traders and in Mauritanian markets. These data indicate that strategies for the procurement of remedies, even for the more common remedies, do not cover all the households and/or all the year(s), and that when a household runs out of remedies it can be difficult to replace them within a short time. These difficulties increase with specific products like the seeds of *Salvia aegyptiaca,* the fruits of *Balanites aegyptiaca,* and the fruits and leaves of *Lycium intricatum,* usually collected in the "liberated territories." In fact, fruits and seeds are available for gathering only during rainy years, where a sequence of years completely without rain is common. The cold season between 2004 and 2005 was characterized by an almost complete absence of rain and was not favorable for plant growing and gathering; hence, some families ran out of stock for specific remedies without being able to replace them. Informants reported some characteristic remedies from the part of Western Sahara under Moroccan control: *Euphorbia officinarum* (*daghmus*) is a common species in the oceanic regions of Western Sahara, while its presence becomes scattered as one moves east toward the Polisario-controlled area. *Daghmus* honey, a product valued by Sahrawis, is produced and sold in the occupied territories, but only in a few cases does it find its way to the camps. Similarly, *enebra* is a product of oceanic origin described by informants as whale's fat, whale's lees, or part of the whale's intestine and probably corresponds with *ambergris.* Sahrawis consider *enebra* as a panacea, and half of the informants reported it as very difficult to procure, while no informants had it in the household during the survey. It can be found sometimes on sale in Tindouf, but at very high prices, whereas in the occupied territories, reportedly, it is easier to procure.

Ostrich fat and gazelle meat are difficult to obtain due to the progressive disappearance of these animals from Western Sahara following their destruction by Spanish colonialists and during the war (see Valverde 1957; Cuzin 1996). Among all the animal fats used by Sahrawi, ostrich fat is the most appreciated, especially for broncho-pulmonary infections (Caro Baroja 1955), although fats that are more easily available, i.e., from goats and camels, progressively have substituted ostrich fat. A common characteristic of non-vegetal products most cited in table 12.5 is their very irregular supply for sale in Tindouf streets and from Mauritanian traders. The high demand in comparison with their availability raises their prices often to unaffordable levels for Sahrawi refugees and makes these remedies subject to faking by individual traders, a fact that refugees are aware of.

Knowledge Transmission and Cultural Identity with Displacement

Sahrawis identify themselves as nomadic people, depending traditionally on their camel herds and on Western Sahara territory and its resources to meet their needs for food, shelter, fuel, medicine, and other necessities of life. Due to the process of becoming sedentary and displacement to refugee camps, Sahrawis found themselves in a context where their traditional lifestyle could not be continued and most of their knowledge could not be applied or transmitted. In the camps, the conservation of traditional medicine and of ties with Western Sahara—along with other practices like the breeding of camels, sheep, and goats—contributed nevertheless to maintaining traditional Sahrawi perceptions of who they are; in other words a collective cultural identity based on shared knowledge and practices, rooted in their traditional lifestyle and nomadic activities. This link contributes to national identity and to the legitimizing of the political referent of refugees—the Polisario. As Chamberlain (2005) notes, the struggle for national liberation is tied to Sahrawi cultural life in a number of ways, and the cultural practices of the Sahrawi are implicitly and explicitly contrasted with those of the Moroccans. As a result, many aspects of everyday Sahrawi existence (including traditional medicine) become cultural markers that distinguish the Sahrawi from other cultures, and this legitimizes the Polisario's role as a defender of Sahrawi culture. In this context, traditional medicine helps to maintain Sahrawi cultural identity by reminding refugees that they are different in the way they categorize and perceive illness and in the remedies used, where these remedies are often resources of the Western Sahara, their "stolen homeland." The importance of the "liberated territories" obtained through a guerrilla war, of Polisario combatants in the procurement of the remedies, as well as of the "campaigns" organized by Polisario for gathering traditional remedies in the "liberated territories" to be distributed subsequently in the camps (Mezzetti 1994): all these factors strengthen the connection between cultural and national identity.

Nevertheless, in spite of the strategies elaborated to obtain the remedies and their cultural significance, sociocultural and productive processes that occurred with displacement have affected refugees' use of traditional medicine. These processes undermined Sahrawi shared knowledge and cultural practices, especially in younger people. Half of the population—everyone under the age of 30 years—was born in the camps, and many have studied or are studying abroad, mainly in Cuba (only primary schools are present in the camps), returning to the camps only after the completion of their studies. They are acculturated into Western biomedicine, adopt Western cultural practices, and often possess only a "narrative" knowledge of Western Sahara. Once they return to the camps they often do not participate in the transmission of traditional knowledge, and thus rupture the generational legacy within Sahrawi culture. Consequently, Sahrawi medicinal knowledge transmission has weakened. Upon our asking informants whether they transmitted their knowledge about traditional medicine and, if so, to whom, 22 percent said that they did not transmit it to anyone. In spite of this trend, however, medicinal knowledge transmission has not weakened as much as might have been supposed: 57 percent of informants reported that they transmitted all or part of their knowledge to daughters, usually the youngest daughter, followed by younger sisters (12 percent) and then by friends or other people who asked for it (9 percent). Daughters are far more likely to receive traditional medicinal knowledge from former generations because the transmission of this knowledge among Sahrawi is gendered; in fact, 89 percent of the informants, or "family experts," were women, and 83 percent of them learned their knowledge from their mothers or grandmothers. In addition, out of four male informants, three did not transmit their knowledge to anybody, indicating that the loss of gender status of popular medicinal knowledge may be related to erosion in knowledge transmission.

To test the hypothesis that the degree of conservation of traditional medicine is related to variations within refugees' generations, we cross-tabulated the age of the informants with the number of different remedies cited by each informant, and present the results in figure 12.3.

Older informants, as expected, cited more remedies. As Zent (2001) argues, traditional ethnobotanical knowledge, under the pressure of external cultures and new living patterns, decreases with the decreasing age of informants. Generally, younger informants store in their tents only a portion of the variety of remedies stored by older informants, mainly those remedies that were most cited (table 12.1). Sahrawis obtain these products mainly from their cultural keystone species, i.e., species that are culturally salient and strongly shape people's cultural

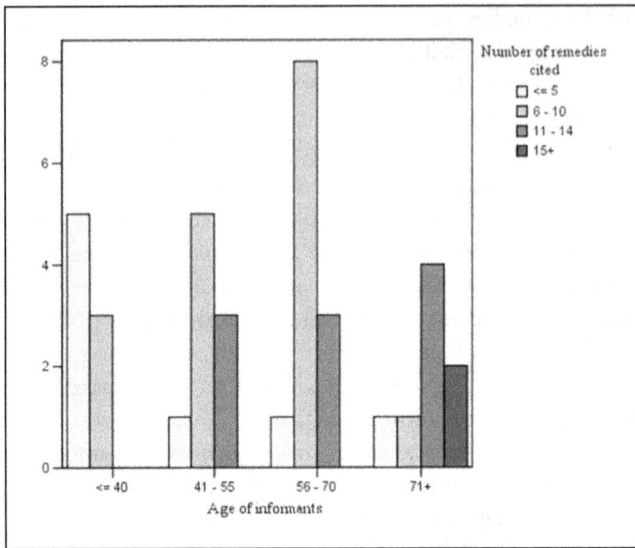

Figure 12.3.
Cross-cultural bar chart between informants' age and number of remedies cited

identity, following the definition by Garibaldi and Turner (2004). For example, Sahrawis use *A. raddiana* in multiple ways: its parts are taxonomically differentiated with specific labels, and it plays a role in Sahrawi traditional narratives, songs, and symbolism. Sahrawi families maintained their use of *A. raddiana* and its products in the camps, where the relation between these remedies and cultural identity is mediated by the possibility to procure them in the Western Sahara. The households that are not integrated into effective procurement networks with the "liberated territories" usually buy culturally important remedies in markets, from traders or obtain them and the related knowledge for free from other tents in the camps. Exchanges of remedies eventually fortify cultural links among refugees, between refugees and the "liberated territories," and between refugees and the Sahrawis living in the "occupied territories."

Conclusions

The data presented in this chapter show that Sahrawi refugees have preserved the use and knowledge of traditional medicinal remedies in the camps, and that they rely on a variety of networks in order to obtain these remedies. Most are wild plants gathered in the part of Western Sahara under control of the Polisario, and soldiers stationed there play an important role in the procurement of these remedies for refugees' families. The conservation of traditional medicine in this context represents a means to maintain cultural identity, and the procurement of remedies from the "liberated territories" in Western Sahara is a means for refugees to maintain ties with their place of origin. The conservation of traditional knowledge and practices also represents resistance against acculturation and despair. Many refugees report feeling that their lives have been wasted, and have abandoned hope of ever returning to Western Sahara. This, as well as the influence of different host cultures for the Sahrawi who have studied abroad, lead slowly to the loss of traditional knowledge and to the ties with Western Sahara for the younger generations, who often know and use only the most culturally relevant remedies.

References

AAA (American Anthropological Association). 1998. *Code of ethics of the American Anthropological Association.* Retrieved 15 March 2006 from http://www.aaanet.org/committees/ethics/ethcode.htm

Alexiades, N.M., and J.W. Sheldon, eds. 1996. *Selected guidelines for ethnobotanical research: A field manual.* New York: New York Botanical Garden.

Berlin, B. 1992. *Ethnobiological classification.* Princeton: Princeton University Press.

Bhatia, M. 2001. "The Western Sahara under polisario control." *Review of African Political Economy* 28: 291–301.

Brainard, J., and A. Zaharlick. 1989. "Changing health beliefs and behaviors of resettled Laotian refugees: Ethnic variation in adaptation." *Social Science and Medicine* 29: 845–852.

Branca, F. 1997. *The health and nutritional status of Saharawi refugees.* Roma: Istituto Nazionale della Nutrizione.

Caro Baroja, J. 1955. *Estudios Saharianos.* Madrid: Ediciones Júcar.

Chamberlain, R. 2005. *Stories of a nation: Historical narratives and visions of the future in Sahrawi refugee camps.* Refugee Studies Centre. RSC Working Paper 29. Oxford: University of Oxford.

Cohen, D. 1963. *Le dialecte Arabe Hassaniya de Mauritanie.* Paris: Librairie C. Klincksieck.

Cotton, C.M. 1996. *Ethnobotany. Principles and applications.* Chichester: Wiley & Sons.

Cozza, N. 2003. "Singing like wood-birds: Refugee camps and exile in the construction of the Saharawi nation." Ph.D. Dissertation. Oxford: University of Oxford.

Cunningham, A.B. 2001. *Applied ethnobotany. People, wild plant use and conservation.* London: Earthscan Publications Ltd, People and Plants Conservation Manual.

Cuzin, F. 1996. "Répartition actuelle et statut des grands mammifères sauvages du Maroc (primates, carnivores, artiodactyles)." *Mammalia* 60: 101–124.

D'Andrade, R. 1995. *The development of cognitive anthropology.* Cambridge: Cambridge University Press.

DNV-RASD. 2005. *Census 2005.* Rabuni, polisario refugee camps: Dirección Nacional de Veterinaria, Ministerio de Salud Pública Sahrawi, RASD—República Árabe Democrática Sahrawi.

Garibaldi, A., and N. Turner. 2004. "Cultural keystone species: Implications for ecological conservation and restoration." Ecology and Society 9:1. Retrieved 10 February 2006 from http://www.ecologyandsociety.org/vol9/iss3/art1

Guinea, A. 1949. *El Sahara Español.* Madrid: Instituto de Estudios Africanos, Consejo Superior de Investigaciones Científicas.

Jacobsen, K., and L. Landau. 2003. *Researching refugees: Some methodological and ethical considerations in social science and forced migration.* Geneva, Switzerland: UNHCR, New Issues in Refugee Research, Working Paper 90.

Kim, Y. 2001. *Becoming intercultural. An integrative theory of communication and cross-cultural adaptation.* Thousand Oaks: Sage Publications.

Lebrun, J. 1998. "Catalogue des plantes vasculaires de la Mauritanie et du Sahara Occidental." *Boissiera* 55: 1–322.

Lippert, A. 1992. "Sahrawi women in the liberation struggle of the Sahrawi people." *Signs: Journal of Women in Culture and Society* 17: 636–651.

Loewenberg, S. 2005. "Displacement is permanent for the Sahrawi refugees." *The Lancet* 365: 1295–1296.

Martin, G.J. 1995. *Ethnobotany.* London: Chapman & Hall.

McDougall, E.A. 1990. "Salts of the Western Sahara: Myths, mysteries, and historical significance." *The International Journal of African Historical Studies* 23: 231–257.

Mezzetti, A. 1994. *L'Organizzazione Sanitaria nei Campi Profughi Sahrawi.* Bologna: Università degli Studi di Bologna.

Monteil, V. and C. Sauvage. 1953. "Contribution á l'étude de la flore du Sahara Occidental. De l'Arganier au Karité. Catalogue des plantes connues des Tekna, des Rguibat et des Maures." *Bullettin del Institute des Hautes Etudes Maroquine, Notes and Documents* 6: 1–147.

Ould Mohamed Baba, A.S. 2004. "Los Berberismos del dialecto Árabe Hassāniyyä de el-Gəblä." *Anaquel de Estudios Árabes* 15: 175–184.

Ozenda, P. 1991. *Flore et végétation du Sahara,* 3d ed. Paris: CNRS Éditions.

OXFAM. 1995. *Western Sahara.* Bruxelles: OXFAM Belgium and Comité Belge de Soutien au Peuple Sahraoui.

Pieroni, A., H. Muenz, M. Akbulut, K. and C. Hüsnü Can Başer, Durmuşkahya. 2005. "Traditional phytotherapy and transcultural pharmacy among Turkish migrants living in Cologne, Germany." *Journal of Ethnopharmacology* 102: 69–88.

Puri, R.K. and C. Vogl. 2004. *A methods manual for ethnobiological research and cultural domain analysis.* Canterbury: Department of Anthropology, University of Kent.

Quitout, M. 1999. *Initiation à l'Arabe Maghrébin. Vocabulaire Bilingue.* París.

San Martín, P. 2004. "Briefing: Western Sahara: Road to perdition?" *African Affairs* 103/413: 651–660.

Seddon, D. 2000. "Western Sahara - Point of no return?" *The Review of African Political Economy* 27: 338–340.

Shelley, T. 2004. *Endgame in Western Sahara: What future for Africa's last colony?* London: Zed Book.

Siouti, J.E. 1994. *La Medecine du Prophete.* Al Bourak.

Spiegel, P.B., and M. Qassim. 2003. *Forgotten refugees and other displaced populations.* The Lancet 362: 72–74.

UNHCR (United Nations High Commissioner for Refugees). 2002. *Anthropometric and micronutrient nutrition survey in Saharawi refugee camps, Tindouf, Algeria.* Tindouf: UNHCR, WFP, Center for International Child Health.

UNHCR (United Nations High Commissioner for Refugees). 2005. *2004 Global refugee trends.* Population and Geographical Data Section, Division of Operational Support, UNHCR, Geneva. Retrieved 24 January 2006 from http://www.unhcr.ch/statistics

Valverde, J.A. 1957. *Aves del Sahara Español. Estudio ecologico del desierto.* Madrid: Instituto de Estudios Africanos, Consejo Superior de Investigaciones Cientificas.

Weller, S.C. 1998. "Structured interviewing and questionnaire construction." In *Handbook of methods in cultural anthropology,* ed. H.R. Bernard. Walnut Creek, CA: Altamira Press, 365–409.

Zent, S. 2001. "Acculturation and ethnobotanical knowledge loss among the Piaroa of Venezuela". In *On Biocultural Diversity,* ed. L. Maffi. Washington: Smithsonian Institution Press, 190–211.

Notes on Contributors

ABDALAHE AHMADI is a veterinarian who graduated in Santa Clara, Cuba. Since 2003 to 2006 he was director in charge for the public veterinary service of the Sahrawi Arabic Democratic Republic (RASD), Algeria. E-mail: abdaahmadi @yahoo.es

MICHAEL J. BALICK, Ph.D., an ethnobotanist, is Director and Philecology Curator, at the Institute of Economic Botany of The New York Botanical Garden. He received his Ph.D. from Harvard University and works with indigenous cultures to document their plant knowledge and understand their traditional management systems. He is a pioneer in urban ethnobotany and has worked for more than a decade with immigrant communities in New York City. His fieldwork elsewhere around the world has focused on tropical and island ecosystems. His current long term projects include understanding the relationship between biodiversity, culture, and human health in Micronesia and Belize.

ALESSANDRO BROGLIA is the coordinator of the project "Animal Health in Sahrawi Refugee Camps—Algeria" in the Sahrawi Arabic Democratic Republic (RASD), Algeria, for SIVtro VSF Italy and Africa70 NGO. His expertise is in animal health in desert and tropical areas, management of veterinary diagnostic laboratories, and veterinary service management in rural areas. E-mail: alessandro.broglia @africa70.org

NEIL CARRIER was awarded his Ph.D. at the University of St. Andrews for a thesis entitled "The Social Life of Miraa: Farming, Trade and Consumption of a Plant Stimulant in Kenya" based on fieldwork in Kenya and the United Kingdom. He furthered his interest in khat as a research assistant on an AHRC / ESRC funded project on the substance, and latterly as part of his ESRC Postdoctoral Research Fellowship at St. Antony's College, Oxford. He currently works as a research officer at the Welcome Unit for the History of Medicine at Oxford on the AHRC funded project "Trauma and Personhood in Late Colonial Kenya." E-mail: Neil.Carrier@wuhmo.ox.ac.uk

MELISSA CEUTERICK holds a. M.Sc. in cultural anthropology and development studies. She is a Ph.D. candidate at the Division of Pharmacy Practice of the University of Bradford, United Kingdom. Her doctoral research deals with the perception and use of plants among the Spanish-speaking Latino communities in the United Kingdom. E-mail: M.T.Ceuterick@Bradford.ac.uk

LINDA F. CUSHMAN, Ph.D. is a sociologist and Associate Clinical Professor at the Mailman School of Public Health, Columbia University. She conducts clinical and community-based research across a variety of public health domains including women's use of alternative medicine and, more recently, evaluations of health programs in Santo Domingo and La Romana, Dominican Republic. E-mail: lfc2@columbia.edu

DOUGLAS C. DALY is a systematic botanist at The New York Botanical Garden, where he is B. A. Krukoff Curator of Amazonian Botany. His work focuses on floristics of Amazonia, systematics and biogeography of the tropical tree family *Burseraceae,* leaf architecture, strategies for identification of tropical trees, and applications of systematics in conservation and economic botany. E-mail: ddaly @nybg.org

SARA DI LELLO works as desk officer for the NGO Africa70—Italy. She is in charge of management in the veterinary sector of the project "Animal Health in Sahrawi Refugee Camps—Algeria." E-mail: sara.dilello@africa70.org

LEVENIA DURÁN holds a B.A. in International Studies from the City College of New York. She has experience conducting surveys and ethnographic interviews, transcribing tape-recorded interviews in Spanish, and translating from English to Spanish. She has worked as a Research Assistant on two different studies investigating Dominican Plants and their uses for the New York Botanical Garden. She is currently a language teacher at Berlitz Language School in Aachen, Germany. E-mail: levenia.duran@gmail.com

LIONEL GERMOSÉN-ROBINEAU, MD, is the coordinator of the TRAMIL (*Traditional Medicine in the Islands*) network that conducts applied research on Caribbean medicinal plants. He is the scientific editor of the Caribbean Herbal Pharmacopoeia. He holds an M.Sc.in Public Health and is a member of UAG University, French West Indies. E-mail: coordina@tramil.net

SARAH KEELER is a Ph.D. candidate in the Department of Anthropology at the University of Kent, UK. Her doctoral research dealt with identity discourses in the context of transnationalism and hybridity among Kurds in London. Her research interests include food anthropology, diaspora, and cultural productions (food, music, and tourism) as sites for the expression of ethnicity, including their changing meanings and functions in a migratory context, with particular focus on the cultures of the Middle East. She currently holds a Marie Curie fellowship at the Centre for Conflict Studies, Utrecht University, The Netherlands, and is a member of a British Academy project on Refugees and Other Forced Migrants in the Middle East. E-mail: keeler_sarah@hotmail.com

FREDI KRONENBERG is Professor of Clinical Physiology and Director of the Richard and Hinda Rosenthal Center for Complementary and Alternative Medicine at Columbia University College of Physicians and Surgeons in New York. Her expertise is in women's health, particularly the physiology of menopause, and in complementary and alternative medicine approaches to maintaining optimal health and treating disease, with a focus on botanical medicine. She works locally to study the ethnomedical practices of immigrant populations in New York City, and internationally to study traditional systems of medicine such as Traditional Chinese medicine and other Asian medical systems. Email: fk11@columbia.edu

SALEH MOHAMED LAMIN is a veterinarian who graduated in Havana, Cuba. He is director in charge for the public veterinary service of the Sahrawi Arabic Democratic Republic (RASD), Algeria.

RAFAEL LANTIGUA is a professor of Clinical Medicine at Columbia University and the Medical Director of the Associates in Internal Medicine Group Practice at the New York Presbyterian Hospital, Columbia University. His research interests are Aging and Quality of Life and The Genetics of Dementia in Hispanics populations. E-mail: ral4@columbia.edu

PRANEE C. LUNDBERG is a registered nurse and Associate Professor of Caring Sciences at the Department of Public Health and Caring Sciences of Uppsala University, Sweden. Her research focuses on cultural care and health beliefs among immigrants and on transcultural nursing. E-mail: Pranee.Lundberg@pubcare.uu.se

MIRIAM MEJIA is the Deputy Director of Alianza Dominicana Inc., a New York-based Dominican community development organization. She holds a BA in Sociology from the Universidad Autónoma de Santo Domingo, Dominican Republic. E-mail: MMejia@alianzadom.org

ANDREANA L. OSOSKI is an ethnobotanist who earned her Ph.D. from The City University of New York. She has done fieldwork in New York City and the Dominican Republic. She currently serves as a Research Consultant for The New York Botanical Garden and as a Science Education Postdoctoral Fellow at Humboldt State University. Her research focuses on medicinal plants, women's health, Caribbean ethnobotany, and Dominican healing traditions and practices. Email: aososki@nybg.org

USHA R. PALANISWAMY holds a Ph.D. in Horticulture from the University of Connecticut. She is a plant physiologst, and phytochemist who takes ethnobotanical approaches to research. Currently, she is the Chair, Division of Natural Sciences and Mathematics, Excelsior College, Albany, NY. She was faculty at the Asian American Studies Institute at the University of Connecticut until 2007. An editor of the *Journal of Herbs, Spices and Medicinal Plants,* she is the North East District delegate and a national board member of The Herb Society of America. Her research focuses on uses of plant species for food and medicine, dietary changes due to migration and related health concerns in immigrant populations. E-mail: usha.palaniswamy@gmail.com

ANDREA PIERONI holds a Ph.D. in Pharmacy from the University of Bonn, Germany. He is Senior Lecturer in Herbal Medicines at the Division of Pharmacy Practice of the University of Bradford, UK He is also Guest Associate Professor in the Department of Social Sciences at the Wageningen University in The Netherlands. His research focuses on Traditional Pharmaceutical Knowledge, gastronomic ethnobotany and related intersections (food medicines, folk functional foods) in the Mediterranean and the Balkan areas (with a special interest in ethnic diasporas), ethnopharmacy, and transcultural health studies among migrant communities in Europe. E-mail: a.pieroni@bradford.ac.uk

CASSANDRA QUAVE is a doctoral candidate in the Department of Biological Sciences, at the Center for Ethnobiology and Natural Products Research at Florida International University, USA. Her research focuses on medical ethnobotany and folk pharmacopoeias in the Mediterranean and the bioactivity of traditional botanic remedies. E-mail: cassy.quave@gmail.com

INGVAR SVANBERG is an ethnobiologist and university lecturer at Södertörn University College and Uppsala University, Sweden. He has published several books and many scientific articles on traditional knowledge and use of animals and plants. His current research interests include Eurasian plant knowledge, traditional ecological knowledge in the Faroes, Estonia, and northern Sweden, European ethnoherpetology, Scandinavian folk ornithology, and cultural history of pets. E-mail: ingvar.svanberg@eurasia.uu.se

Bren Torry is a lecturer in the Division of Pharmacy Practice, School of Life Sciences, University of Bradford. Her current research interests include health psychology and cultural diversity within veterinary medicine and transcultural health and the elderly. Email: b.torry@bradford.ac.uk

Tinde van Andel is an ethnobotanist at the National Herbarium of the Netherlands at Utrecht University. Her research focuses on non-timber forest products of the Guianas and especially on Surinamese medicinal plants and the use of traditional medicine among Surinamese residents and migrants. E-mail: T.R.vanAndel @uu.nl

Ina Vandebroek is a Postdoctoral Research Associate at the Institute of Economic Botany of The New York Botanical Garden. She holds a Ph.D. in Medical Sciences from Ghent University, Belgium. Her current research focuses on traditional health care knowledge, beliefs and practices of Dominican immigrants in New York City and aims at making a transnational comparison of medicinal plant use for common health conditions between New York City and the Dominican Republic. The results of this applied urban ethnobotany project funded by the National Institutes of Health/National Center for Complementary and Alternative Medicine (NIH/NCCAM) will serve as a basis for a medicinal plant guidebook for the Dominican community in New York City. E-mail: ivandebroek@nybg.org

Charlotte van 't Klooster graduated as a Biologist at the Free University of Amsterdam with a study of medicinal plants sold by a Saramaccan shop in Amsterdam. After that she published a book on vernacular plant names of Suriname. She now works as a study enlightener at the Free University. E-mail: charlotte .vant.klooster@falw.vu.nl

Anahi Viladrich is a sociologist and medical anthropologist of Argentine origin, and currently serves as a faculty member at the Urban Public Health Program, Hunter College of the City University of New York, where she directs the Immigration and Health Initiative aimed at developing research, training, and advocacy on the health issues affecting immigrant populations worldwide. She has published extensively on reproductive health and gender both in Argentina and in the United States, and more recently has written on tango enclaves, Argentine migrants, and Latino immigrants' alternative healing practices in the United States, including the use of herbs and plants. E-mail: aviladri@hunter .cuny.edu

Gabriele Volpato is an ethnobotanist and Ph.D. candidate in the Department of Social Sciences of Wageningen University, The Netherlands. He conducted fieldwork in Italy, Cuba, Algeria, and Western Sahara. His research focuses on food ethnobotany and folk pharmacopoeias, as well as on ethnoveterinary. E-mail: gabriele.volpato@wur.nl

CHRISTINE WADE, MPH. is the Associate Director of Research and Administration, the Richard and Hinda Rosenthal Center for Complementary and Alternative Medicine and has coordinated research and education activities there since 1995. She has considerable expertise in minority women's health and traditional systems of medicine. She was the project director of an NICHD and NIH/NC-CAM funded national survey of complementary and alternative medicine use for women's health conditions in four race/ethnic groups. Ms. Wade has been collaborating internationally on research on Traditional Chinese Medicine since 2000 when she helped develop a pilot of acu-injection therapy for dysmenorrhea in Shanghai and Milan. E-mail: wade@columbia.edu

JOLENE YUKES is a medical anthropologist and clinical herbalist in New York City. She compiled *Dominican Medicinal Plants: A Guide for Health Care Providers,* coordinated the "Urban Ethnobotany Project: Dominican Herbal Remedies for Women's Health in New York City," and designed and implemented the community health initiative "Dominican Herbal Medicine and Culturally Effective Health Care" at The New York Botanical Garden's Institute of Economic Botany. Her primary research interests include applied ethnobiology, Latin American and Caribbean healing traditions, the sociology of scientific knowledge, and bioregional herbalism. E-mail: jolene_yukes@yahoo.com

Index